Principles of Inductive Near Field Communications for Internet of Things

RIVER PUBLISHERS SERIES IN COMMUNICATIONS

Volume 18

Consulting Series Editors

MARINA RUGGIERI   HOMAYOUN NIKOOKAR
University of Roma "Tor Vergata"   *Delft University of Technology*
Italy   *The Netherlands*

This series focuses on communications science and technology. This includes the theory and use of systems involving all terminals, computers, and information processors; wired and wireless networks; and network layouts, procontentsols, architectures, and implementations.

Furthermore, developments toward new market demands in systems, products, and technologies such as personal communications services, multimedia systems, enterprise networks, and optical communications systems.

- Wireless Communications
- Networks
- Security
- Antennas & Propagation
- Microwaves
- Software Defined Radio

For a list of other books in this series, please visit www.riverpublishers.com

Principles of Inductive Near Field Communications for Internet of Things

Johnson Ihyeh Agbinya (PhD)

La Trobe University, Melbourne, Australia

Aalborg

ISBN: 978-87-92329-52-3 (hardback)

Published, sold and distributed by:
River Publishers
PO Box 1657
Algade 42
9000 Aalborg
Denmark

Tel.: +4536953197
www.riverpublishers.com

MATLAB Codes available for download:
http://riverpublishers.com/river_publisher/book_details.php?book_id=90

Dedication

Dedicated to my sweetheart and wife Jemimah who has refocused me towards higher goals and excellence.

Contents

Preface

Over the past five years renewed interest and rapid progress has been made in the application of inductive methods to telecommunications in difficult terrains where radiating radio waves are prone to interference, in near field communications, power transfer and data telemetry from embedded medical devices, wireless power transfer, sensing and in the recent two years, Internet of things (IoT). All of these emerging technologies operate over short range and mostly in the near field region of antennas where energy is transferred by coupling magnetic flux from one antenna to the other. The book covers detailed summary of the electronic principles which support these technologies as well as reporting on unpublished latest advances in the area from our research laboratory at La Trobe University. The contents of the first three chapters, including 6, 7 are available here and there in current literature waiting to be organised as in the book. However chapters 4, 5, 15 and 16 are unpublished research undertaken during the author's sabbatical at the National Information and Communication Australia at Redfern in Sydney. Apart from the research chapters, the fundamental principles covered in the book lend themselves to easy understanding by good undergraduate and postgraduate students in electronics, electrical, computer engineering and Physics.

Objectives and Prerequisites

Communicating with transceivers buried in dielectric materials is an emerging focus for mine communication and tunnels, remote sensing and through biological tissues. Adding to these the need to network things in close proximity and provide secure near field communication requires an air interface able to resist interception from a remote probe. Inductive communication provides these advantages. Unfortunately the limited range, rapid power decline of orders many times what obtains in radio wave transmissions requires understanding of inductive near field methods to design close contact data transfer. The book has therefore covered the basic principles of inductive communications from the foundations up to applications. This includes:

- Distinguishing between near-field and far-field communication and propagation
- Circuit models of inductive antennas
- Resonant inductive communications
- Narrow band system capacity
- Broadband methods including Inductive multiple-input-multiple-output systems
- Modulation and Coding methods
- Inductive link and channel characteristics
- Magneto Inductive Link budgets
- Limitations of inductive Systems (misalignment, crosstalk and parasitic losses)
- Magnetic waveguides devices and range extension methods based on
- Application and effects of trans-impedance
- Application of near-field methods
- Wireless power transfer

A rich set of Matlab codes for simulating various inductive communication components is included as a chapter in the book and also made available online at the Publisher's site.

Organisation of the Book

The book is organised into twenty chapters, starting from chapters 1 and 2 which cover the essential principles of near-field communications and more so near-field inductive communication. The first two chapters provide clear distinctions in the definitions of near-field edge as used in various research fields. The distinctions are mostly based on the wavelength of radiation and the size of antenna.

Chapter three provide answers to the question on how to use inductive coupling for normal data communication between two end points. It highlights the value and efficacy of maximum power transfer at resonance and derives an expression for maximum system capacity which is based on the quality factors of the transceiver components. This idea is further developed in chapter four new ideas on how to determine the size of the magnetic bubble over which secure inductive communications can be established, the so-called sphere of silence. Normally the capacity single-input-single-output systems can be increased significantly by using multiple transceivers. This is the subject of chapter five where inductive multiple-input-single-output and multiple-input-multiple-output systems are introduced and discussed in great details. To quantify the performance of inductive communication systems, chapter six provides the circuit models and topologies of inductive communication systems which is used to discuss the effects of coil misalignment on system performance.

Chapter seven may be seen as a standalone chapter. It is a summary of various types of inductive coils and their design equations. The objective in the chapter is to provide a snapshot of the system equations that could be used to derive the inductances of coils. It can be read in isolation without missing much from the earlier and later chapters of the book.

Circuit models of various inductive system topologies are discussed in chapter eight. The models are used to quantify the impact of parasitic resistances and capacitances in the transmitter and receiver circuits on system performance in chapter nine. Chapter nine also covers trans-impedance, inductive links and efficiencies of the links based on various system topologies. The question of performance is carried forward to chapter ten where crosstalk is discussed. Like radio wave communication where interference is a great issue, crosstalk is the major interference in inductive systems. Since it reduces the performance of the systems, a reasonable analysis of crosstalk is presented in chapter ten and fourteen. The next three chapters are dedicated to signal conditioning with the aim of sustained high data rate communications. The chapters discuss modulation techniques and coding. They are followed by a thorough analysis of inductive channel models in chapter fifteen and then magneto-inductive devices such as mirrors, two-port devices, Fabry Perot resonators and resonant cavities, tapers, splitters and Bragg gratings. The Physics of these devices is covered. The remaining chapters deal with the emerging applications of inductive methods including hearing aids, wireless power transfer, embedded medical systems, health monitoring, near field communication devices and energy harvesting. An invaluable set of Matlab codes is given as the concluding chapter. The codes may be used to simulate various aspects of inductive systems and for their analysis. This code is available as a free online resource.

Acknowledgment

When I started conducting research on inductive communication and sensing, the intention was never to write a book on the subject. However finding out how perfect the educational foundations I received are suited to the task and a period of six months spent at NICTA Redfern in Sydney provided a perfect climate for not only research but also to document research and perform a thorough review of the subject area. This has given birth to this book. For that I am mostly grateful to the Professor Aruna Seneviratne, the laboratory director of NICTA Redfern for providing a place for undertaking sabbatical at NICTA. The quiet office space and the rich but free coffee provided at the NICTA office kept me on my desk during this period. I am also grateful to my students who have contributed to this book. Notably Mehrnoush Masihpour undertook the re-drawing of most figures in the book, adding colour and artistic flavours to them. She also ran most of the simulations reproduced in chapter twenty. La Trobe University tends to bring the best out of me and the opportunity to return to my alma mater after many years in industry to work again under Professor John Devlin has helped a great deal to finish and progress the research upon which some sections of this book are based. For this opportunity I am grateful.

About the Author

Johnson graduated from La Trobe University with a PhD (1994) in microwave radar remote sensing (MSc Research University of Strathclyde, Glasgow, Scotland (1982) in microprocessor techniques in digital control systems) and BSc (Electronic/Electrical Engineering, Obafemi Awolowo University, Ife, Nigeria (1977). He is currently Associate Professor (remote sensing systems engineering) in the department of electronic engineering and Extraordinary Professor in telecommunications at Tshwane University of Technology/French South African Technical Institute in Pretoria, South Africa. He is also Extraordinary Professor in computer science at the University of the Western Cape in South Africa, Cape Town. He was a Senior Research Scientist at CSIRO Telecommunications and Industrial Physics (1994–2000; renamed CSIRO ICT) in biometrics and remote sensing and Principal Engineer Research at Vodafone Australia research from 2000 to 2003. He is the author of six technical books in electronic communications including Principles of Inductive Near Field Communications for Internet of Things (River Publishers, Aalborg Postkontor, Denmark, 2011); IP Communications and Services for NGN (Auerbach Publications, Taylor & Francis Group, USA, 2010) and Planning and Optimisation of 3G and 4G Wireless Networks (River Publishers, Aalborg Postkontor, Denmark, 2009).

He is Consulting Editor for River Publishers Denmark on new areas in Telecommunications and Science. He is the founder and editor-in-chief of the

African Journal of ICT (AJICT) and founder of the International Conference on Broadband Communications and Biomedical Applications (IB2COM). His current research interests include remote and short range sensing, inductive communications and wireless power transfer, Internet of Things , wireless and mobile communications and biometric systems.

Dr. Agbinya is a member of IEEE and African Institute of Mathematics (AIMS). He has published extensivelyoin broadband wireless communications, biometrics, vehicular networks, video and speech compression and coding, contributing to the development of voice over IP, intelligent multimedia sub-system and design and optimisation of 3G networks. He was recipient of research and best paper awards and has held several advisory roles including the Nigerian National ICT Policy initiative.

1

Introduction to Near Field Communications

Traditionally, the emphasis in telecommunications has been to reach long distances (far field). Even though the first telephone the first telephone call on March 10 1876 by Alexander Graham Bell and his assistant Thomas A Watson was a short distance call in nearly by offices, the emphasis has been long distance communications. Thirty nine years after the first short range telephone call was made, on the 25th of January 1915, Alexander Graham Bell and Thomas Watson also made the first long distance voice call between New York and San Francisco. Recently new focus has been placed on short range communications due to its apparent advantages of easier achievement of wireless large capacity. The most commercially prevalent communication systems have therefore been those which operate in the higher microwave frequencies and they include mobile telephones and short range data communication systems. The shift in increasing capacity is also directly related to the shift to higher and higher frequencies where large spectrums exist to be dedicated to capacity hungry applications such as imaging and multimedia. These higher frequency ranges also mean shorter wavelengths which further facilitate the design of more efficient and small sized antennas. Of course there are applications which do not need large spectrum outlays such as hands free wireless communication, real time location systems, low data rate voice communications and low data rate communication systems. Practically, low frequencies (long wavelengths) penetrate objects more and are subject to diffraction around objects which would normally have blocked high frequency transmissions. Low frequencies are also known to be less susceptible to multipath effects. Hence low frequency systems have for a long time enjoyed commercial patronage.

Short range communications (a.k.a near field communications) have for long been neglected by engineers and most researchers in the field. This focus is intentional to take advantage of far field communications where near field negative effects are less pronounced.

J. Ihyeh Agbinya (PhD), Principles of Inductive Near Field Communications for Internet of Things, 1–11.
© 2011 *River Publishers. All rights reserved.*

1.1 Near Field or Far Field

The terms "near field" and "far field" are defined in relation to the communication properties of a radiation source or a transmitting antenna. Unfortunately in current literature there are numerous definitions of these terms and to the uniformed there appears to be no clear distinction between where near field ends and where far field starts (Table 1.1). In reality, there is an intermediate zone between the near and far fields so that there are two boundaries between the regions.

This chapter distinguishes between the two region models (near field and far field) and the three region models [1]. The two region model consists of the near and far field regions only. The three region model consists of the near field, transition and far field regions.

Table 1.1 Definitions of the Near-Field / Far-Field Boundary [1]

Definition for shielding	Remarks	Reference
$\lambda/2\pi$	$1/r$ terms dominant	Ott, White
$5\lambda/2\pi$	Wave impedance $= 377\ \Omega$	Kaiser
For antennas		
$\lambda/2\pi$	$1/r$ terms dominant	Krause
3λ	D not $\gg \lambda$	Fricitti, White, Mil-STD-449C
$\lambda/16$	Measurement error < 0.1 dB	Krause, White
$\lambda/8$	Measurement error < 0.3 dB	Krause, White
$\lambda/4$	Measurement error < 1 dB	Krause, White
$\lambda/2\pi$	Satisfies the Rayleigh criteria	Berkowits
$\lambda/2\pi$	For antennas with $D \ll \lambda$ and printed-wiring-board trace	White, Mardiguian
$2D^2/\lambda$	For antennas with $D \gg \lambda$	White, Mardiguian
$2D^2/\lambda$	If transmitting antenna has less than 0.4D of the receiving antenna	MIL-STD 462
$(d + D)^2/\lambda$	If $d > 0.4\ D$	MIL-STD 462
$4D^2/\lambda$	For high accuracy antennas	Kaiser
$50\ D^2/\lambda$	For high accuracy antennas	Kaiser
$3\lambda/16$	For dipoles	White
$(d^2 + D^2)/\lambda$	If transmitting antenna is 10 times more powerful than receiving antenna, D	MIL-STD-449D

We will distinguish between the fields produced by an elemental dipole antenna and a magnetic loop and provide expressions also for the radial and angular fields for each. We thus use an algebraic methods for distinguishing between near and far fields. Many of the equations given in this chapter were originally derived by S. K. Schelkunoff from Maxwell's equations. With the equations to be given, three approaches are therefore adopted for defining near field and far field. The first relies on the range of transmission, the second on the wave impedance and the third on the phase variations of the radiations. In radio communication the main interest is in how the power in the transmitted signal attenuates with distance. Antenna designers are traditionally interested in how the phase of the radiation is affected by distance, whereas in shielding of radiations, the interest is on how the wave impedance changes with distance. Hence these three perspectives are pursued in terms of the definitions of near field.

Question 1. Explain why the definition of near field boundary is not as clear cut as expected.

Question 2. Explain three approaches for defining the near field boundary.

1.1.1 Fields from a dipole antenna

An electric dipole produces two fields, an electric field and a magnetic field. Electric and magnetic fields behave differently in the near field and far fields and hence require different equations to model them in free space and in practical situations. Also two sets of equations describing these fields at angular directions and radial directions need to be given. We first provide the angular behaviours of the fields.

The induced voltage (per meter) in a dipole antenna excited by a current source at angular frequency w along an angular direction θ is given by the expression:

$$E_\theta = \frac{I * l * \beta^3}{4\pi\omega\varepsilon_\theta} \left[\frac{j}{\beta * r} + \frac{1}{(\beta * r)^2} - \frac{j}{(\beta * r)^3} \right] \sin(\theta) * e^{-j\beta r} \; V/m \quad (1.1)$$

In this expression

l is the physical length of the dipole antenna in meters
$\beta = \omega/c = 2\pi/\lambda$ is its electrical length per meter of wavelength of the antenna
$\omega = 2\pi f$ is the angular frequency in radians per second

f is frequency of the radiation in Hertz

λ is the wavelength in meters

c the speed of light ($3 * 10^8 m/\sec$)

r is the distance from the source of radiation to the point of observation and $\varepsilon_0 = \frac{1}{36\pi * 10^{+9}} F/m$ is the permittivity of free space

The magnetic field induced by the current I along the angle ϕ in the dipole antenna is also given by the expression:

$$H_\phi = \frac{I * l * \beta^2}{4\pi} \left[\frac{-1}{j\beta * r} + \frac{1}{(\beta * r)^2} \right] \sin(\theta) * e^{-j\beta r} \quad A/m \qquad (1.2)$$

The range characteristic of the electric field in a dipole antenna is given by the expression:

$$E_r = \frac{I * l * \beta^3}{4\pi \omega \varepsilon_\theta} \left[\frac{j}{\beta * r} + \frac{1}{(\beta * r)^2} - \frac{j}{(\beta * r)^3} \right] \cos(\theta) * e^{-j\beta r} \quad V/m \quad (1.3)$$

While the electric field has units of voltage per meter, the magnetic field has units of current per meter. Hence their ratio provides a measure of the impedance of the space in which the dipole operates.

1.1.2 Fields from a magnetic loop antenna

Apart from a dipole antenna, a magnetic loop antenna has also been used for near field magnetic induction communications. When a magnetic loop antenna is excited with a current I at frequency f, the electric field at direction ϕ is

$$E_\phi = -j \frac{\omega * \mu_0 * m * \beta^2}{4\pi} \left[\frac{-1}{j\beta * r} + \frac{1}{(\beta * r)^2} \right] \sin(\theta) * e^{-j\beta * r} \quad V/m \quad (1.4)$$

The magnetic field at distance r is

$$H_r = 2j \frac{\omega * \mu_0 * m * \beta^2}{4\pi \eta_0} \left[\frac{1}{(\beta * r)^2} - \frac{j}{(\beta * r)^3} \right] \cos(\theta) * e^{-j\beta * r} \quad A/m$$

$$(1.5)$$

and

$$H_\theta = j \frac{\omega * \mu_0 * m * \beta^2}{4\pi \eta_0} \left[\frac{1}{\beta * r} + \frac{1}{(\beta * r)^2} - \frac{j}{(\beta * r)^3} \right] \sin(\theta) * e^{-j\beta * r} \quad A/m$$

$$(1.6)$$

In this expression

$\mu_0 = 4\pi * 10^{-7} H/m$ is the permeability of free space.

$\eta_0 = 376.7\Omega$ is the free space impedance

1.1.2.1 Range-based near field boundary

Equations (1.1) to (1.6) depend on range in the form $\frac{1}{r}$, $\frac{1}{r^2}$ *and* $\frac{1}{r^3}$. Close to the antenna (near field), when r is small the $\frac{1}{r^3}$ term dominates. Farther away from the antenna the terms $\frac{1}{r^2}$ *and* $\frac{1}{r^3}$ attenuate very fast and are close to zero and the term inversely proportional to range is the most important. This is the far field region. Hence the boundary between the near and far field can be estimated to be where the last two terms in each of the equation vanish. For example in equation (1.2) we set

$$\frac{1}{\beta * r} = \frac{1}{(\beta * r)^2} \tag{1.7}$$

and

$$r = \frac{\lambda}{2\pi} \tag{1.8}$$

Therefore the boundary between the near and far fields is where the range is equal to the inverse of the electrical length of the antenna. Thus if we know the electrical length the antenna, we also know the near field boundary. Of course we do know the electrical length through the frequency of the radiation. By defining the boundary of the near field in terms of the wavelength, we have a moving boundary in space. Thus the boundary is not fixed but changes with the frequency of the excitation. Hence at very high frequencies, the boundary is very near the source of radiation and at low frequencies it is further away.

Question 3. a) If the electric field from a dipole antenna is given by the equation

$$E_r = \frac{I * l * \beta^3}{4\pi\omega\varepsilon_\theta} \left[\frac{j}{\beta * r} + \frac{1}{(\beta * r)^2} - \frac{j}{(\beta * r)^3} \right] \cos(\theta) * e^{-j\beta r} \ V/m$$

Derive an expression for the near field boundary.

b) What is the value of the near field boundary if the transmitting frequency is 10 MHz?

Question 4. The magnetic field from a loop antenna is given by the expression

$$H_r = 2j \frac{\omega * \mu_0 * m * \beta^2}{4\pi \eta_0} \left[\frac{1}{(\beta * r)^2} - \frac{j}{(\beta * r)^3} \right] \cos(\theta) * e^{-j\beta * r} \quad A/m$$

Use it to derive an expression for the near field boundary.

Question 5. Explain why the near field communication is not a fixed point in space.

1.1.2.2 Near field boundary from free-space impedance

At the near field boundary, the free space impedance is constant or the electromagnetic wave remains constant. We do know that the ratio of the electric field and the magnetic fields for a dipole antenna or a loop antenna is called the free space impedance. Hence at the near field boundary, their ratio approaches 377 ohms. Although the exact value may not be obtained in most circumstances, the ratio is relatively close to this value at the near field boundary. The free space impedance can be computed from the ratios of equations (1.1) to (1.2) and the ratio of equations (1.4) and (1.5). This gives the following two expressions for free space impedance as functions of distance from the radiating source:

$$Z_E(r) = \frac{\eta_0 \left| 1 + \frac{1}{j.\beta.r} + \frac{-1}{(\beta.r)^2} \right|}{\left| \left(1 + \frac{1}{j.\beta.r} \right) \right|} \tag{1.9}$$

and

$$Z_H(r) = \left| \frac{\eta_0 \left(1 + \frac{1}{j.\beta.r} \right)}{1 + \frac{1}{j.\beta.r} + \frac{-1}{(\beta.r)^2}} \right| \tag{1.10}$$

In this expression $j = \sqrt{-1}$ and $\eta_0 = 377$. When $\beta.r \ll 1$ in equation (1.9) it reduces to

$$Z_E(r) = \frac{-j.\eta_0}{\beta.r} \tag{1.11}$$

As the range increases in (1.11) the free space impedance approaches 377 ohms and that value of r is the near field boundary. Usually because of the difficulty in finding the exact near field point using this method, a transition region is defined with the following expressions:

$$\text{Near field at } r < 0.1 * \frac{\lambda}{2\pi}$$

$$\text{Transition region } 0.1 * \frac{\lambda}{2\pi} < r < 0.0 * \frac{\lambda}{2\pi}$$

$$\text{Far Field region } r > 0.8 * \frac{\lambda}{2\pi}$$

The exact near field boundary is therefore left to the engineer to pin point and confirm based on the evaluation of these equations. The near field boundary point is therefore a problem that is a bit subjective.

1.1.2.3 Near field boundary from phase front

Antenna designers are interested in the phase front of the wave radiated by the antenna. The starting point is to consider a dipole antenna. From the analyses in the previous sections, the parameters of the antenna play a significant role in the determination of the boundary between the near and far fields.

The parameters are used further to estimate the boundary. Consider Figure 1.1. The length of the antenna is $2z$. The distances to the receiving antenna at point P from the middle of the antenna r' and from the top r are nearly equal if the receiver point is close. The near field point occurs somewhere along the distance r'. Using the cosine rule, the relationship between them is:

$$r' = \sqrt{r^2 + z^2 - 2rz * \cos(\theta)} \tag{1.12}$$

When $r \gg z$, the relationship between r and r' is:

$$r' = \sqrt{r^2 - 2rz * \cos(\theta)} \tag{1.13}$$

Under this condition the receiver is too far away from the transmitting antenna. A second approximation is necessary to reveal the hidden relationships that lead to the near field point. For that we use the binomial expansion method to have

$$r' = r - z * \cos(\theta) + \frac{1}{r} * \left(\frac{z^2 * \sin^2(\theta)}{2} \right) + \frac{1}{r^2} * \left(\frac{z^3 * \cos(\theta) \sin^2(\theta)}{2} \right) + \ldots \tag{1.14}$$

When the angle $\theta = \pi/2$, ie the receiver lies on the y axes on a line perpendicular to the antenna length, the second and the 4th terms in the above

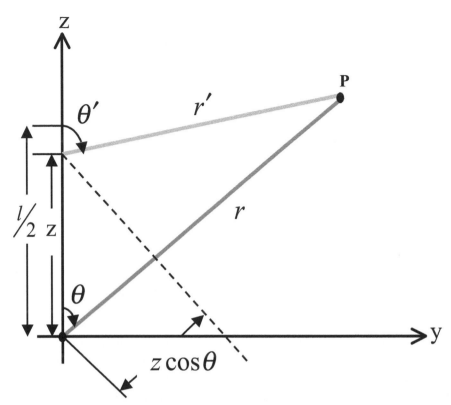

Figure 1.1 Receiving and Transmitting Antennas Close to Each Other [1]

expression reduce to zero and

$$r' = r - \frac{1}{r} * \frac{z^2}{2} \tag{1.15}$$

Hence the path difference is only a maximum of $\frac{z^2}{2r}$. Generally phase differences of up to $\pi/8$ produces acceptable measurement errors. Hence the phase difference must be less than $\pi/8$ or $\frac{\beta z^2}{2r} \leq \pi/8$.

When the angle θ is small, the third and fourth and subsequent terms in equation (1.14) are negligible and the path difference is only $z * \cos(\theta)$. Thus the wave from r arrives first at point P and the wave that took path r' arrives later (lags) and with a phase difference proportional to the path difference and at reduced amplitude. Therefore the last two terms in the wave equation (1.3) are equal and it reduces to

$$E_r = \frac{jI * l * \beta^2}{4\pi \omega \varepsilon_\theta r} \cos(\theta) * e^{-j\beta(r'-z\cos(\theta))} \ V/m \tag{1.16}$$

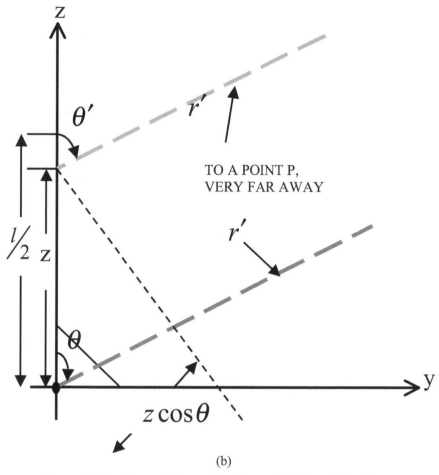

TO A POINT P,
VERY FAR AWAY

(b)

Figure 1.2 Receiving and Transmitting Antennas Far From Each Other

Figure 1.3 shows two transmitting antennas and one receiving antenna. At a far point when r is very large, the big circle has a larger contact circumference with the plane wave front line compared with the smaller circle (smaller wave front). From the diagram

•There are no citation for Figure 1.2

$$(r + \Delta r)^2 = z^2 + r^2$$

Or

$$2r\,\Delta r + (\Delta r)^2 = z^2$$

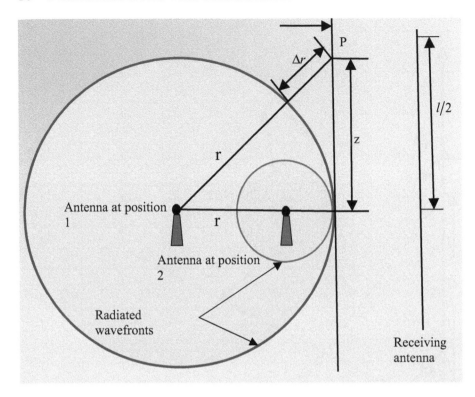

**PLANE
WAVEFRONT**

Figure 1.3 Plane Wave Front at Far Field [1]

and

$$r = \frac{z^2}{2\Delta r} - \frac{\Delta r}{2} \quad \text{and when } (\Delta r \gg z), \Delta r \approx \frac{z^2}{2r}$$

In this expression Δr is the path difference at point P on the wave front and the wave that took the direct route perpendicular to the line of the plane wave front.

As shown in Table 1, the locations of near and far field boundaries are defined in relation also to the application of interest.

$$\frac{\beta z^2}{2r} \leq \pi/8 \quad \text{and } r \approx \frac{8z^2}{\lambda} = \frac{8l^2}{\lambda}$$

Where $z = l$ is assumed in the final approximation. Note this derivation is similar to [1] with corrections for the derivations in [1].

References

[1] Charles Capps, "Near Field or Far Field", EDN, August 2001, pp. 95–102.

2

Near Field Propagation in Free Space

Near field communication is a non contact wireless form of short range communications which use either near field electric fields or near field electromagnetic fields for transporting information. It has applications in near field biomedical monitoring, ranging, RFID, personal area networks and in mobile phones and devices including payment cards. While the technologies behind NFC have been around for decades, their applications and characteristics are now being properly studied. The range of great interest in NFC is over short distances of a couple of meters to about 5 meters. Over such ranges communications can be achieved by having the transmitter and receiver placed close to each other.

Far field channel models have been well studied. There are however fundamental differences between far field channel models and near field channel models. Far field channel models were first proposed by Harald Friis [1] and have become the fundamental guide for the estimation of signal propagation and degradation in a radio frequency channel. Friis law states the relationship between the transmitted and received signal power with distance as a function of the antenna gains. In the far field, the power loss as a function of distance and frequency is given by the expression:

$$P_L(r, f) = \frac{P_R}{P_T} = \frac{G_T G_R \lambda^2}{(4\pi)^2 r^2} = \frac{G_T G_R}{4} \frac{1}{(\beta.r)^2} \qquad (2.1)$$

In this expression, r is the distance between the transmitter and receiver G_T and G_R are the gains of the transmitting and receiving antennas, $\beta = 2\pi/\lambda$ and λ is the wavelength of transmission in meters. Friis expression shows that far field free space power loss is proportional to the inverse power of distance squared. This law does not apply directly to near field communications. The path loss is a description of the relationship between the transmitted power and the received power and is subject to further degradation depending on the existing terrain in the communication.

J. Ihyeh Agbinya (PhD), *Principles of Inductive Near Field Communications for Internet of Things*, 13–20.

The path loss in the near field is proportional to the inverse fourth power of distance $(1/r^4)$ or higher [2, 3]. The consequences of this is that the signal power at the near field is likely going to be higher than expected from the expression for the far field, thereby giving a better signal to noise ratio and better comparative link quality and capacity as well. The higher power roll off in the near field suggests that the range of communication is shorter and hence less prone to interception from long distances and hence relatively more secure. A furthermore advantage of near field communications therefore is that they should interfere less with other radio frequency communication systems around the same spectral range and outside the operating environment as well.

2.1 Link Equations in Near Field Communications

The behaviour of electric and magnetic fields are markedly different in the near field region. Different equations are therefore required for the transmission link. A magnetic antenna (eg. a loop antenna) is necessary for receiving magnetic field signals and an electric antenna (eg. a dipole) is required to receive electric field signals.

2.1.1 Link equations from a dipole antenna

The link equation depicts the received signal power at the receiver antenna and it is proportional to the square of the time average of the electric field intensity, where

$$P_R(E) \propto \langle |E|^2 \rangle \propto \left(\frac{1}{(\beta r)^2} - \frac{1}{(\beta r)^4} + \frac{1}{(\beta r)^6} \right) \qquad (2.2)$$

Therefore the path loss expression at the near field region is

$$P_L(r, f) = \frac{P_R(E)}{P_T} = \frac{G_T G_R(E)}{4} \left(\frac{1}{(\beta r)^2} - \frac{1}{(\beta r)^4} + \frac{1}{(\beta r)^6} \right) \qquad (2.3)$$

The received electric field signal is proportional to the product of the gains of the transmitter and receiver antenna gains. This equation depicts the situation for like antennas for transmission and reception when an electric field receiver antenna receives from an electric field transmitting antenna (electric to electric) and also the received signal with a magnetic field antenna from a transmitting magnetic field antenna (magnetic to magnetic).

2.1.2 Link equations from a magnetic loop antenna

Similar equations can be written for the received magnetic field signals from a co-polarised transmitting antenna. The received magnetic field signal is proportional also to the square of the time average of the magnetic field intensity as follows:

$$P_R(H) \propto \langle |H|^2 \rangle \propto \left(\frac{1}{(\beta r)^2} + \frac{1}{(\beta r)^4} \right) \tag{2.4}$$

Consequently the path loss expression for the link is

$$P_L(r, f) = \frac{P_R(H)}{P_T} = \frac{G_T G_R(H)}{4} \left(\frac{1}{(\beta r)^2} + \frac{1}{(\beta r)^4} \right) \tag{2.5}$$

This equation models the received signal from two unlike antennas (electric to magnetic and magnetic to electric).

Clearly from Figure 2.1, when the range $r < 0.1\lambda$, the received signal decays at a rate of about +60dB/decade. There appears to be a path gain in this situation or the received signal appears to be larger than the transmitted signal. However conservation of energy should apply to the link. Therefore the gain of the receiving antenna cannot be arbitrarily large.

This is a path gain. Above the intermediate range (ie, $r > 0.1\lambda$), the decay in the received signal or the path loss is about -6dB/decade and at large range $(r \gg 0.1\lambda)$ the path loss is about -18dB per decade [2,3].

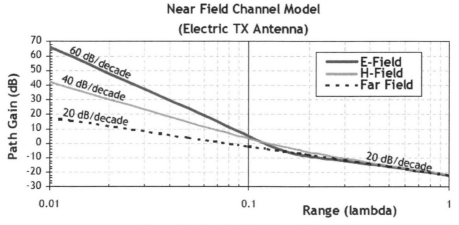

Figure 2.1 Near Field Path Loss [3]

From equation (2.3), when $\beta r = 1$; $r = \frac{\lambda}{2\pi}$ the near field path loss approximates to

$$P_L(r, f) \approx \frac{G_T G_R(H)}{4} \tag{2.6}$$

Hence the path loss for a typical near field channel is approximately -6dB. A normal approximation for matched antennas (electric to electric or magnetic to magnetic) is:

$$P_L = 10 \log_{10}(P_L(r, f)) = -6 + 20 \log_{10}(G) \, dB \tag{2.7}$$

When unlike antennas are used (electric to magnetic, magnetic to electric and $\beta r = 1$; $r = \frac{\lambda}{2\pi}$), the near field path loss approximates to:

$$P_L = 10 \log_{10}(P_L(r, f)) = -3 + 10 \log_{10}(G_T G_R(H)) \, dB \tag{2.8}$$

2.1.3 Limits of antenna sizes versus gain

Bearing in mind that the gain of an antenna cannot be arbitrarily large, what then is the limit? The answer to this question is provided by Wheeler [5] and Chu [6]. They used the concept of a boundary sphere. A boundary sphere is the smallest sphere that encloses an antenna. Thus two antennas placed side by side fits into the bounding sphere and the radius of the sphere R is therefore characteristics of the antenna (Figure 2.2).

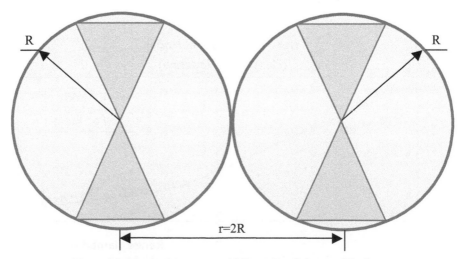

Figure 2.2 Matched Antennas with Boundary Spheres of Radius R

The closest distance from end to end of two matched antennas without overlapping is $2R$ where R is the radius of the boundary sphere. With this as limit, the near field propagation equation provides the means to estimate the limit of antenna gain versus size. For like antennas this is:

$$P_E(r, f) = \frac{P_R(E)}{P_T} \leq 1 \geq \frac{G_T G_R(E)}{4} \left(\frac{1}{(2\beta R)^2} - \frac{1}{(2\beta R)^4} + \frac{1}{(2\beta R)^6} \right)$$
(2.9)

Since $r = 2R$ and for like antennas $G_T = G_R = G$, we have

$$G \leq \sqrt{\frac{4}{\left(\frac{1}{(2\beta R)^2} - \frac{1}{(2\beta R)^4} + \frac{1}{(2\beta R)^6} \right)}} = \frac{2(2\beta R)^3}{\sqrt{1 - (2\beta R)^2 + (2\beta R)^4}}$$

$$= \frac{2(4\pi R_\lambda)^3}{\sqrt{1 - (4\pi R_\lambda)^2 + (4\pi R_\lambda)^4}}$$
(2.10)

In this expression $R_\lambda = R/\lambda$ and the radius is in units of the system wavelength. Figure 2.3 shows that there is a gain in the path loss equation when the boundary sphere has radius in wavelengths greater than one tenth ($R_\lambda \geq 0.1$).

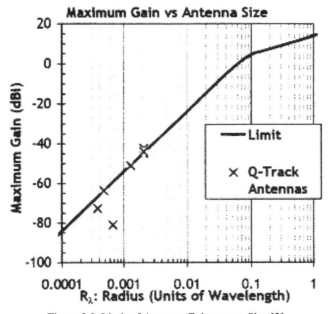

Figure 2.3 Limit of Antenna Gain versus Size [2]

2.1.4 Near field phase change, attenuation and delay spread

The near field phase change when an electric antenna is used for transmission is given in [7]. The electric field phase change is

$$\varphi_E = \frac{180}{\pi}\left\{\beta.r + \left[\cot^{-1}\left(\beta.r - \frac{1}{\beta.r}\right) + n\pi\right]\right\} \tag{2.11}$$

The magnetic field phase change is also given by the expression

$$\varphi_H = \frac{180}{\pi}\{\beta.r + \cot^{-1}(\beta.r) + n\pi\} \tag{2.12}$$

The phase difference between the electric and magnetic waves therefore is expressed as

$$\varphi = \varphi_E - \varphi_H = \frac{180}{\pi}\left\{\left[\cot^{-1}\left(\beta.r - \frac{1}{\beta.r}\right) - \cot^{-1}(\beta.r)\right]\right\} \tag{2.13}$$

This phase difference is a function of range and frequency and not dependent on the antenna parameters. It has a one-to-one mapping to range and therefore suitable for ranging. In fact this idea will be used in our latter chapter on near field ranging.

Free space propagation is only an approximation and in practice the operating environment affects the link quality and behaviour. The environment can lead to attenuation or enhancement of the signal and this change is limited to about 20 dB. Near field indoor propagation is well approximated by the free space model. The major impact of indoor environment is phase distortion caused by multipath delay distortion. The phase distortion is caused by the echo response of objects in the environment. The echo response is largely not sensitive to frequency. Typical delay distortions indoors lie within the 30 to 50 ns range which is consistent with expectations. The worst near field delay distortion measured to date at 1MHz is around 100 ns. Delay spread is a function of distance [8] given by the expression

$$\tau_{RMS} = \tau_0\sqrt{\frac{r}{r_0}} \tag{2.14}$$

In this expression $r_0 = 1m$ is the reference distance and the delay spread parameter $\tau_0 = 5.5$ ns [9]. When the RMS delay spread in the limit is smaller than the period of the signal, the delay spread is given by the expression

$$\varphi_{RMS} = \omega\tau_{RMS} = 2\pi f\tau_{RMS}$$

The phase behaviour of the signal at near field therefore needs a correction. The correction accounts for the delay spread experienced by the signal. For the electric wave the expression of the phase change therefore becomes

$$\varphi_E = \frac{180}{\pi} \left\{ \beta.r + \left[\cot^{-1} \left(\beta.r - \frac{1}{\beta.r} \right) + n\pi \right] \right\} + Norm(0, \varphi_{RMS}) \quad (2.15)$$

And for the magnetic wave, the expression also becomes

$$\varphi_H = \frac{180}{\pi} \{ \beta.r + \cot^{-1}(\beta.r) + n\pi \} + Norm(0, \varphi_{RMS}) \quad (2.16)$$

References

[1] Harald Friis, "A Note on a Simple Transmission Formula," Proc. IRE, 34, 1946, pp. 254–256.

[2] Hans Gregory Schantz, "A Near Field Propagation Law & A Novel Fundamental Limit to Antenna Gain Versus Size", pp. 237–240.

[3] Hans Schantz, Near Field Channel Model, IEEE P802.15-04/0417r2, Oct. 27, 2004, pp. 1–13.

[4] Charles Capps, "Near Field or Far Field", EDN, Aug. 2001, pp. 95–102.

[5] Harold A. Wheeler, "Fundamental Limitations of Small Antennas," Proc. IRE, 35, (1947), pp. 1479–1484.

[6] L.J. Chu, "Physical Limitations of Omni-Directional Antennas," Journal of Applied Physics, 19, Dec. 1948, pp. 1163–1175.

[7] Hans Schantz, "*Near Field Ranging Algorithm*," IEEE802.15-04/0438r0, 17 Aug. 2004.

[8] Kai Siwiak et al, "On the relation between multipath and wave propagation attenuation," Electronic Letters, 9 Jan. 2003 Vol. 39, No. 1, pp. 142–143.

[9] Kai Siwiak, "UWB Channel Model for under 1 GHz," IEEE 802.15-04/505r0, 10 Oct., 2004.

3

Near Field Magnetic Induction Communications

Most of modern telecommunications is due mainly to propagation of electromagnetic waves. Electromagnetic wave is not a single wave but rather a superposition of an electric and a magnetic wave which co-propagate. The two waves propagate together and they support long distance communications. Short range communications however is not wholly dependent on a propagating electromagnetic wave. For this reason near field communications has followed two traditional lines. The first is near field electric field communications, a lot well known. Its techniques and characteristics are used in RFID, short range communications and Bluetooth fall into this category and they rely on conventional transmission of RF signals between two antennas – the transmitter and receiver. The system uses an electric antenna which generates and transmits a propagated electromagnetic wave.

In electromagnetic wave propagation, the electric and magnetic fields are perpendicular to each other and to the direction of propagation (Figure 3.1).

The second approach to short range communications is based on magnetic induction. Near field magnetic induction (NFMI) communications relies on coupling of magnetic energy between two non-contact magnetic antennas. No RF signal propagates between the antennas except that the transmitter and receiver resonate at the same angular frequency ω_0. Magnetic Inductive communication technique is less known but is however generating a lot of interest due to its seemingly secure communication potential and reduced interference with other communications systems at near field. Its ability to support relatively significant system capacity over short distances to enable transport of data between two devices adds to the interest it is generating.

3.1 Real and Imaginary Power

To further understand the distinction between the two forms of short range communications, it is essential to understand the notion of near field and far

J. Ihyeh Agbinya (PhD), Principles of Inductive Near Field Communications for Internet of Things, 21–34.

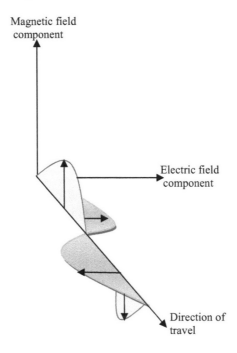

Figure 3.1 Propagating Electromagnetic Electric and Magnetic Fields

field and "real power" and "reactive power". As aptly explained by Schantz [4], "a simple thought experiment involving electromagnetic energy flow establishes why the fields are in quadrature close to an electrically small antenna and in phase far away. The Poynting vector (S = ExH) is the measure of the energy flux around the hypothetical small antenna

If the electric and magnetic fields are phase synchronous, then when one is positive, the other is positive and when one is negative the other is negative. In either case, the Poynting flux is always positive and there is always an outflow of energy. This is the radiation (or "real power") case. If the electric and magnetic fields are in phase quadrature, then half the time the fields have the same sign and half the time the fields have opposite signs. Thus, half the time the Poynting vector is positive and represents outward energy flow and half the time the Poynting vector is negative and represents inward energy flow. This is the reactive (or "imaginary power") case. Thus, fields in phase are associated with far field radiation and fields in quadrature are associated with near field quadrature" [4]. There is in practical terms a "gradual transition from near field phase quadrature to far field phase synchronicity" [4].

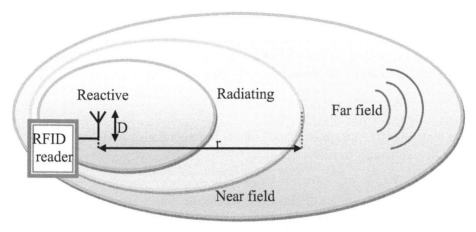

Figure 3.2 Near and Far Fields of an RFIDRFID Reader [2]

3.1.1 Near field and far field regions

The space around an antenna is normally divided into two main regions as illustrated in Figure 3.2: far field and near field. In the far field, electric and magnetic fields propagate outward as an electromagnetic wave and the two fields are perpendicular to each other and to the direction of propagation (Figure 3.1). This is a well known phenomenon in telecommunications and well described in microwave theory and Maxwell equations.

The angular field distribution does not depend on the distance from the antenna. The fields are uniquely related to each other via free-space impedance and decay as $1/r$. In the near field, the field components have different angular and radial dependence (e.g. $1/r3$). The near field region includes two sub-regions: radiating, where the angular field distribution depends on distance and then reactive where the energy is stored but not radiated [2].

Authors continue to differ in the definition of where near field ends and far field begins. From practical terms, there is no unique point where near field ends and far field begins. Near field therefore has to be defined in many circumstances from the application point of view. Conventionally, near field is defined using equation (3.2). This definition is practical and used for stored energy (reactive or inductive) near-field communications when the wavelength of communication is short (ie, high frequency communication) or the size of the antenna is small. For electrically large antennas, the boundary between near field and far field may further be split into two, the radiating near field and the radiating far field regions. In the radiating near field, the angular field distribution depends on distance from the RF source. In the far field it does not. Energy is radiated as well as exchanged between the source and a reactive near field. Using this argument, from [2 and 3] near field edge is defined in terms of the largest dimension of the antenna and is

$$r = 2D^2/\lambda \qquad (3.1)$$

For small electric antennas, many authors have continued to define reactive near field boundary to be at

$$r = \lambda/2\pi \qquad (3.2)$$

This expression is suited to situations where the dimensions of the antenna are much smaller than the wavelength. For electrically large antennas, the reactive near field boundary is better described by the expression

$$r = 0.62\sqrt{D^3/\lambda} \qquad (3.3)$$

D is the largest dimension of the antenna [3]. "Near field magnetic induction system is a short range wireless physical layer that communicates by coupling a tight, low-power, non-propagating magnetic field between devices. The concept is for a transmitter coil in one device to modulate a magnetic field which is measured by means of a receiver coil in another device" [1].

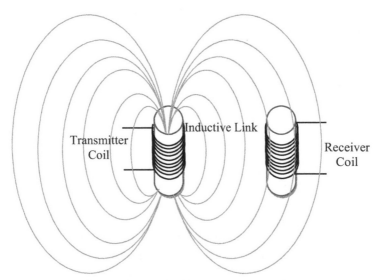

Figure 3.3 Inductive Transmitter and Receiver

3.1.2 Energy coupling in inductive communication

To understand the concept of inductive communications an understanding of the energy coupling principles in inductive networks is essential. While these basic principles may be found in many text books in physics or electronics, a brief summary of the key principles are given in this section. First and foremost we will go back to the basics and provide the relevant expressions without showing how they are derived in most cases.

3.1.2.1 Series resonant circuit

Figure 3.4 represents a series resonant circuit with an operating input signal at a typical NFMI frequency of 13.56 MHz. the total impedance of the circuit is easily found to be

$$Z(jw) = R + j(X_L - X_C)(\Omega) \qquad (3.4)$$
$$X_L = \omega L = 2\pi f L(\Omega)$$
$$and\ X_C = 1/\omega C = 1/2\pi f C(\Omega)$$

X_L and X_C are the inductive and capacitive reactance of the inductor and capacitor respectively and f (Hertz) is frequency. The resistance R represents the sum of the ohmic source resistance of the input, and any parasitic resistances of the inductor and capacitors. The impedance of the circuit is purely resistive when $X_L = X_C$. This is known as the resonant condition of the circuit at which point we have $\omega_0 = 1/\sqrt{LC}$. Therefore by knowing the resonance frequency and choosing one of the components for the circuit, the other can be computed quite easily. An important measure of the circuit

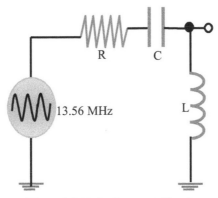

Figure 3.4 Series Resonant Circuit

performance is its half power bandwidth or the so-called -3 dB point which can easily be derived to be given by the expression

$$B = \frac{R}{2\pi L}(Hz) \tag{3.5}$$

Another measure of performance of the circuit is its quality factor, Q. The Q factor is defined by the ratio of the energy stored in the circuit to the energy dissipated by the circuit.

$$Q = \frac{Energy\ stored\ in\ the\ circuit\ per\ cycle}{Energy\ dissipated\ by\ the\ circuit\ per\ cycle} = \frac{Reactance}{Resistance} \tag{3.6}$$

The quality factor refers more to the qualities of the inductors and capacitors in the circuit and pertains to their ability to transfer energy from the source to the load or their ability to dissipate very little energy in the series resistor. Hence

$$Q = \frac{f_0}{B} = \frac{\omega_0 L}{R} = \frac{1}{R\omega_0 C} \tag{3.7}$$

The Q factor has no units. The output voltage measured across the inductor is obtained using a voltage divider principle and

$$V_0 = \frac{jX_L}{R + j(X_L - X_C)} V_{in} \tag{3.8}$$

Optimum voltage is dropped across the inductor at the output at resonance and hence

$$V_0 = \frac{jX_L}{R} V_{in} = jQV_{in} \tag{3.9}$$

For example if the inductor has a $Q = 50$, the output voltage across it is 50 times higher than the input voltage. Hence this implies that while the input voltage was spread across a large spectrum, the output voltage is compressed to be in a very narrow band around where the resonance frequency is.

Example. The resistance at the input in Figure 3.4 is 7.5 ohms and the resonance frequency $f = 13.56$ MHz. If $Q = 50$, what are the values of L and C?

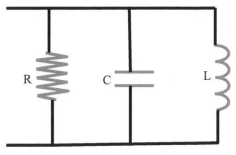

Figure 3.5 Parallel Resonant Circuit

Solution

$$Q = \frac{\omega_0 L}{R} \text{ or } X_L = QR = 50 * 7.5(\Omega) = 375(\Omega)$$

$$L = \frac{X_L}{2\pi f} = \frac{375}{2\pi(13.56 * 10^6)} = 4.401(\mu H)$$

$$C = \frac{1}{2\pi f X_L} = \frac{1}{2\pi(13.56 * 10^6)(375)} = 31.299(pF)$$

3.1.2.2 Parallel resonant circuit

The total impedance of the parallel resonant circuit can be shown to be equal to

$$Z(j\omega) = \frac{j\omega L}{(1 - \omega^2 LC) + \dfrac{jwL}{R}}(\Omega) \qquad (3.10)$$

The circuit resonates when the impedance of the circuit is purely resistive ($Z = R$) or when the denominator of this equation is minimized and that occurs when

$$1 - \omega^2 LC = 0$$

and

$$f_0 = \frac{1}{2\pi\sqrt{LC}}(Hz) \qquad (3.11)$$

This is the resonant condition. The 3 dB bandwidth of the circuit is determined by the parallel L and C tank circuit and is given by

$$B = \frac{1}{2\pi RC}(Hz) \tag{3.12}$$

$$Q = \frac{\text{Reactance}}{\text{Resistance}} = \frac{\omega L}{R} = \frac{1}{\omega RC} = \frac{f_0}{B} \tag{3.13}$$

Or

$$Q = R\sqrt{\frac{C}{L}} \tag{3.14}$$

Thus the Q factor for this circuit is proportional to the product of its resistance and the square root of the ratio of its capacitive and inductive values.

3.1.3 Principles of power transfer

To appreciate NFMI communications, understanding the basic principles of power transfer using inductors is essential. Consider two inductors placed side by side. One has an input voltage (transmitter) which excites the inductor and the second (receiver) merely receives some current from the transmitter. This principle is similar to heat transfer in which we have a source of heat and a second body merely absorbs the heat from the other body. The energy coupling is described as weak because only a small proportion of the magnetic field produced actually is intercepted by the receiver antenna.

The coupling coefficient k used to qualitatively estimate the efficiency of energy transfer from the transmitter coil to the receiver coil is given by the expression

$$k = \frac{M}{\sqrt{L_1 L_2}} \tag{3.15}$$

In this expression M is the mutual inductance and L_1 and L_2 are the self inductances of the transmitter and receiver coils respectively.

Kirchoff's voltage law is applied to the two inductive circuits in this figure resulting to the equations

$$V_1 = j\omega L_1 I_1 + j\omega M I_2$$
$$V_2 = j\omega M I_1 + j\omega L_2 I_2 \tag{3.16}$$

The power coupling is seen in the transmitter as a product of the mutual inductance M and the current I_2 flowing in the receiver and at the receiver,

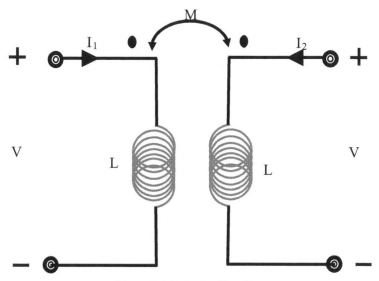

Figure 3.6 Inductive Coupling

power coupling is due to the current I_1 flowing in the transmitter being coupled through the mutual inductance as well.

Let the transmitter and receiver be tuned by inserting capacitors and resistors in their circuits. In practice the inductors in the transmitter and receiver will have inherent DC resistances which are augmented by the inserted resistors.

The resulting voltage equations become

$$V_1 = [R_1 + j\omega L_1 + (1/j\omega C_1)]I_1 + j\omega M I_2$$
$$0 = j\omega M I_1 + [R_2 + j\omega L_2 + (1/j\omega C_2)]I_2 \qquad (3.17)$$

Since there is no voltage source in the receiver, the sum of its voltages is zero and they result from induced current I_2 and the mutual inductance M coupling the transmitter. The induced coupling voltages are small compared to the other voltages and in the transmitter and receiver. Also, we will assume that the two circuits are at resonance making the imaginary parts of the equations zero. Hence

$$I_1 = \frac{V_1}{R_1} \quad \text{and} \quad P_1 = \frac{0.5 * |V_1|^2}{R_1} \qquad (3.18)$$

Knowing the value of the current I_1 makes it easy to solve the simultaneous equation so that

$$I_2 = \frac{-j\omega M I_1}{R_2} = \frac{-j\omega M V_1}{R_1 R_2} \qquad (3.19)$$

Therefore the input power P_1 can now be given in terms of the parameters of the transmitter and receiver coils as

$$P_2 = \frac{\omega^2 M^2 |V_1|^2 R_2}{2(R_1 R_2)^2} \qquad (3.20)$$

At the receiver, the inductive voltage is really the only source of power to it, therefore we keep that component in our analysis to obtain the power ratio

$$\frac{P_2}{P_1} = \frac{\omega^2 M^2 R_1 R_2}{(R_1 R_2)^2} = \frac{\omega^2 k^2 L_1 L_2}{R_1 R_2} = k^2 Q_1 Q_2 \qquad (3.21)$$

This equation is the power coupling equation desired and is fundamental to NFMI communications. It shows that the coupled power can be estimated if we know the Q factors of the transmitter and receiver coils and the coupling coefficient k. It also provides insight into how we could estimate the capacity of such systems using Shannon's equation. The expression however fails to show how we could compute the coupling coefficient k. In the subsequent sections this expression is developed further to include the efficiencies of the coil and the coupling coefficient as a function of distance.

3.1.3.1 Power transfer with inductive antennas

Inductive coils behave like antennas by transferring magnetic energy between themselves. The power transfer between two inductive coils is a function of their dimensions, the distance between the coils, their resonance frequency and their orientation with respect to each other. Understanding the concept of mutual inductance and the role of coupling coefficients is central to a practical understanding of energy transfer in inductive communications.

When two coils are brought together either wound on top of each other or individually would and isolated but in close proximity to each other, varying current in one of the coils induces a voltage (current) in another. This is due to mutual inductance. In this section the inductive energy characteristics of different types of coils arranged or misaligned with respect to each other in various ways are summarized. Since it is impossible to be exhaustive in this summary, only the most prevalent arrangements in current literature and the ones that have been used widely in certain applications are summarised.

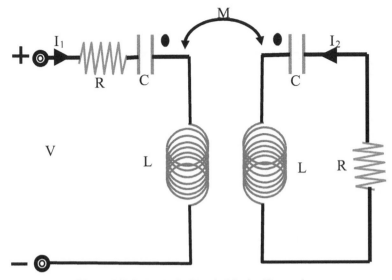

Figure 3.7 Inductively Coupled Series Transceiver

3.1.3.2 Two parallel coils centered on a single axis [5]

The two coils in this analysis are arranged in parallel and separated from each other by a distance x with radii r_1 and r_2 respectively and $r_2 \leq r_1$. The two coils are modelled with lumped circuits as in Figure 3.7.

The coupling coefficient as a function of distance x has been given in [5] as

$$k(x) = \frac{\pi^2 r_1^2 r_2^2}{\sqrt{r_1 r_2}\left(\sqrt{x^2 + r_1^2}\right)^3} \tag{3.22}$$

Maximum energy is transferred between the coils when $k(x) = 1$. In Figure 3.9, R_{L1} and R_{L2} are the ohmic resistances of the transmitter and receiver coils respectively. The capacitances attached to them to ensure series resonances are given by C_1 and C_2 respectively. Therefore the resonance

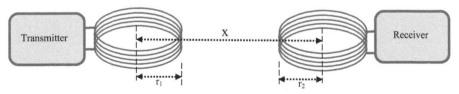

Figure 3.8 Two Parallel coils centred on an axis [5]

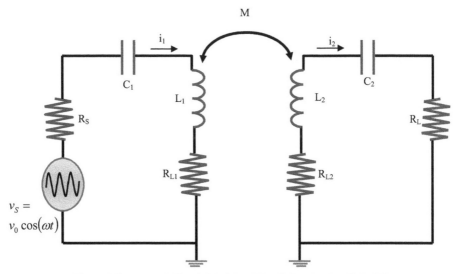

Figure 3.9 Lumped Circuit Models of Parallel Inductive Coils [5]

frequency is $\omega_0 = \frac{1}{\sqrt{L_1 C_1}} = \frac{1}{\sqrt{L_2 C_2}}$. The efficiencies of the coils and their quality factors at resonance are therefore

$$\eta_1 = \frac{R_S}{R_{L1} + R_S}; \quad \eta_2 = \frac{R_L}{R_{L2} + R_L} \tag{3.23}$$

and

$$Q_1 = \frac{\omega_0 L_1}{R_{L1} + R_S}; \quad Q_2 = \frac{\omega_0 L_2}{R_{L2} + R_L} \tag{3.24}$$

The currents near the resonant frequency are

$$i_1(\omega) = \frac{v_0}{(R_{L1} + R_S)\left(1 + j\dfrac{2Q_1 \Delta \omega}{\omega_0}\right)} \tag{3.25}$$

$$i_2(\omega) = \frac{j\omega k \sqrt{L_1 L_2} i_1}{(R_{L2} + R_L)\left(1 + j\dfrac{2Q_2 \Delta \omega}{\omega_0}\right)} \tag{3.26}$$

and $\Delta\omega = \omega - \omega_0$. Therefore the power delivered to the load at the receiver is

$$P(\omega)_L = \frac{|i_2|^2 R_L}{2} = \frac{P_S Q_1 Q_2 \eta_1 \eta_2 k^2}{\left(1 + Q_1^2 \frac{(2\Delta\omega)^2}{\omega_0^2}\right)\left(1 + Q_2^2 \frac{(2\Delta\omega)^2}{\omega_0^2}\right)} \tag{3.27}$$

The power delivered to the load is proportional to the square of the distance based coupling factor $k(x)$, where: $P_S = \frac{1}{2}\frac{v_0^2}{R_S}$.

The link efficiency in a coupled system is defined as the product of the transmitter and receiver coil efficiencies. This is

$$\eta = \eta_1 \eta_2 = \frac{R_S}{R_{L1} + R_S} * \frac{R_L}{R_{L2} + R_L} \tag{3.28}$$

Near resonance if $R_S \gg R_{L1}$ and $R_L \gg R_{L2}$; $\eta \approx 1$

References

[1] Freelinc, "FreeLinc Near-Field Magnetic Induction Technology", pp. 1–5.
[2] Pavel V. Nikitin, K. V. S. Rao and Steve Lazar, "An Overview of Near Field UHF RFID", in Proc. IEEE International Conference on RFID, Gaylord Texan Resort, Grapevine, TX, USA, March 26–28, 2007, pp. 167–174.
[3] T. Lecklider, "The world of the near field", Evaluation Engineering, October 2005, available at http://archive.evaluationengineering.com/archive/articles/1005/1005the_world.asp
[4] Hans Gregory Schantz, "Near Field Phase Behaviour", 2005, pp. 134–137.
[5] Hengzhen C. Jiang and Yuanxun E. Wang, "Capacity Performance of an Inductively Coupled Near Field Communication System", in Proc. IEEE International Symposium of Antenna and Propagation Society, July 5–11, 2008, pp. 1–4.

4

Capacity of Near Field Magnetic Induction Communications

Near field magnetic induction (NFMI) communication as opposed to near field radiative communication has recently started to gain grounds as a means for providing future medium capacity personal area networks [1, 2] and close proximity condition monitoring.

4.1 Near Field Magnetically Coupled MIMO Systems

The capacity of near field magnetically coupled single input single output communication system in free space is analyzed in this chapter based on the equivalent circuit model estimated. A proof has been given for the radial dependence of power losses as a function of distance to power 6. The extent of a NFMI communication a crucial measure of its security performance is measured as we also propose and define three bubble factors which provide evidence of the extent of the magnetic communication bubble for required receiver sensitivity. The capacity of near field magnetically coupled MIMO communication system in free space is analyzed based on the equivalent circuit model estimated for the two by two Alamouti scheme. The analysis provides a basis for assessing the general inductively coupled MIMO communication system.

4.1.1 Capacity of multiple antenna near field magnetic induction communications

The capacity performance of a single input single output inductively coupled communication system was recently discussed [3] and related to the Q-factors of the inductors in the application. The block diagram of a single input single output (SISO) magnetically inductive communication system is shown in Figure 4.1. The transmitter and receiver are composed of two coils

J. Ihyeh Agbinya (PhD), Principles of Inductive Near Field Communications for Internet of Things, 35–47.
© 2011 *River Publishers. All rights reserved.*

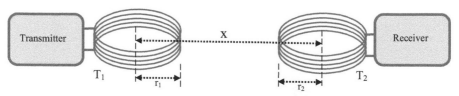

Figure 4.1 Inductively Coupled Near Field System

Figure 4.2 System Implementation

each of radius r_1 and r_2 and separated from each other by a distance x. the communications link between them is an inductive coupling k where based on the lump circuit model of Figure 4.3, the coupling between the two magnetic antennas can be estimated using the equation

$$k = \frac{M}{\sqrt{L_1 L_2}} \qquad (4.1)$$

And M is the mutual coupling between the two inductive circuits and L_1 and L_2 are the inductive values of the transmitter and receiver coils respectively.

The coupling coefficient between the transmitter and receiver as a function of distance provides an estimate of the energy transfer at each point along the separation between the transmitter and receiver. This is derived with a few approximations in this section. We start this by using the expression for the coupling volume of a single turn coil given by the expression:

$$V_C = \frac{\mu_0 A^2}{L} \qquad (4.2)$$

In this expression A is the area of the coil that really collects some of the magnetic flux created by the source and L is the self-inductance of the coil. Antennas create power which spreads or propagates within a volume of space. Given this we also require to estimate the reactive power density per unit volume created at the receiver site. The reactive power density at the receiver is:

$$V_D = \frac{\textit{Reactive power flowing in transmitter coil}}{\textit{Volume density of reactive power at the receiver created by transmitter}} \qquad (4.3)$$

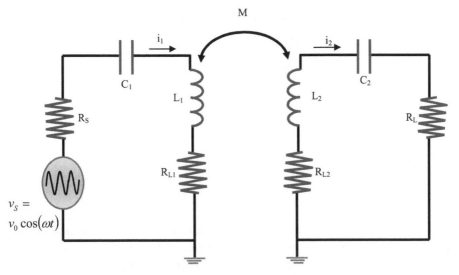

Figure 4.3 Equivalent Circuit of a Pair of Antennas [3]

From Figure 4.3, the following relationships can be established for the currents flowing in the transmitter and the inductive current in the receiver. Let $\Delta\omega = \omega - \omega_0$ and

$$i_1(\omega) = \frac{v_0}{(R_{L1} + R_S)\left(1 + j\dfrac{2Q_1\Delta\omega}{\omega_0}\right)} \qquad (4.4)$$

$$i_2(\omega) = \frac{j\omega k\sqrt{L_1 L_2}\, i_1}{(R_{L2} + R_L)\left(1 + j\dfrac{2Q_2\Delta\omega}{\omega_0}\right)} \qquad (4.5)$$

The efficiencies of the coils are also given to describe how effective they are in transferring power and by definition

$$\eta_1 = \frac{R_S}{R_{L1} + R_S}; \quad \eta_2 = \frac{R_L}{R_{L2} + R_L} \qquad (4.6)$$

and the quality factors are also given by definition to be

$$Q_1 = \frac{\omega_0 L_1}{R_{L1} + R_S}; \quad Q_2 = \frac{\omega_0 L_2}{R_{L2} + R_L} \qquad (4.7)$$

The power delivered to the receiver load is given by the expression:

$$\frac{P_L(\omega)}{P_S} = \frac{V_C}{V_D}\eta_1\eta_2 Q_1 Q_2 = \eta_1\eta_2 Q_1 Q_2 k^2(x) \qquad (4.8)$$

$$k^2(x) = \frac{V_C}{V_D}$$

Where

$$P_L(\omega) = \frac{|i_2|^2 R_L}{2} = \frac{P_S Q_1 Q_2 \eta_1 \eta_2 k^2}{\left(1 + Q_1^2\frac{(2\Delta\omega)^2}{\omega_0^2}\right)\left(1 + Q_2^2\frac{(2\Delta\omega)^2}{\omega_0^2}\right)} \qquad (4.9)$$

To complete the proof for $k(x)$ we need to illustrate the physical arrangements and show distances and the engaging magnetic fields as in Figure 4.4.

The magnetic field at point x along the axis of a single turn coil of diameter $D = 2r$ carrying a peak phasor current I_1 is given by the expression

$$H_z(0, 0, x) = \frac{I_1 r^2}{2[(r^2 + x^2)]^{\frac{3}{2}}} \qquad (4.10)$$

The reactive power density is also given by the expression

$$P_r = \frac{\omega\mu_0|H|^2}{2}(VAm^{-3}) \qquad (4.11)$$

Hence

$$P_r = \frac{0.5\omega\mu_0 I_1^2 r^4}{4[(r^2 + x^2)]^3}(VAm^{-3}) \qquad (4.12)$$

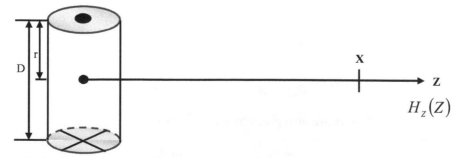

Figure 4.4 Field Measurement Geometry

Since the reactive power from the source magnetic coil is given by

$$P_S = \frac{|I_1|^2 \omega L_1}{2} \tag{4.13}$$

The reactive power at the load becomes

$$k^2(x) = \frac{V_C}{V_D} = \frac{\mu_0 A_2^2 \mu_0 r_1^4}{4 L_1 L_2 [(r^2 + x^2)]^3} \tag{4.14}$$

The self inductances themselves can be estimated with the expressions

$$L = \frac{\mu_0 \pi r^2 N^2}{l + 0.9r} = \frac{\mu_0 \pi r^2}{l + 0.9r}; \quad N = 1 \tag{4.15}$$

We make one further approximation to reign in the formula for mutual coupling. We assume that the radius of the coil is far less than the length: $r \ll l$ and $l = 2\pi r$.

Hence

$$L_1 \approx \frac{\mu_0 r_1}{2}; \quad \text{and } L_2 \approx \frac{\mu_0 r_2}{2}; \quad A_2 = \pi \cdot r_2^2 \quad N = 1 \tag{4.16}$$

Finally we obtain the expression for the coupling coefficient to be

$$k^2(x) = \frac{\pi^2 r_1^3 r_2^3}{(x^2 + r_1^2)^3} \tag{4.17}$$

The ratio of the two values is a metric of performance for the coils. The

$$k(x) = \frac{\pi \cdot r_1^2 r_2^2}{\sqrt{r_1 r_2} \left(\sqrt{x^2 + r_1^2} \right)^3} \tag{4.18}$$

Where $r_2 \leq r_1$. From the equivalent circuit model of Figure 4.3, the following relationship is assumed between the antennas that $\omega_0 = \frac{1}{\sqrt{L_1 C_1}} = \frac{1}{\sqrt{L_2 C_2}}$. Hence the two coils are chosen such that they resonate at the same frequency and hence permit optimum mutual communication to be established between them.

4.1.2 Definition of bubble factors

NFMI communication is promising in its ability to create a so-called secure communication 'bubble'. However, current literature is devoid of the characteristics, the size and extent of the magnetic bubble. By inference, most authors appear to assume that the size of the NFMI bubble is the same as the edge of near field. The two are not the same. The signal level at the near field edge is however still too high and easily available for interception. In this section we define the size of the magnetic communication bubble.

4.1.3 Size of NFMI bubble

Intuitively, we define the magnetic bubble in terms of the sensitivity of the receiver. Let d be the distance at which the received signal power is equal to the sensitivity of the receiver $P_r = P_S$, where P_S is the sensitivity of the receiver. The size of the magnetic bubble is defined as the distance d where the received power is just equal to the sensitivity of the receiver. With this definition, the size or the extent of the bubble is not fixed but rather a function of the capability of the receiver. If the receiver is highly sensitive, the bubble it sees has a large radius. We may also define the size of the bubble in terms of the signal to noise ratio (SNR) of the system. With this abstraction, we define the radius of the magnetic bubble as the distance where the received signal power is just equal to the noise power ($P_r = N$ or SNR $= 1$. Therefore the system capacity at the edge of the NFMI bubble is given by

$$C = B_f f_0 \log_2 \left(1 + \frac{P_{rd}(\omega = \omega_0)}{N} \right) = B_f f_0 \log_2 2 = B_f f_0 \qquad (4.19)$$

At the edge of the bubble, no signal amplification will help in detecting the signal as noise is also amplified equivalently so the bubble remains secure and silent to someone outside it. Bearing in mind that the 3dB fractional bandwidth B is defined purely by the Q of the coils and the centre frequency, where

$$B_f = \frac{B}{f_0} = \frac{\sqrt{-(Q_1^2 + Q_2^2)^2 + \sqrt{(Q_1^2 + Q_2^2)^2 + 4Q_1^2 Q_2^2}}}{\sqrt{2}Q_1 Q_2} \qquad (4.20)$$

By letting $Q_1 = Q_2 = Q$, this expression reduces to $B_f = \frac{0.644}{Q}$ and the capacity at the edge of the bubble

$$C = B_f f_0 = \frac{0.644 f_0}{Q} \qquad (4.21)$$

is determined exclusively by the Q-factors of the coils and the resonance centre frequency. For a resonance frequency of 13.56 MHz and Q $= 40$, this capacity is approximately 218 Kbps. The capacity at the edge of the bubble is directly proportional to the resonant frequency. The higher Q is, the lower the capacity at the edge of the bubble. Thus a high transmitting Q or a high receiving Q do not automatically lead to high capacity. This paradox is the major benefit of NFMI.

4.1.4 Bubble factors

The performance of NFMI communication systems should be evaluated in terms of the efficiency of the communication bubble. This section therefore introduces new terms called bubble factors which can be used to assess the performance of the transmitter with respect to keeping its communication within a required bubble size and extent. We propose three bubble factors which are required to estimate the extent of the communication bubble in body area networks and also the level of inherent security and the degree of difficulty for interception of a communication using near field magnetic induction.

The power transferred to the receiver load resistance R_L in NFMI communication (Figure 4.1) is proportional to the transmitted power, the quality factors of the coils, the coupling coefficient and their efficiencies. When the radius of the transmitting coil is far smaller than the distance of coverage, we can approximate the received power at x by the expression:

$$P_r(\omega = \omega_0) = P_t Q_1 Q_2 \eta_1 \eta_2 \frac{\pi^2 r_1^3 r_2^3}{(x^2 + r_1^2)^3} \cong \frac{\sigma}{x^6}; \quad r_1 \ll x \qquad (4.22)$$

From this expression we observe that the received power at any point in space is a decreasing function of distance to power six (6) provided the radius of the transmitting coil in the transmitter is far smaller than the distance at which power is measured away from it ($r_1 \ll x$). We define σ as the ***distance bubble factor*** for near field magnetic induction communications.

$$\sigma = \pi^2 P_t Q_1 Q_2 \eta_1 \eta_2 r_1^3 r_2^3 \qquad (4.23)$$

We define also a second bubble factor, called the ***resonance bubble factor***. We substitute for Q in the previous equation. Coupling of energy from the transmitter to the receiver is optimum at the resonant frequency. Hence the performance of the system as a function of the resonant frequency of the transmitting and receiving coils is of interest. From equations (4.22) and

(4.23) and Figure 4.3 we can show that:

$$P_r(\omega = \omega_0) \cong \frac{\mu \omega_0^2}{x^6}; \text{ where} \tag{4.24}$$

$$\mu = \frac{\pi^2 P_t L_1 L_2 \eta_1 \eta_2 r_1^3 r_2^3}{(R_{L1} + R_S)(R_{L2} + R_L)}; \; r_1 \ll x$$

We define μ as the **resonance bubble factor** for near field magnetic induction communications. Received signal power at any distance is high if the resonant frequency is high. Hence for small magnetic "bubble" communications, the resonant frequencies of the coils should be chosen to be small and appropriate. Coils with small radii, small efficiencies, low inductances, high source resistance and high load resistance are most suitable for this purpose.

A third bubble factor ε is also defined and called the **receiver load bubble factor**. We introduce a further approximation that the load resistance in the receiver is much greater than the self resistance of receiver inductor $((R_{L2} \ll R_L)$ to obtain the expression:

$$P_r(\omega = \omega_0) \cong \frac{\varepsilon \omega_0^2}{R_L x^6}; \text{ where} \tag{4.25}$$

$$\varepsilon = \frac{\pi^2 P_t L_1 L_2 \eta_1 \eta_2 r_1^3 r_2^3}{(R_{L1} + R_S)}; \; R_{L2} r_1 \ll R_L$$

The receiver load bubble factor shows that high source and load resistances lead to smaller bubbles. Hence when an INTERCEPTOR tries to maximize reception by increasing a receiver load resistance, there is no obvious advantages in doing so. This expression also shows that we can make the communication system directly proportional to the resonant frequency and distance by equating the receiver load bubble factor to the receiver resistance, or set:

$$P_r(\omega = \omega_0) \cong \frac{\omega_0^2}{x^6}; \text{ when} \tag{4.26}$$

$$\frac{\varepsilon}{R_L} = 1 \Rightarrow \frac{\pi^2 P_t L_1 L_2 \eta_1 \eta_2 r_1^3 r_2^3}{(R_{L1} + R_S)} = R_L; \; R_{L2} r_1 \ll R_L$$

This is an essential system design expression for selection of system components or parameters.

4.1.5 Inductively coupled SIMO system

The capacity of a SISO system (Figure 4.1) is limited by the number of channels. To take advantage of the multiple paths which inductive links provide, in this section we propose and analyse the capacities of multiple input multiple output systems starting with a single input multiple output near field communications (Figure 4.4).

4.1.6 NFMI SIMO system

A SIMO communication system normally consists of multiple receiving antennas and a single transmitting antenna as in Figure 4.5.

In Figure 4.5 we assume that the receiver R_2 is at the boresight of the transmitter and R_3 is at an angle θ, so that the distance d is

$$P_{r1}(\omega = \omega_0) = P_t Q_1 Q_2 \eta_1 \eta_2 k_1^2(x) \tag{4.27}$$

Where $k_1(x)$ is as in the SISO system.

$$P_{r2}(\omega = \omega_0) = P_t Q_1 Q_2 \eta_1 \eta_3 k_2^2(x) \tag{4.28}$$

and

$$k_2(x) = \frac{\pi \cdot r_1^2 r_3^2}{\sqrt{r_1 r_3} \left(\sqrt{(x/\cos\theta)^2 + r_1^2} \right)^3} \tag{4.29}$$

This is the general expression for the coupling coefficient when many receiving antennas are deployed. At boresight, the angle is zero and we have the first receiving antenna.

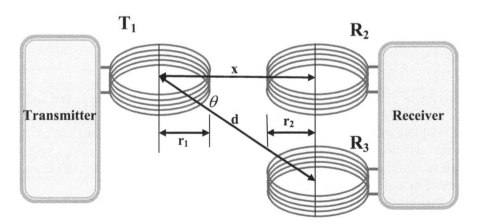

Figure 4.5 NFMI SIMO System

With two receiving antennas the total induced power is given by the expression:

$$P_r(\omega = \omega_0) = P_{r1}(\omega = \omega_0) + P_{r2}(\omega = \omega_0) = P_t Q_1 Q_2 \eta_1 \eta_2 k_1^2(x) \quad (4.30)$$
$$+ P_t Q_1 Q_3 \eta_1 \eta_3 k_2^2(x) = P_t Q_1 \eta_1$$
$$\times (Q_2 \eta_2 k_1^2(x) + Q_3 \eta_3 k_2^2(x))$$

Assume that the two receiving antennas resonate with the transmitter at w_0 and have identical Q so that $Q_1 = Q_2 = Q_3 = Q$; We have previously given the expression for Q_2 and

$$Q_3 = \frac{\omega_0 L_3}{R_{L3} + R_S}$$

Therefore

$$P_r(\omega = \omega_0) = P_t Q^2 \eta_1 \left(\eta_2 k_1^2(x) + \eta_3 k_2^2(x) \right) \quad (4.31)$$

For the general case when the radii of the receiving antennas are not equal and less than that of the transmitting antenna we have $r_3, r_2 \leq r_1$ and

$$P_r(\omega = \omega_0) \cong \frac{\pi^2 P_t Q^2 \eta_1 r_1^3}{x^6} \left(\eta_2 r_2^3 + \eta_3 r_3^3 \cos^6 \theta \right); \quad r_1^2/x^2 \ll 1 \quad (4.32)$$

and

$$r_1^2 \cos^2 \theta / x^2 \ll 1$$

When the receiving antennas have equal radius $r_2 = r_3$

$$P_r(\omega = \omega_0) \cong \frac{\pi^2 P_t Q^2 \eta_1 r_1^3 r_2^3}{x^6} (\eta_2 + \eta_3 \cos^6 \theta); \quad r_1^2/x^2 \ll 1 \quad (4.33)$$

and

$$r_1^2 \cos^2 \theta / x^2 \ll 1$$

Thus the receive power is higher, decreases with distance as a power of 6 and is a function of the efficiencies of the inductors of the receivers. To maximize the coupling in the second non-boresight receiver, we must maximize its efficiency and also locate it at a small angle off the boresight. This result comes with an inherent understanding that the two receivers are well isolated from each other and there is no magnetic coupling between them.

This result is significant for two reasons. Firstly it provides a basis for intercepting NFMI communications based on the following fact. Clearly, the *distance bubble factor* is larger and is a function of the efficiencies of the receiver coils and their displacement angles from boresight. Furthermore, the security of the transmission is reduced when an interceptor deploys an array of receivers. An interceptor could deploy an array of receivers closely spaced from each other with high inductive efficiency and large radii individually. The received power from an array of N inductive sensors therefore would be

$$P_r(\omega = \omega_0) \cong \frac{\pi^2 P_t Q^2 \eta_1 r_1^3 r_2^3}{x^6} \left(\eta_2 + \frac{1}{r_2^3} \sum_{k=1}^{N} r_k^3 \eta_k \cos^6 \theta_k \right); \quad (4.34)$$

$$r_1^2/x^2 \ll 1; r_k^2 \cos^2 \theta_k / x^2 \ll 1$$

For example, if the radius of each of the inductive coils is N times as large as that of the receiver at the boresight, the received power is maximized to

$$P_r(\omega = \omega_0) \cong \frac{\pi^2 P_t Q^2 \eta_1 r_1^3 r_2^3}{x^6} \left(\eta_2 + N^3 \sum_{k=1}^{N} \eta_k \cos^6 \theta_k \right); \quad (4.35)$$

$$r_1^2/x^2 \ll 1; r_k^2 \cos^2 \theta_k / x^2 \ll 1$$

This received power can be very significant at larger ranges if a large array is deployed.

Secondly in commercial non-military applications, the objective is neither the security of the personal area network nor the extent of the bubble, but the capacity of the system is essential and crucial. The capacity of a 1×2 NFMI SIMO system therefore can be estimated and is:

$$C = B_f f_0 \log_2 \left(1 + \frac{P_r(\omega_0)}{N} \right) \quad (4.36)$$

The capacity by deploying one more receiver is given by the expression:

$$C_s = \frac{P_{rs}(\omega = \omega_0)}{P_r(\omega = \omega_0)} = \frac{\eta_2 k_1^2(x) + \eta_3 k_2^2(x)}{\eta_2 k_1^2(x)} \quad (4.37)$$

Therefore

$$C_s = 1 + \frac{\eta_3 \cos^6 \theta \left(1 + \dfrac{r_1^2}{x^2} \right)^3}{\eta_2 \left(1 + \dfrac{r_1^2 \cos^2 \theta}{x^2} \right)^3} \quad (4.38)$$

The capacity change when one more receiver is deployed is given by the expression:

$$C_\Delta \cong B_f f_0 \log_2 \left[\frac{(\eta_3 \cos^6 \theta) \left(1 + \dfrac{r_1^2}{x^2} \right)^3}{\eta_2 \left(1 + \dfrac{r_1^2 \cos^2 \theta}{x^2} \right)^3} \right]; \qquad (4.39)$$

$$r_1^2/x^2 \ll 1; r_2 = r_3; r_1^2 \cos^2 \theta / x^2 \ll 1$$

By making the radius of the receiving antenna at off boresight larger than the one directly opposite the transmitter, capacity increase can be improved by:

$$C_\Delta \cong B_f f_0 \log_2 \left[\frac{r_3^3 (\eta_3 \cos^6 \theta) \left(1 + \dfrac{r_1^2}{x^2} \right)^3}{r_2^3 \eta_2 \left(1 + \dfrac{r_1^2 \cos^2 \theta}{x^2} \right)^3} \right]; \qquad (4.40)$$

$$r_1^2/x^2 \ll 1; \; r_2 < r_3; r_1^2 \cos^2 \theta / x^2 \ll 1$$

Clearly, the increase in capacity is a function of the distance x, the radii of the coils, the efficiencies of the coils and the angular displacement of the receivers. By trading off these variables higher capacity increases can be achieved. For example, by making $r_3 \gg r_2$ and/or making $\eta_3 \gg \eta_2$ this change can be significant. It is evident from the equation that small angular displacements are preferred for higher capacity increases.

References

[1] Christine Evans-Pughe, "Close encounters of the magnetic kind", IEE Review, May 2005, pp. 38–42.

[2] Rajeev Bansal, "Near Field Magnetic Communications", IEEE Antennas and Propagation Magazine, Vol. 46, No. 2, Apr. 2004, pp. 114–115.

[3] Hengzhen C. Jiang and Yuanxun E. Wang, "Capacity Performance of an Inductively Coupled Near Field Communication System", in Proc. IEEE International Symposium of Antenna and Propagation Society, Jul. 5–11, 2008, pp. 1–4.

5

Near Field Magnetic Induction MISO Communication Systems

As in the previous chapter, this chapter demonstrates the concept of Near field magnetic induction (NFMI) multiple-input single-output (NFMI-MISO) communication system.

5.1 Near Field Magnetically Coupled MISO Systems

A NFMI MISO system consists of multiple antennas at the transmitter and a single receiver. The system provides transmission diversity by using more than one transmitter as sources of energy coupling to the receiver.

We can showthat the coupling coefficient between the transmitter T_1 and Receiver R_1 and between the transmitter T_2 and also R_1 are given by the following expressions. The derivation of these equations follow the approach used in chapter 4.

$$P_{R1}(\omega = \omega_0) = P_t Q_r Q_1 \eta_1 \eta_r k_1^2(x) \tag{5.1}$$

Where $k_1(x)$ is as in the SISO system. The received power from transmitter T_2 at R_1 is also given by the expression:

$$P_{rd}(\omega = \omega_0) = P_t Q_r Q_2 \eta_r \eta_2 k_d^2(x) \tag{5.2}$$

The diagonal coupling coefficient between T_2 and R_1 is given by the expression:

$$k_d(x) = \frac{r_2^2 r_r^2}{\sqrt{r_2 r_r} \left(\sqrt{(x/\cos\theta)^2 + r_2^2} \right)^3} \tag{5.3}$$

With a single receiving antenna the total induced power from the two transmitters assuming that the receiver is able to distinguish between the two

J. Ihyeh Agbinya (PhD), Principles of Inductive Near Field Communications for Internet of Things, 49–64.

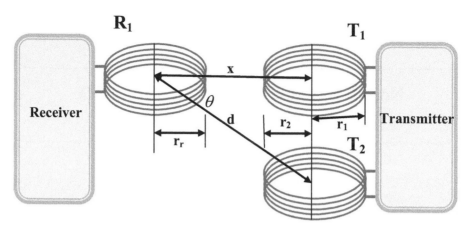

Figure 5.1 NFMI MISO System

transmissions is given by the expression:

$$
\begin{aligned}
P_{rd}(\omega = \omega_0) &= P_{r1}(\omega = \omega_0) + P_{rd}(\omega = \omega_0) \\
&= P_t Q_1 Q_r \eta_1 \eta_r k_1^2(x) + P_t Q_2 Q_r \eta_r \eta_2 k_d^2(x) \qquad (5.4) \\
&= P_t Q_r \eta_r \left(Q_2 \eta_2 k_d^2(x) + Q_1 \eta_1 k_1^2(x) \right)
\end{aligned}
$$

Assume that the receiving antenna resonates with the transmitters at w_0 and they have identical Q so that $Q_1 = Q_2 = Q_r = Q$; We have previously given the expression for Q_2 and

$$
Q_r = \frac{\omega_0 L_r}{R_{Lr} + R_L}
$$

Therefore

$$
P_{rd}(\omega = \omega_0) = P_t Q^2 \eta_r \left(\eta_1 k_1^2(x) + \eta_2 k_d^2(x) \right) \qquad (5.5)
$$

For the general case when the radii of the receiving antenna is less than those of the transmitting antennas $r_1,\ r_2 \geq r_r$ and $r_1,\ r_2, r_r \ll x$ we have

$$
P_{rd}(\omega_0) \cong \frac{P_t Q^2 \eta_r r_r^3}{x^6} \left(\eta_1 r_1^3 + \eta_2 r_2^3 \cos^6 \theta \right); \qquad (5.6)
$$

$$
r_1^2/x^2 \ll 1 \quad \text{and } r_2^2 \cos^2 \theta / x^2 \ll 1
$$

When the transmitting antennas have equal radii $r_2 = r_1$

$$
P_{rd}(\omega_0) \cong \frac{P_t Q^2 \eta_r r_r^3 r_1^3}{x^6}(\eta_1 + \eta_2 \cos^6 \theta); \qquad (5.7)
$$

$$
r_1^2/x^2 \ll 1;\ r_1^2 \cos^2 \theta / x^2 \ll 1
$$

Thus the receiver power is higher, decreases with distance as a power of 6 and is a function of the efficiencies of the inductors of the transmitters. To maximize the coupling in the receiver, we must maximize its efficiency and also locate it at a small angle off the boresight. This result comes with an inherent understanding that the two transmitters are well isolated from each other and there is no magnetic coupling between them. Although the dependencies in this expression differ from the ones for SIMO, the expressions are similar. In SIMO, the capacity is a function of the efficiencies of the receiver coils with the second term (due to the second receiver) decreasing as the cosine to the power six of its displacement from boresight. In the MISO case, the capacity is a function of the efficiencies of the transmitter coils with the second term (due to the second transmitter) decreasing as a function of the cosine to the power six of its displacement from boresight. If there is coupling between the two transmitter coils, it serves noise power to reduce the power coupled by them to the receiver and hence will decrease the capacity of the system.

5.1.1 Capacity of NFMI MISO system

Form equation (5.7) the capacity of a 2×1 NFMI MISO system can be estimated and is:

$$
\begin{aligned}
C &= B_f f_0 \log_2 \left(1 + \frac{P_{rd}(w = \omega_0)}{N} \right) \\
&= B_f f_0 \log_2 \left(1 + \frac{P_t Q^2 \eta_r r_r^3 r_1^3}{N \cdot x^6} (\eta_1 + \eta_2 \cos^6 \theta) \right)
\end{aligned}
\tag{5.8}
$$

The improved capacity by deploying one more transmitter is given by the increased received signal power by the expression:

$$
\begin{aligned}
C_s &= \frac{P_{rd}(\omega = \omega_0)}{P_r(\omega = \omega_0)} = \frac{\eta_1 k_1^2(x) + \eta_2 k_d^2(x)}{\eta_1 k_1^2(x)} \\
&= 1 + \frac{\eta_2}{\eta_1} \cdot \frac{k_d^2(x)}{k_1^2(x)}
\end{aligned}
\tag{5.9}
$$

Clearly, by using two transmitting coils we have improved the system performance in delivering power to the receiver and hence the system capacity as well. The change in capacity is approximately

$$C_\Delta \cong B_f f_0 \log_2 \left[\frac{\left(r_2^3 \eta_2 \cos^6 \theta\right)\left(1 + \frac{r_1^2}{x^2}\right)^3}{r_1^3 \eta_1 \left(1 + \frac{r_2^2 \cos^2 \theta}{x^2}\right)^3} \right]$$

$$\cong 3 B_f f_0 \log_2 \left(\frac{r_2 \eta_2}{r_1 \eta_1}\right) + B_f f_0 \log_2 \left[\frac{\left(\cos^6 \theta\right)\left(1 + \frac{r_1^2}{x^2}\right)^3}{\left(1 + \frac{r_2^2 \cos^2 \theta}{x^2}\right)^3} \right] \quad (5.10)$$

When the displacement angle is very small the increase in capacity is approximately equal to

$$C_\Delta \cong B_f f_0 \log_2 \left(\frac{r_2^3}{r_1^3} \frac{\eta_2}{\eta_1} \cos^6 \theta \right) ; \quad (5.11)$$

$$r_1^2/x^2 \ll 1; \quad r_2^2 \cos^2 \theta / x^2 \ll 1$$

Clearly from equation (5.11), there is advantage in making the efficiency and radius of the transmitter off boresight much larger than the efficiency and radius of the one at boresight.

Clearly, the increase in capacity is a function of the distance x (equation (5.10)), the radii of the coils, the efficiencies of the coils and the angular displacement of the transmitters. By trading off these variables higher capacity increases can be achieved. For example, by making $r_2 \gg r_1$ and/or making $\eta_2 \gg \eta_1$ this change can be significant. It is evident from the equation that small angular displacements are preferred for higher capacity increases.

5.1.2 Inductively coupled MIMO system

Clearly from previous analyses, it has been demonstrated that there is advantage in deploying either more transmitters or more receivers. Doing so has augmented the system capacity compared with that of SISO systems. To take further advantage of the multiple paths which inductive links provide, in this section we propose and analyze the capacities of multiple input multiple output systems starting with a two input and two output near field inductive communication system (Figure 5.2).

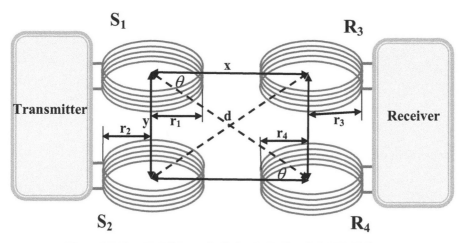

Figure 5.2 Near Field Magnetic (Inductively Coupled) MIMO System

5.1.3 NFMI MIMO system

Figure 5.2 describes a near field inductively coupled MIMO communication system. To the author's knowledge, this innovative step has never been proposed by any other author in the form presented here.

A general form is presented in which each coil has its distinct radius, efficiency and quality factor Q. We however relax this rigidity and allow coils to resonate at the same resonant frequency. As this represents the more optimum form, it is a welcome choice of system architecture. Depending on the circuitry that exists in the transmitter and receiver and the distances separating the inductive antennas, six mutual inductances and coupling coefficients can be defined and they are: M_{12}; M_{14}; M_{13}; M_{32}; M_{34} and M_{24} and also k_{12}; k_{14}; k_{13}; k_{32}; k_{34} and k_{24}. Both k_{12} and k_{34} are couplings between the antenna coils in the transmitter and receiver respectively and exist if the isolation between the antennas are inadequate. We assume that for the coils the following holds $k_{ij} \neq k_{ji}$. Equality holds only when the radii of the solenoids are equal. We define and categorize three types of coupling. Horizontal (x) coupling refers to the coupling between coils whose axis line on a line (k_{12} and k_{34}). Diagonal (d) coupling takes between two coils with centers that are diagonally opposite each other (k_{23} and k_{14}). Finally, vertical (y) coupling takes place between two coils whose centers lie on a vertical line (k_{13} and k_{24}). We derive the following six coupling coefficients as functions of the distance x (the remaining six coupling coefficients are obtained by exchanging the radius in the denominators of the expressions with the second radius in the expressions):

$$k_{13}^2(x) = \frac{r_1^3 r_3^3}{\left(x^2 + r_1^2\right)^3} \tag{5.12}$$

$$k_{34}^2(x) = \frac{r_3^3 r_4^3}{\left((x \tan \theta)^2 + r_3^2\right)^3} \tag{5.13}$$

The vertical distance (y) is $y = x \tan(\theta)$. Therefore

$$k_{12}^2(x) = \frac{r_1^3 r_2^3}{\left[(x \tan \theta)^2 + r_1^2\right]^3} \tag{5.14}$$

and

$$k_{14}(x) = \frac{r_1^3 r_4^3}{[(x/\cos \theta)^2 + r_1^2]^3} \tag{5.15}$$

It is also observed that the diagonal distances are equal but are at different angles. The remaining two coupling coefficients are:

$$k_{23}(x) = \frac{r_2^3 r_3^3}{\left[(x/\cos \theta)^2 + r_2^2\right]^3} \tag{5.16}$$

and

$$k_{23}(x) = \frac{r_2^3 r_3^3}{\left[(x/\sin \theta)^2 + r_2^2\right]^3} \tag{5.17}$$

In this chapter we will consider the coupling coefficients between the transmitters and receivers as either

i) zero and therefore they are considered in the analysis and then
ii) as sources of noise in the transmitting and receiving ends and have influence on the signal to noise ratios at each receiver and hence the system capacity. In a latter chapter we will demonstrate how the vertical mutual inductances could be used to aid transmission and hence be useful sources of communications and hence help to increase system capacity by efficient signal processing techniques.

5.1.4 Zero coupling between transmitters (receivers)

In the following expressions, the first index on the variables refers to the receiver and the second index to the transmitter. The power received by receiver j due to transmission from transmitter i is:

$$P_{ji} = P_{t1} Q_i Q_j \eta_i \eta_j k_{ji}^2(x) \tag{5.18}$$

The power received by receiver 3 from due to transmission from transmitter 1 is

$$P_{31} = P_{t1} Q_1 Q_3 \eta_1 \eta_3 k_{31}^2(x) \tag{5.18a}$$

Similarly, the remaining received powers are

$$P_{32} = P_{t2} Q_2 Q_3 \eta_2 \eta_3 k_{32}^2(d) \tag{5.18b}$$

$$P_{41} = P_{t1} Q_1 Q_4 \eta_1 \eta_4 k_{41}^2(d) \tag{5.18c}$$

and

$$P_{42} = P_{t2} Q_2 Q_4 \eta_2 \eta_4 k_{42}^2(x) \tag{5.18d}$$

The received power matrix therefore is

$$P_r = \begin{bmatrix} P_{31} & P_{32} \\ P_{41} & P_{42} \end{bmatrix} \tag{5.19}$$

By combining the received power, improvement in system capacity can be achieved. This chapter does not yet consider methods of combining the received power nor methods of transmissions which facilitate separation of transmissions.

5.1.5 Non-zero coupling between transmitters (receivers)

Normally, except in efficiently isolated sources and receivers, energy coupling between transmitters (receivers) takes place. This undesirable coupling becomes noise sources which limit the performance of the transmission system. In this section an analysis of the coupling between transmitters (receivers) is undertaken. Consider the coupling between transmitters or receivers. Assume the radii of the transmitters (receivers) are not equal and that they are not placed orthogonally or axially collinear. Hence the coupling between transmitters (receivers) is not symmetrical. Hence the noise seen by receiver 3 due to the power coupled by receiver 4 is:

$$N_{34} = P_{41} Q_3 Q_4 \eta_3 \eta_4 k_{34}^2 + P_{42} Q_3 Q_4 \eta_3 \eta_4 k_{34}^2 = Q_3 Q_4 \eta_3 \eta_4 k_{34}^2 (P_{41} + P_{42}) \tag{5.20}$$

and

$$N_{43} = Q_3 Q_4 \eta_3 \eta_4 k_{43}^2 (P_{31} + P_{32}) \tag{5.21}$$

Hence the differences in the amount of interference introduced by each source on the other are due to the different coupling coefficients and the different power sources as shown in the above equations. If

$$k_{43}^2(x) = \frac{r_4^3 r_3^3}{\left(x^2 + r_3^2\right)^3} \neq k_{34}^2(x) = \frac{r_4^3 r_3^3}{\left(x^2 + r_4^2\right)^3} \tag{5.22}$$

$$N_{43} = Q_3^2 \eta_3^2 Q_4 \eta_4 \left(P_{t1} Q_1 \eta_1 k_{31}^2 + P_{t2} Q_2 \eta_2 k_{32}^2\right) k_{43}^2 \tag{5.23}$$

$$N_{34} = Q_4^2 \eta_4^2 Q_3 \eta_3 \left(P_{t1} Q_1 \eta_1 k_{41}^2 + P_{t2} Q_2 \eta_2 k_{42}^2\right) k_{34}^2 \tag{5.24}$$

Let

$$a = Q_3 \eta_3 (Q_4 \eta_4 k_{34})^2 \tag{5.25a}$$

and

$$b = Q_4 \eta_4 (Q_3 \eta_3 k_{43})^2 \tag{5.25b}$$

Then the noise interference for this system is

$$\begin{bmatrix} N_{34}/a \\ (N_{43}/b) \end{bmatrix} = \begin{bmatrix} k_{31}^2 & k_{32}^2 \\ k_{41}^2 & k_{42}^2 \end{bmatrix} * \begin{bmatrix} P_{t1} Q_1 \eta_1 \\ P_{t2} Q_2 \eta_2 \end{bmatrix} \tag{5.26}$$

The system interference matrix is a product of the coupling coefficients matrix and amplified power vector. The power from each transmitter is first amplified by the product of its efficiency and Q. Also as the coupling matrix shows, all the coupling coefficients contribute to interference in some manner.

The interference matrix system when current receivers become transmitters can be obtained from the above expression by swapping indices. For the 2×2 system, the swapping procedure is (index 1 becomes 3 and index 3 becomes 1) $1 \leftrightarrow 3$ and index $2 \leftrightarrow 4$. This also transposes the interference matrix and the noise vector indexes. The two by two index swapping can be made recursive.

5.1.6 General case

The previous section provided an analysis of a 2×2 MIMO system using NFMIC. This section extends the analysis as a general framework for an M (transmitters) $\times N$ (receivers) case. Each transmitter transmits power $P_n a \leq n \leq M$.

In Figure 5.3, we have M transmitters and N receivers. We assume that each of the receiver is able to hear the transmission from all the transmitters in the system who couple energies to them. We also assume that vertical

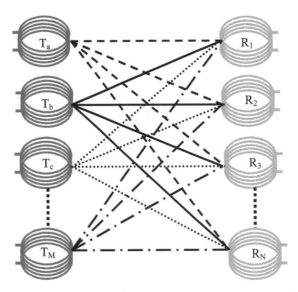

Figure 5.3 MIMO – The General Case of $M \times N$ Transmitters and Receivers

coupling between neighbouring transmitters (receivers) exist and the minimum number of interference sources to each transmitter (receiver) is two and the maximum interference sources is M (for transmitters) and N (for receivers). We will undertake the analysis of the noise at each receiver, first for the minimum case and then for the maximum case.

5.1.6.1 Minimum case (two interferers)

The power received by the first receiver R_1 is

$$P_1 = Q_1\eta_1(P_a Q_a \eta_a k_{1a}^2 + P_b Q_b \eta_b k_{1b}^2 + P_c Q_c \eta_c k_{1c}^2 + \cdots \\ + P_M Q_M \eta_M k_{1M}^2) \tag{5.27}$$

The interference seen by receiver 2 as a result of coupling from receiver 1 is

$$N_{21} = P_1 Q_1 \eta_1 Q_2 \eta_2 k_{21}^2 = Q_2 \eta_2 (Q_1 \eta_1 k_{21})^2 (P_a Q_a \eta_a k_{1a}^2 + P_b Q_b \eta_b k_{1b}^2 \\ + P_c Q_c \eta_c k_{1c}^2 + \cdots + P_M Q_M \eta_M k_{1M}^2) \tag{5.28}$$

$$P_2 = Q_2 \eta_2 (P_a Q_a \eta_a k_{2a}^2 + P_b Q_b \eta_b k_{2b}^2 \\ + P_c Q_c \eta_c k_{2c}^2 + \cdots + P_M Q_M \eta_M k_{2M}^2) \tag{5.29}$$

With this framework, the expressions for the interference seen at receivers 1, 2 and 3 are:

Interference at the receiver 1 due to receiver 2

$$N_{12} = P_2 Q_1 \eta_1 Q_2 \eta_2 k_{12}^2 = Q_1 \eta_1 (Q_2 \eta_2 k_{12})^2 (P_a Q_a \eta_a k_{2a}^2 + P_b Q_b \eta_b k_{2b}^2$$
$$+ P_c Q_c \eta_c k_{2c}^2 + \cdots + P_M Q_M \eta_M k_{1M}^2) \tag{5.30}$$

Interference at the receiver 2 due to receiver 3

$$N_{23} = Q_2 \eta_2 (Q_3 \eta_3 k_{23})^2 (P_a Q_a \eta_a k_{3a}^2 + P_b Q_b \eta_b k_{3b}^2$$
$$+ P_c Q_c \eta_c k_{3c}^2 + \cdots + P_M Q_M \eta_M k_{3M}^2) \tag{5.31}$$

Interference at the receiver 3 due to receiver 2

$$N_{32} = Q_3 \eta_3 (Q_2 \eta_2 k_{32})^2 (P_a Q_a \eta_a k_{2a}^2 + P_b Q_b \eta_b k_{2b}^2$$
$$+ P_c Q_c \eta_c k_{2c}^2 + \cdots + P_M Q_M \eta_M k_{2M}^2) \tag{5.32}$$

The total interference at receiver 2 is due to receiver 1 and 3: $N_2 = N_{21} + N_{23}$ or

$$N_2 = P_a Q_a \eta_a \left(\sigma_1^2 k_{1a}^2 + \sigma_3^2 k_{3a}^2\right) + P_b Q_b \eta_b \left(\sigma_1^2 k_{1b}^2 + \sigma_3^2 k_{3b}^2\right) + \cdots$$
$$+ P_M Q_M \eta_M \left(\sigma_1^2 k_{1M}^2 + \sigma_3^2 k_{3M}^2\right) \tag{5.33}$$
$$\sigma_1^2 = Q_1^2 \eta_1^2 Q_2 \eta_2 k_{21}^2; \quad \sigma_3^2 = Q_2^2 \eta_3^2 Q_2 \eta_2 k_{23}^2 \tag{5.34}$$

Some of these noise contributions are likely to be very small. In the same manner the interference at receiver 3 is given by $N_3 = N_{32} + N_{34}$ and at receiver 4 is $N_4 = N_{43} + N_{45}$. The signal to noise ratio at node 2 P_2/N_2 is given by the expression:

$$SNR = \cfrac{\begin{array}{c} Q_2 \eta_2 (P_a Q_a \eta_a k_{2a}^2 + P_b Q_b \eta_b k_{2b}^2 \\ + P_c Q_c \eta_c k_{2c}^2 + \cdots + P_M Q_M \eta_M k_{2M}^2) \end{array}}{\begin{array}{c} P_a Q_a \eta_a (\sigma_1^2 k_{1a}^2 + \sigma_3^2 k_{3a}^2) + P_b Q_b \eta_b (\sigma_1^2 k_{1b}^2 + \sigma_3^2 k_{3b}^2) \\ + \cdots + P_M Q_M \eta_M (\sigma_1^2 k_{1M}^2 + \sigma_3^2 k_{3M}^2) \end{array}} \tag{5.35}$$

This can be decoupled into a ratio of power from the source and the noise components from the two interferers as:

$$SNR = \frac{f\left(k_{2n}^2\right)}{(Q_1 \eta_1 k_{21})^2 \, f\left(k_{1n}^2\right) + (Q_3 \eta_3 k_{23})^2 \, f\left(k_{3n}^2\right)}; \quad n \in (a, b, c, \ldots, M) \tag{5.36}$$

When $P = P_a = P_b = \cdots = P_M$; $Q = Q_a = Q_b = \cdots = Q_M$ and $\eta = \eta_a = \eta_b = \cdots = \eta_M$

$$SNR = \frac{\displaystyle\sum_{n=a}^{M} k_{2n}^2}{(Q_1\eta_1 k_{21})^2 \displaystyle\sum_{n=a}^{M} k_{1n}^2 + (Q_3\eta_3 k_{23})^2 \displaystyle\sum_{n=a}^{M} k_{3n}^2}$$

$$= \frac{\displaystyle\sum_{n=a}^{M} k_{2n}^2}{\beta_1^2 \displaystyle\sum_{n=a}^{M} k_{1n}^2 + \beta_3^2 \displaystyle\sum_{n=a}^{M} k_{3n}^2}; \quad n \in (a, b, c, \ldots, M) \qquad (5.37)$$

Thus the noise at receiver 2 is the sum of the contributions of energy couplings to receiver 1 and also of coupling of energy through receiver 3. For this case, if we know the coupling coefficients at each receiver, we can easily compute the signal to noise ratios at each receiver. This is a very fundamental equation when identical solenoids are used for transmission of information. In general we can write

$$SNR = \frac{\displaystyle\sum_{n=a}^{M} k_{jn}^2}{\beta_{j-1}^2 \displaystyle\sum_{n=a}^{M} k_{(j-1)n}^2 + \beta_{j+1}^2 \displaystyle\sum_{n=a}^{M} k_{(j+1)n}^2}; \quad n \in (a, b, c, \ldots, M) \quad (5.38)$$

This expression is valid if the solenoids are collinear and they have identical Q, equal efficiency η and transmitting at the same power P_t. We can show that in general when a receiver is able to listen to N interferers, the signal to noise ratio is

$$SNR_r = \frac{\displaystyle\sum_{n=a}^{M} k_{rn}^2}{\displaystyle\sum_{s=1}^{N} \left(\beta_{rs}^2 \sum k_{rs}^2 \right) + N_T} \qquad (5.39)$$

Here N_T is thermal noise. In practice some of the interfering sources will have little effect on the overall noise heard by a receiver and only a few sources dominate production of noise. This is a desirable situation. For this to be the case, several conditions are desirable:

i) isolation of receivers is necessary to block noise being coupled to them
ii) orthogonal orientation of receivers to limit interference
iii) orthogonal orientation of transmission from the transmitters
iv) locating the receivers far away from each other to limit the coupling coefficients.

For example when the radii of the transmitting and receiving solenoids are equal the coupling coefficients reduce to $1/(1 + (d_{ij}/r)^2)^3$ and d_{ij} is the distance between receiver i and transmitter j.

5.1.6.2 Maximum case (S interferers)

There is a more serious interference problem if the receivers are not collinearly located with respect to each other or when they are placed randomly and or along two collinear axes as in Figure 5.4. In practice, in a combat situation or in a field operation, the receivers are located randomly with respect to each other. Consider Figure 5.4 below. In that scenario too many noise sources interfere with the main signal.

When the interferers are not a priori located linearly or planned, the coupling coefficients are functions of distance and angle.

The interference seen by receiver j as a result of coupling from receiver i is

$$
\begin{aligned}
N_{ji} &= P_i Q_i \eta_i Q_j \eta_2 k_{ji}^2 \\
&= Q_j \eta_j (Q_i \eta_i k_{ji})^2 (P_a Q_a \eta_a k_{ia}^2 + P_b Q_b \eta_b k_{ib}^2 \\
&\quad + P_c Q_c \eta_c k_{ic}^2 + \cdots + P_M Q_M \eta_M k_{iM}^2)
\end{aligned}
\tag{5.40}
$$

The signal to noise ratio at any receiver r is therefore given by the expression

$$
SNR_r = \frac{\displaystyle\sum_{n=a}^{M} k_{rn}^2}{(Q_1 \eta_1 k_{r1})^2 \displaystyle\sum_{n=a}^{M} k_{1n}^2 + (Q_2 \eta_2 k_{r2})^2 \displaystyle\sum_{n=a}^{M} k_{2n}^2 \\ + \cdots + (Q_m \eta_m k_{rm})^2 \displaystyle\sum_{n=a}^{M} k_{mn}^2}
\tag{5.41}
$$

$$= \frac{\displaystyle\sum_{n=a}^{M} k_{rn}^2}{\displaystyle\beta_1^2 \sum_{n=a}^{M} k_{1n}^2 + \beta_2^2 \sum_{n=a}^{M} k_{2n}^2 + \cdots + \beta_m^2 \sum_{n=a}^{M} k_{mn}^2}$$

This expression simplifies to

$$SNR_r = \frac{\displaystyle\sum_{n=a}^{M} k_{rn}^2}{\displaystyle\sum_{m=1}^{R} \sum_{n=a}^{M} \beta_m^2 k_{mn}^2}; \quad n \in (a, b, c, \ldots, M); \ m \neq r \tag{5.42}$$

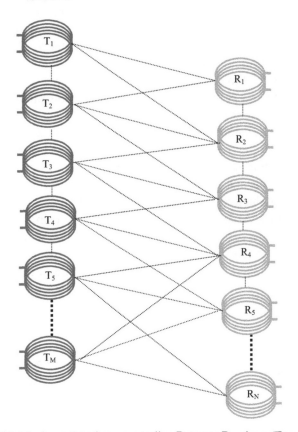

Figure 5.4 Maximum Interference coupling Between Receivers (Transmitters)

Fortunately this equation applies to the transmitters as well. Interference is compounded by the coupling coefficients if inherent in the system.

5.1.7 Relay coupling

While it looks like mutual coupling between neighbouring transmitting and receiving antennas could be a problem, we propose the use of relay coupling. Relay coupling is defined as when two antennas of the same function type (transmitter or receiver) aid each other in relaying of information for each other in such a manner that they alternate in either transmission or reception. In relay magnetic inductive transmission using the 2×2 MIMO system, during time slot T_1, source S_1 transmits inductively to receivers R_2 and R_4 and transmitter S_3 and they couple the information received. During the second time slot T_2 the transmitter S_3 relays the received data to receivers R_2, R_4 and S_1 rests. During this time slot, S_1 does not transmit any of its own data but receives only. In the third time slot T_3, the transmitter S_3 now to receivers R_2 and R_4 and transmitter S_1 receives. In the fourth time slot S_1 relays its data and S_3 rests. In the 5th time slot the process repeats. The receivers are always busy receiving data except that during alternate time slots the received data is relayed data. The process is depicted in Table 5.1.

The efficiency of system utilization is 75% for each transmitter and 100% for the receivers. We can increase the efficiency of system utilization if it were possible for the two transmitters to co-resonate at two different frequencies and thus can transmit and receive simultaneously. This can be achieved by altering either the inductive and or the capacitive components of the transmitting lumped circuits. Assume that we achieve this by using two capacitors, the first connected in series with the source resistance in the transmitters and the second in series with the load resistance in the receiver, so that when the capacitor is switched in, the second resonant mode is used as well and $\omega_{01} = \frac{1}{\sqrt{L_1 C_1}} = \frac{1}{\sqrt{L_2 C_2}}$ and $\omega_{02} = \frac{1}{\sqrt{L_1 C_1'}} = \frac{1}{\sqrt{L_2 C_2'}}$. Hence transmitting and receiving can take place simultaneously on ω_{01} and ω_{02}.

Table 5.1 NFMI Transmission and Reception Diversity Scheme

Time Slot	S_1	S_3	R_2	R_4
1	Transmit	Receive	Receive	Receive
2	Rest	Relay	Receive	Receive
3	Receive	Transmit	Receive	Receive
4	Relay	Rest	Receive	Receive

Table 5.2 NFMI Channel Sounding Scheme

Time Slot	S_1	S_3	R_2	R_4
1	Transmit	Receive	Receive	Receive
2	Receive	Relay	Receive	Receive
3	Relay	Rest	Receive	Receive
4	Rest	Transmit	Receive	Receive

5.1.8 NFMI MIMO channel matrix

The channel matrix of a MIMO system is normally used to estimate the capacity of the system. Both ergodic and outage probabilities depend on the characteristics of the MIMO channel matrix. To derive the channel matrix, we first derive expressions for the mutual coupling that exists between the antennas. The derivation is for the 2×2 channel matrix as depicted in the system of Figure 5.3.

From Table 5.1, each transmitter transmits, receives, relay and rest for the 4th time slot. The receivers are busy for all the time slots, but must buffer data received the previous time slot to be used to enhance capacity this time slot. The transmitters waste one time slot in every 4. However, there is an inherent advantage that when they transmit their energies are coupled to three receivers.

References

[1] Christine Evans-Pughe, "Close encounters of the magnetic kind", IEE Review, May 2005, pp. 38–42.

[2] Rajeev Bansal, "Near Field Magnetic Communications", IEEE Antennas and Propagation Magazine, Vol. 46, No. 2, Apr. 2004, pp. 114–115.

[3] Hengzhen C. Jiang and Yuanxun E. Wang, "Capacity Performance of an Inductively Coupled Near Field Communication System", in Proc. IEEE International Symposium of Antenna and Propagation Society, Jul. 5–11, 2008, pp. 1–4.

6

Circuit Models and Power Estimates of Antennas in Inductive Near Field Communications Links

This chapter presents a systematic model of inductive transmission systems and the power estimates for different types of coils and their spatial orientations with respect to each other.

6.1 Lumped Circuit Models of Inductive Links

The objective of an inductive communication system is to induce a voltage in the receiver load that is proportional to the load resistance and the current induced in the receiver coil. Of course if the current in the transmitter is modulated with data or information, this information is therefore inherent in the current induced in the receiver. Optimum current can be induced but it is a function of the coil dimensions, number of turns, their distances from each other and the orientations. Optimum power is also induced if the coils resonate at the same frequency. These concepts by themselves are not new because they are mostly well covered in undergraduate text books on circuit theory. The next sections provide the fundamental principles behind the use of loop antenna for applications in near field communications. As an objective, the chapter evaluates the characteristics and power delivery capabilities of different orientations of loop antennas.

6.1.1 Circuit model of loop antenna

The loop antenna has a special place in short range communication in application where wireless delivery of power is required such as in biomedical implants and difficult terrain. Ability to deliver power could save lives in out-door applications including bush walking, mountaineering, mining, caves and tunnels. It provides a means to charge dwindling battery power remotely. Furthermore, it is unique for short range communication where no output

J. Ihyeh Agbinya (PhD), Principles of Inductive Near Field Communications for Internet of Things, 65–77.

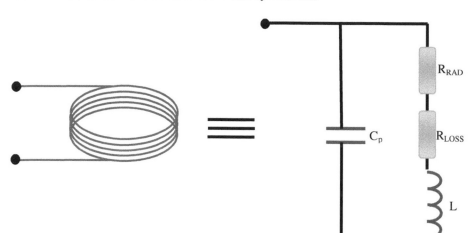

Figure 6.1 Electric Circuit Model of Loop Antenna

power amplification is required. The loop antenna also has better efficiency compared to the monopole and dipole antennas.

In addition to the series radiation and loss resistances, the loop antenna model has a parallel capacitor as shown in Figure 6.1 [2].

The orientation of the loop transmitting and receiving antennas with respect to each other determines the amount of energy coupled to the receiver by the transmitter. In this section, the energy coupling properties of the loop antenna are presented. From Figure 6.1, the loop antenna is modeled with a tank circuit consisting of an inductor in series with the loss and radiation resistances of the antenna. These are in parallel with its inherent capacitive value.

In the reactive near-field region, the radiation resistance is zero. However, in the radiating near field an inductive loop antenna is modelled as series resistors representing the *radiation* resistance of the antenna and a *loss* resistance, its *inductive* value and the mutual inductance M. The radiation resistance of the antenna is a *virtual* consideration and is equal to the value of the resistor that would dissipate the same amount of power if it is inserted in replace the antenna. In Figure 6.2 both the transmitter and receiver models are shown with each having an extra component (mutual inductance, M) resulting from the coupling.

The inductive link consists of a transmitting part (Reader as in RFID) and a receiver part (tag as in RFID). The transmitter has at its input a source of

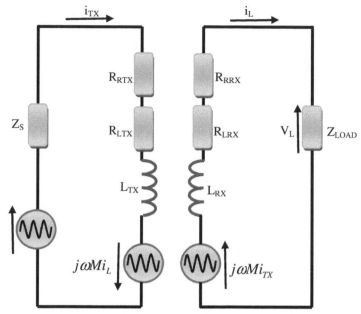

Figure 6.2 Lumped Circuit model of an Inductive Link

current i_{TX} and input impedance Z_s. Due to mutual inductance, the current flowing in the transmitter coil induces a current i_{RX} in the receiver. The receiver coil also has a self inductance L_{RX}, radiation resistance and a loss resistance. At its output is the load impedance (resistance) Z_L. Normally the radiation resistances are very small, a few tens of Ohms. In this chapter, both the radiation and loss resistors will be added and be considered as a single resistor. Hence

$$R_{TX} = R_{RTX} + R_{LTX} \tag{6.1}$$

$$R_{RX} = R_{RRX} + R_{LRX} \tag{6.2}$$

We apply Kirchoff's voltage law to the circuits in Figure 6.2 to sum the voltages in the transmitter and receiver circuits to have:

$$v_{TX} = (Z_S + R_{TX})i_{TX} + j\omega L_{TX}i_{TX} - j\omega M i_L \tag{6.3}$$

$$v_L = Z_L i_L = j\omega M i_{TX} - j\omega L_{RX}i_L - R_{RX}i_L \tag{6.4}$$

These equations reveal that at the transmitter the voltage resulting from the mutual inductance is a loss whereas at the receiver it is a gain. Hence these equations provide a means for assessing the level of induced voltages. The load voltage at the receiver is directly proportional to the mutual inductance from which we can derive some estimation of the effects of the coupling coefficient $k(x)$ on the communication link. This effect will now be quantified formally. Before we delve into that, a careful observation of equations (6.2) and (6.3) show that in the reactive near-field region when the radiation resistance is zero, more energy is coupled to the load because ($R_{RTX} = R_{RRX} = 0$). Observe also from equation (6.1) that the radiation resistance increases with the radius r of the loop antenna. Hence in the radiating far field region, less power is transferred to the receiver load because it appears as a loss in equations (6.3). The radiation efficiency of the loop antenna is therefore given by the ratio

$$\eta_{rad} = \frac{R_{RAD}}{R_{RAD} + R_{LOSS}} \tag{6.5}$$

6.1.2 Link equations: radiating near field

Friis equation applies only to communication in the radiating far field in which the receiver is located. In the radiating near field signal propagation has been shown to not follow the Friis equation [2]. For this case Yates [2] showed that when the input impedance Z_S of the transmitter is chosen to resonate with the transmitting antenna inductor L_{TX}, maximum power is transferred to the receiver load. Hence when

$$Z_S = \frac{1}{j\omega C} \quad \text{where } C = \frac{1}{\omega^2 L_{TX}} \tag{6.6}$$

The transmitter power therefore is given by the expression

$$P_{TX} = I_{TX_{rms}}^2 R_{TX} \tag{6.7}$$

The resulting magnetic field induced to the receiver antenna is given by the Biot-Savart law as

$$H = \frac{I_{TX} N_{TX} r_{TX}^2}{2 \left(\sqrt{r_{TX}^2 + x^2} \right)^3} \tag{6.8}$$

In this expression x is the distance separating the transmitter and receiver and aligned on a common axis. The transmitting coil has radius r and N_{TX} turns.

Using Faraday's law the induced voltage in the receiver coil (antenna) is given by the expression

$$V = N_{RX} \mu_0 A_{RX} \cdot j\omega H \tag{6.9}$$

This equation assumes uniform flux linkage to the receiver coil and the distance between the transmitter and receiver is much larger than the coil radii. In the equation, N_{RX} is the number of turns of the receiver antenna coil A_{RX} is the area of the receiver antenna, and $\phi_{RX} = \mu_0 A_{RX} \cdot j\omega H$ is the permeability. By substituting equation (6.8) in (6.9), the induced voltage at the receiver antenna is

$$V = \frac{j\mu_0 \omega I_{TX} N_{TX} N_{RX} A_{RX} r_{TX}^2}{2 \left(\sqrt{r_{TX}^2 + x^2} \right)^3} \tag{6.10}$$

Maximum power transfer is achieved at the receiver by matching the receiver load impedance as a conjugate match of the receiver antenna impedance. Therefore

$$Z_L = R_{RX} - j\omega L_{RX} \tag{6.11}$$

Therefore the power delivered to the load is maximum and given by the expression

$$P_{RX} = \frac{V_{RMS}^2}{4R_{RX}} = \frac{\left(\mu_0 \omega I_{TX(RMS)} N_{TX} N_{RX} A_{RX} r_{TX}^2 \right)^2}{16 R_{RX} \left(r_{TX}^2 + x^2 \right)^3} \tag{6.12}$$

The power transfer ratio given by the ratio of equations (6.12) and (6.7) is

$$\frac{P_{RX}}{P_{TX}} = \frac{\left(\mu_0 \omega N_{TX} N_{RX} A_{RX} r_{TX}^2 \right)^2}{16 R_{TX} R_{RX} \left(r_{TX}^2 + x^2 \right)^3} \tag{6.13}$$

When the distance between the receiver and transmitter is much larger than the radius of the transmitting coil, the power transfer ratio reduces to

$$\frac{P_{RX}}{P_{TX}} \cong p \cdot \frac{\left(\pi \mu_0 \omega N_{TX} N_{RX} r_{RX}^2 r_{TX}^2 \right)^2}{16 R_{TX} R_{RX} x^6} \tag{6.14}$$

$A_{RX} = \pi r_{RX}^2$ and p represents a factor for misaligning the antennas. Hence the power transfer increases with the radii of the coils, frequency and the number of turns of the coils in the transmitter and receiver antennas.

Table 6.1 System Variables

Factors	Symbol
Coil Radii	a, b
Coil Spacing	D
Lateral Misalignment	Δ
Angular Misalignment	γ
Tx, Rx coils	C_{Tx}, C_{Rx}
Tx, Rx Number of Turns	N_{Tx}, N_{Rx}
Tx, Rx Excitation Current	I_{Tx}
Ohmic Losses in the Tx, Rx Coils ($4\pi \times 10^{-7} H/m$)	R_{Tx}, R_{Rx}
Free Space Permeability	μ_0
Magnetic Permeability of Ferrite Core	μ_r

6.1.3 Effects of misalignments

Let us re-write equation (6.14) to account for the effects of misaligned coils. The new expression is:

$$\frac{P_{RX}}{P_{TX}} = \frac{(\mu_0 \omega N_{TX} N_{RX} A_{RX})^2}{16\pi^2 R_{TX} R_{RX} x^6} \cdot H_{INT}^2 \tag{6.15}$$

$$H_{INT} = \int_0^\pi \frac{dl_{TX} x \, r}{r^3}$$

These variables are defined in [1] and used in the Figures 6.2 to 6.4.

Figure 6.3 represents an ideal orientation of the coils along a common axis.

6.1.3.1 Lateral misalignment of coils

In the lateral misalignment [1] between the coils (Figure 6.4), the x and y components of the magnetic field intensity are parallel to the receiver plane and do not contribute to the lines of flux that cut through the receiver coil.

The z-component of the magnetic field intensity is the most prominent and given by the expression:

$$H_z = \int_0^\pi \frac{a^2 - a\Delta \cos\varphi}{\left(\sqrt{a^2 + D^2 + \Delta^2 - 2a\Delta \cos\varphi}\right)^3} d\varphi \tag{6.16}$$

By evaluating the above elliptic integral we obtain

$$H_z = \frac{a\sqrt{2m}}{(2a\Delta)^{3/2}} \cdot \left(\Delta K + \frac{am - (2-m) \cdot \Delta}{2 - 2m} \cdot E\right) \tag{6.17}$$

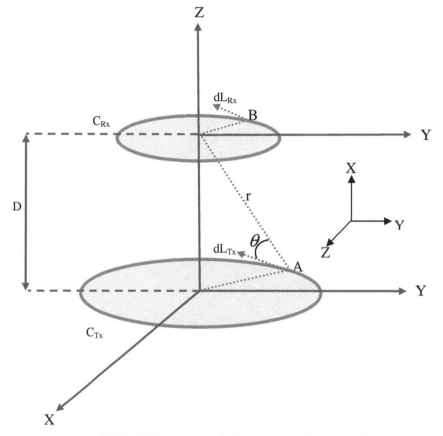

Figure 6.3 Ideal Orientation for the Transmitter and Receiver Coils

$K(m)$ and $E(m)$ are the complete elliptic integrals of the first and second kinds respectively and m is the modulus ($0 \leq m \leq 1$).

$$m = \left[\frac{4a\Delta}{(a + \Delta)^2 + D^2} \right] \tag{6.18}$$

The power transfer ratio for this case is

$$H_z = \frac{(m\mu_0\omega N_{TX}N_{RX}b^2)^2}{64a\,R_{TX}R_{RX}\Delta^3} \cdot \left(\Delta K + \frac{am - (2 - m) \cdot \Delta}{2 - 2m} \cdot E \right) \tag{6.19}$$

The power transferred is thus a function of the coil dimensions, the material of the core, the frequency of transmission and the misalignment of the coils with respect to each other.

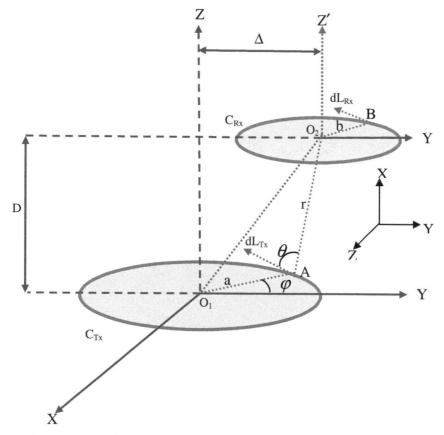

Figure 6.4 Lateral Misalignment Between the Transmitter and Receiver Coils [1]

6.1.3.2 Angular misalignment [1]

The x and y components of the magnetic field vectors cancel out because the of the circular symmetry of the centre of the receiver coil. Therefore the integration of the magnetic field intensity yields the value:

$$H_z = \frac{a^2 \cdot \pi}{\left(\sqrt{a^2 + D^2}\right)^3} \qquad (6.20)$$

The component of the magnetic field vertical to the plane of the coil is a function of the tilt angle and is obtained with the expression

$$H_z(\gamma) = n \cdot H_z \qquad (6.21)$$

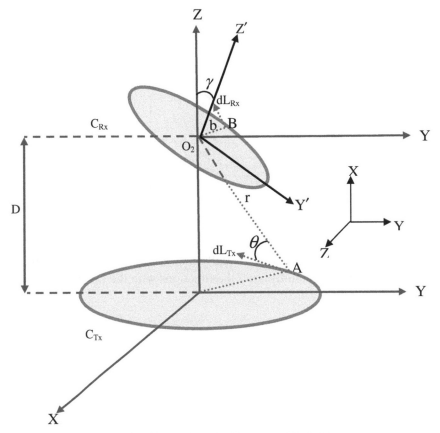

Figure 6.5 Angular Misalignment of Coils [1]

By defining $\boldsymbol{n} = (0, \sin\gamma, \cos\gamma)$ we obtain the relation

$$H_z = \frac{a^2 \cdot \pi \cdot \cos\gamma}{\left(\sqrt{a^2 + D^2}\right)^3} \tag{6.22}$$

This component of the magnetic field is vertical to the receiver coil. The power transfer ratio for this case from equation 6.15 is

$$\frac{P_{RX}}{P_{TX}} = \frac{\left(\pi\mu_0\omega N_{TX}N_{RX}a^2b^2\right)^2 \cos^2\gamma}{16R_{TX}R_{RX}\left(a^2 + D^2\right)^3} \tag{6.23}$$

It is evident from this power transfer ratio that the efficiency of the link reduces as a function of the angular misalignment between the coils increases.

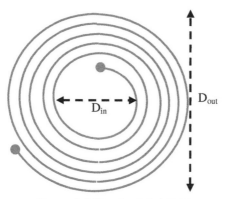

Figure 6.6 Circular Spiral Coil

The power transfer functions apply equally to non air cores or ferromagnetic cores and ferrite rods by multiplying the functions with the permeability μ_r.

6.1.4 Coil orientation and geometry models

In the previous section we paid attention to how the coils are misalinged with respect to each other.

The material and type of coil affects the performance of the link. A spiral coil can be approximated by closed concentric circular coils. The current flows through them in the same direction. The induced magnetic field at the centre of the receiver coils in Figure 6.7 is therefore the sum of the fields induced by the N turns of the transmitter coil. Thus in the ideal case the field in the z direction is

$$H_{z(ideal)} = \sum_{i=1}^{n} \frac{I_{TX} a_i^2}{2 \left(\sqrt{a_i^2 + D^2} \right)^3} \tag{6.24}$$

When the coils are laterally misaligned, the resulting H field is given by the expression

$$H_z = \frac{I_{TX}}{2\pi} \sum_{i=1}^{n} \frac{a_i \sqrt{2m_i}}{(2a_i \Delta)^{3/2}} \cdot \left(\Delta K + \frac{a_i m_i - (2 - m_i) \cdot \Delta}{2 - 2m_i} \cdot E \right) \tag{6.25}$$

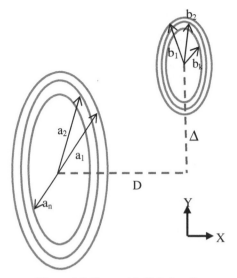

Figure 6.7 Concentric Spiral coils

In this expression, $K(m_i)$, and $E(m_i)$ are the complete elliptic integrals of the first and second kinds respectively and m is the modulus ($0 \leq m_i \leq 1$); and

$$m_i = \left[\frac{4a_i \Delta}{(a_i + \Delta)^2 + D^2} \right] \tag{6.26}$$

When the coils have angular misalignment with respect to each other

$$H_{z(angular)} = \sum_{i=1}^{n} \frac{I_{TX} a_i^2 \cos \gamma}{2 \left(\sqrt{a_i^2 + D^2} \right)^3} \tag{6.27}$$

For K turn spiral when each one has a radius b, the induced voltage from Faraday's law is

$$V_{cs} = j\mu_0 \omega H \cdot \sum_{j=1}^{K} \pi b_j^2 \tag{6.28}$$

Therefore the power transfer ratio for the circular spiral is given by the expression

$$\frac{P_{RX}}{P_{TX}} = \frac{(\mu_0 \pi \omega)^2}{16 R_{RX} R_{TX}} \left(\sum_{i=1}^{n} \frac{a_i^2}{\left(\sqrt{a_i^2 + D^2} \right)^3} \right)^2 \left(\sum_{j=1}^{K} \pi b_j^2 \right)^2 \tag{6.29}$$

Generally for the lateral displacement case, we have

$$\frac{P_{RX}}{P_{TX}} = \frac{(\mu_0 \omega)^2}{16 R_{RX} R_{TX}} \sum_{i=1}^{n} \frac{a_i \sqrt{2m_i}}{(2a_i \Delta)^{3/2}}$$

$$\cdot \left(\Delta K + \frac{a_i m_i - (2 - m_i) \cdot \Delta}{2 - 2m_i} \cdot E \right) * \left(\sum_{j=1}^{K} b_j^2 \right)^2 \tag{6.30}$$

For the angular misalignment case we have

$$\frac{P_{RX}}{P_{TX}} = \frac{(I_{TX} \mu_0 \pi \omega)^2}{4 R_{RX} R_{TX}} \left(\sum_{i=1}^{n} \frac{a_i \pi \cdot \cos \gamma}{\left(\sqrt{a_i^2 + D^2} \right)^3} \right)^2 \left(\sum_{j=1}^{K} b_j^2 \right)^2 \tag{6.31}$$

References

[1] Kyriaki Fotopoulou and Brian W. Flynn, "Optimum Antenna Coil Structure for Inductive Powering of Passive RFID Tags", in Proc. 2007 IEEE International Conference on RFID, March 26–28, 2007, Gaylord Texan Resort, Grapevine, Texas, USA, pp. 71–77.

[2] David C Yates, Andrew S Holmes and Alison J Burdett, "Optimal Transmission Frequency for Ultralow-Power Short Range Radio Links", IEEE Trans. On Circuits and Systems–I: Regular Papers, Vol. 51, No. 7, Jul. 2004, pp. 1405–1413.

7

Resistive and Inductive Properties of Near Field Magnetic Inductive Communication Links

This chapter presents selected coil types and the formulae required to estimate their self inductances. It also presents a systematic model of inductive coupling transmission systems and the estimates of the cross talk that exists in the system. We also discuss the inductive link in general and its characteristics based on different types of coils and their spatial orientations with respect to each other.

7.1 Ohmic Resistance of a Conductor

No matter how pure the material and processes used to produce a conductive wire are, a wireless possesses Ohmic resistance. The Ohmic (DC) resistance of a wire is a function of its length, conductivity and cross sectional area given by the expression:

$$R_{DC} = \frac{l}{\sigma S} = \frac{l}{\sigma \pi r^2} (\Omega) \tag{7.1}$$

The area S of most wires is circular and hence the DC resistance becomes a function of the radius r of the wire. Here σ (mho/m) is the conductivity, l is the total length of the wire. In NF communication, a large Q is essential for a high system capacity and hence the resistance of the wire must be small or its radius must be large or the length short.

The loss resistance is typically a DC resistance. The expression for the Ohmic resistance of a solenoid of N turns using a material of conductivity σ, radius r and cross sectional area S is

$$R_{LOSS(Solenoid)} = \frac{N \cdot 2\pi \cdot r}{\sigma \cdot S} \tag{7.2}$$

J. Ihyeh Agbinya (PhD), Principles of Inductive Near Field Communications for Internet of Things, 79–91.
© 2011 River Publishers. All rights reserved.

The loss resistance for a spiral of n turns and radii r_i is

$$R_{LOSS(Solenoid)} = \frac{2\pi}{\sigma \cdot S} \sum_{i=1}^{n} r_i \tag{7.3}$$

A conductor antenna also possesses a radiation resistance which needs to be accounted for in the near field radiation regime. The radiation resistance of a loop antenna is given by the expression:

$$R_R = \frac{\pi N^2 Z_0 (\beta \cdot r)^4}{6} \tag{7.4}$$

Z_0 is the impedance of free space and $\beta = 2\pi/\lambda$. The radiation resistance is thus a function of the radius of the dimensions of the coil and the frequency of operation.

7.1.1 AC resistance of a conductor

At DC, charge carriers are distributed evenly through out the cross section of the wire. However as the frequency of the signal the conductor carries is raised, the magnetic field at the centre of the conductor increases as well. This means that the reactance near the centre of the conductor also increases resulting in higher impedance and also higher current density. Hence electric charge move away from the centre to the edge or surface of the conductor making the current density to decrease towards the centre of the conductor and increase towards the edge or surface. This phenomenon is well known in Physics and referred to as skin effect. The skin depth therefore for the conductor is the depth at which the current density in it reduces to 1/e or 37% of its value along the surface. Skin depth is a function of the frequency f of radiation, the conductivity σ (mho/m) and also the permeability μ of the medium. The skin effect results in a net decrease in the area of cross section of the conductor and a rise in its AC resistance. Skin depth can be estimated with the expression:

$$\delta = \frac{1}{\sqrt{\mu\pi\sigma f}} \tag{7.5}$$

The resistance of a conductor due to skin depth is called AC resistance. In Table 7.1, typical values for the conductivities of popular conductors are given.

Brass as shown has very low conductivity is non-magnetic and hence is also used for magnetic shielding of materials and instruments.

Table 7.1 Magnetic Properties of Selected Metals

Conductor	Conductivity $\sigma\,(mho/m)$	Relative Permeability (μ_r) $\mu_0 = 4\pi \times 10^{-7}(H/m); \mu = \mu_0\mu_r$
Aluminum	3.82×10^7	1
Brass	1.5×10^7	1
Cobalt		60
Copper	$5,8 \times 10^7$	1
Gold	4.1×10^7	1
Iron (pure)		4000 (8000)
Nickel		50
Silver	6.1×10^7	0.999 999 81

The AC resistance of a conductor is given by the expression:

$$R_{ac} = \frac{l}{\sigma A_{active}} \approx \frac{l}{2\pi r \delta \sigma}\,(\Omega)$$

$$= \frac{l}{2r}\sqrt{\frac{f\mu}{\pi\sigma}} = R_{DC}\frac{r}{2\delta}\,(\Omega) \tag{7.6}$$

$A_{active} \approx 2\pi r\delta$ is the skin depth area on the conductor. Thus the AC resistance increases with the square root of the operating frequency and permeability and inversely with the conductivity of the wire. If the conductor is etched onto a dielectric substrate, the AC resistance is modified as a function of the width w and thickness t of the conductor and given by the expression

$$R_{ac} = \frac{l}{\sigma(w+t)\delta} = \frac{l}{(w+t)}\sqrt{\frac{\pi f\mu}{\sigma}}\,(\Omega) \tag{7.7}$$

A low frequency approximation of this resistance is used for low frequency applications. The equation applies when the skin depth is comparable to the radius of the conductor and is

$$R_{low\,freq} \approx \frac{l}{\sigma\pi \cdot r^2}\left[1 + \frac{1}{48}\left(\frac{r}{\delta}\right)^2\right] = R_{DC} \cdot \left[1 + \frac{1}{48}\left(\frac{r}{\delta}\right)^2\right]\,(\Omega) \tag{7.8}$$

$$R_{ac} = \left[1 + \frac{1}{48}\left(\frac{r}{\delta}\right)^2\right]\,(\Omega)$$

Thus the low frequency approximation is the product of the DC resistance and an AC resistance.

7.1.2 Inductive wireless link

When an electric current flows through a conductor, it creates a magnetic field. Maxwell's equations demonstrate this. If the magnetic field is time varying, it is capable of creating a flow of current in a neighbouring coil or conductor as well. This is called inductance (a current is induced in the other conductor). The inductance is a function of the physical characteristics of the conductor. The inductance L of the conductor is defined as the ratio of the total magnetic flux Ψ linking it to the current I (amps) flowing through it.

$$L = \frac{N\Psi}{I} \tag{7.9}$$

In this expression N is the total number of turns of the conductor. Hence the more the number of turns of the coil the greater its self inductance and the inductance increases the more compact (close) the turns of the coil are. Over the years scientists have developed expressions for approximating the inductance of wires and coils of various shapes, orientations and formations. Usually because of the distributed capacitance of wires, the values obtained are purely approximations and used as starting points for estimating the inductances. Different types of coil shapes can be used to create the near link antennas and hence expressions are given in this section to describe how to compute their inductances.

An inductive link is created by the use of a coil carrying currents and exciting a second coil in its close proximity. While this technique is well known since the early days of electrical engineering, we are now coming to terms with the real benefits of using the well known circuit theory for wireless communications. A few definitions and equations will help to place in context the underlying principles which permit the creation of inductive links.

7.1.3 Self inductance

In other to characterize an inductive link, the **self inductance** of the coil needs to be known. The self inductance is an inherent property of a coil and is a function of its dimensions and parameters. This section provides a few formulae that are essential for estimating the self inductances of several coil configurations.

Straight wire: The self inductance of a straight wound wire of length l (cm) and radius r (cm) is given by the expression

$$L = 0.002l \cdot \left(\log_e \frac{2l}{r} - 0.75 \right) \quad (\mu H) \tag{7.10}$$

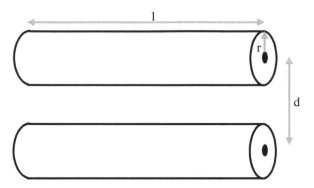

Figure 7.1 Two Parallel Conductors

Two Parallel Wires: For two parallel line wires of length l each of radius r and separated from each other by a distance d, the self-inductance is

$$L \approx \frac{\mu_0 l}{\pi} \left[\ln \left(\frac{d}{r} \right) + \frac{1}{4} - \frac{d}{l} \right] \quad (\mu H) \tag{7.11}$$

Circular Coil: The self inductance of a circular coil is given by the expression [1]

$$L = \mu_0 r \cdot \ln \left(\frac{2r}{d} \right) \quad (\mu H) \tag{7.12}$$

The diameter of the wire is d, the radius of the coil is r and μ_0 is the permeability of free space. This equation is valid only if the diameter of the wire is much smaller than the diameter of the coil and will be used when $d/2r \leq 0.001$. This approximation ensures that only small errors are made in the value of the self-inductance. A general approximation for self-inductance that can be used as well is given by the expression

$$L = 1.257 \cdot r \cdot \left[2.303 * \log_{10} \left(\frac{16r}{d} - 2 \right) \right] * 10^{-6} \quad (\mu H) \tag{7.13}$$

In practice most coils are composed of many turns N in close proximity to each other and hence the self inductance becomes $L_0 = N^2 L$. The above expressions are not valid for spiral and rectangular coils. We use them for illustration on how to compute the parameters required for implementing an inductive link.

Coil of N turns: The self inductance of a thin wire solenoid coil of N turns that is wound over a length of l of a former of diameter $2r$ can be

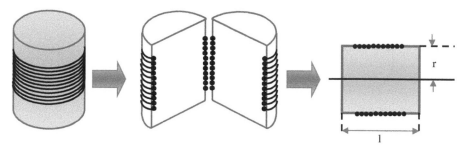

Figure 7.2 Single Layer Circular Coil

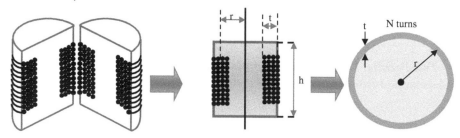

Figure 7.3 Multiple Layer Circular Coil

approximated also with the expression

$$L = \frac{\mu_0 \pi r^2 N^2}{l + 0.9r} \tag{7.14}$$

N-Turn Single Layer Circular Coil: The inductance of an N-turn single layer circular coil of radius r (cm) and length l (cm) is given by the expression

$$L = \frac{(rN)^2}{22.9r + 25.4l} \quad (\mu H) \tag{7.15}$$

N-Turn Multiple Layer Circular Coil: The inductance of an N-turn multiple layer circular coil of average radius r (cm), winding thickness t (cm) and winding height h (cm) is

$$L = \frac{0.31(rN)^2}{6r + 9h + 10t} \quad (\mu H) \tag{7.16}$$

Spiral Coil: For spiral coils the self inductance is

$$L = \frac{0.3937(rN)^2}{8r + 11t} \quad (\mu H) \tag{7.17}$$

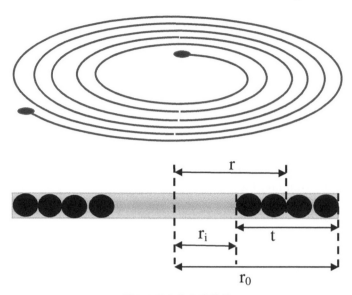

Figure 7.4 Spiral Coil

The following relations hold in the Figure 7.4: $r = (r_0 + r_i)/2$; $t = (r_0 - r_i)$; r_i and r_0 are the inner and outer radii of the spiral respectively. The inductance of a spiral coil was also estimated with the following expression in [3] considering the mutual inductance between the loops of various radii.

$$L = \sum L(r_{1i}, r) + \sum_{i=1}^{N} \sum_{i=1}^{N} M(r_{1i}, r_{1j}, d = 0, \rho = 0)(1 - \alpha_{ij}) \quad (\mu H)$$

(7.18)

Where $\alpha_{ij} = 1$ when $i = j$ and $\alpha_{ij} = 0$ otherwise [2], ρ is the alignment of coils and M the mutual inductance of two neighbouring coils is

$$M = \mu_0 \sqrt{r_1 r_2} \left[\left(\frac{2}{s} - s \right) K(s) - \frac{2}{s} E(s) \right]$$

(7.19)

$$s = \sqrt{\left(\frac{4 r_1 r_2}{(r_1 + r_2)^2 + d^2} \right)}$$

$K(s)$ and $E(s)$ are the complete elliptic integrals of the first and second kind respectively.

Multiple Layer N-Turn Square Loop: the inductance of an N-turn square loop of multiple layers is approximated from the expression [4]:

$$L = 0.008aN^2 \left\{ 2.303 \log_{10}\left(\frac{a}{b+c}\right) + 0.2235\frac{b+c}{a} + 0.726 \right\} \quad (\mu H)$$

(7.20)

The dimensions a, b and c represent one half the side of the square measured to the centre of the rectangular cross section of the winding, the length of the winding and depth of the winding respectively (as shown in Figure 7.5).

Multi Layer N-turn Rectangular Coil: rectangular coils lend themselves to easy construction and their dimensions can be easily measured. The self inductance of such coils is given by the expression

$$L = \frac{0.0276(CN)^2}{1.908C + 9b + 10h} \quad (\mu H)$$

(7.21)

Rectangular Thin Film Inductor: this inductor is formed as a rectangular thin film of length l, width w and thickness t. The inductance is given by the

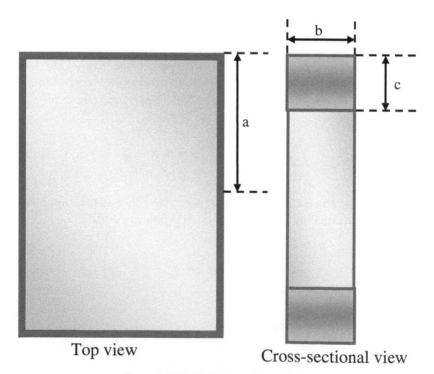

Top view   Cross-sectional view

Figure 7.5 Multiple Layer Square Loop

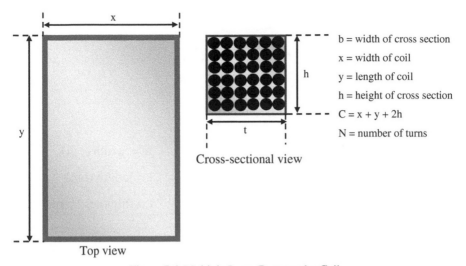

Cross-sectional view

Top view

Figure 7.6 Multiple Layer Rectangular Coil

expression

$$L = 0.008l \left\{ \ln\left(\frac{2l}{w+t}\right) + 0.50049 + \frac{w+t}{3l} \right\} \quad (\mu H) \qquad (7.22)$$

A Flat Square Coil: The inductance of a flat square coil of thickness t, width w and length a is given by the expression:

$$L = 0.0467aN^2 \left\{ \begin{array}{l} \log_{10}\left(\frac{2a^2}{w+t}\right) - \log_{10}(2.414a) \\[2mm] +0.02032aN^2 \left[0.914 + \left(\frac{0.2235}{a}(w+t)\right)\right] \end{array} \right\} \quad (\mu H)$$

(7.23)

Observe that the dimensions in this formula are given in inches.

Rectangular loop or flat plate element: Many tag antennas are of the planar rectangular loop types. The inductance of such antennas can be derived as follows

Let a be the radius of the wire, $l_c = \sqrt{(l_a^2 + l_b^2)}$ and $A = l_a * l_b$. Then

$$L = 4 \left\{ l_b \ln\left(\frac{2A}{a(l_b+l_c)}\right) + l_a \ln\left(\frac{2A}{a(l_a+l_c)}\right) + a[a + l_c - (l_a + l_b)] \right\}$$

(7.24)

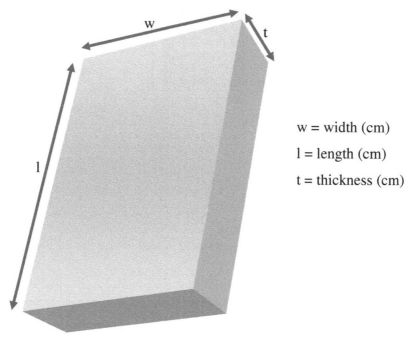

w = width (cm)

l = length (cm)

t = thickness (cm)

Figure 7.7 Rectangular Film Inductor

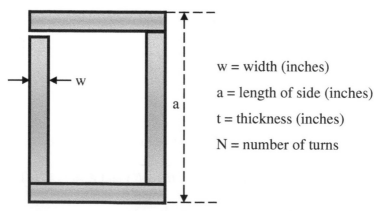

w = width (inches)

a = length of side (inches)

t = thickness (inches)

N = number of turns

Figure 7.8 Flat Square Coil

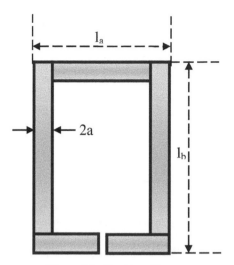

Figure 7.9 Flat Plate Element

N-Turn Planar Spiral Coil: the inductance of a coil is often impacted by the mutual inductances of its constituent components. The self inductance of a planar spiral coil is one such example. The expression for estimating it is

$$L = L_0 - M_- - M_+ \quad (\mu H) \tag{7.25}$$

The magnetic fields of adjacent conductors create mutual inductances. When the directions of currents in the conductors are the same, the mutual inductances are positive. They are negative when the current directions in adjacent conductors are in opposite directions. These mutual inductances are functions of the geometric mean d of distances and of their lengths.

In many applications, the antennas are formed by just two parallel conductors and for such cases the mutual inductance is given by the expression

$$M = 2lF \quad (nH) \tag{7.26}$$

F is the mutual inductance factor and l is the length of each conductor and F is given by

$$F = \ln\left\{\left(\frac{l}{d}\right) + \sqrt{\left[1 + \left(\frac{l}{d}\right)^2\right]}\right\} - \sqrt{\left[1 + \left(\frac{l}{d}\right)^2\right]} + \frac{d}{l} \tag{7.27}$$

Consider Figure 7.10, let j and k be the indices of the conductors and p and q the indices of the lengths of the difference of the conductors, the mutual

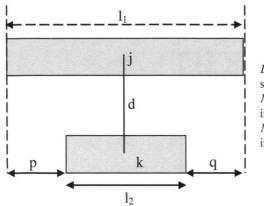

L_0 = sum of self inductances of all straight segments

M_- = sum of negative mutual inductances

M_+ = sum of positive mutual inductances

Figure 7.10 Parallel Conductors

inductance of conductors j and k is given by the expression

$$
\begin{aligned}
M_{j,k} &= \frac{1}{2}\left\{\left(M_{k+p} + M_{k+q}\right) - \left(M_p + M_q\right)\right\} \\
&= M_k \quad \text{for } p = q = 0 \quad (a) \\
&= M_{k+p} - M_p \quad \text{for } p = q \quad (b) \\
&= \frac{1}{2}\left\{\left(M_j + M_k\right) - M_p\right\} \quad \text{for } q = 0 \quad (c) \\
&= \frac{1}{2}\left\{\left(M_j + M_k\right) - M_q\right\} \quad \text{for } p = 0 \quad (d)
\end{aligned}
\tag{7.28}
$$

When ($l_1 = l_2$), equation (a) in the above expression is used and the mutual inductance factor is also used in association with it to obtain the expressions

$$
M_{k+p} = 2l_{k+p} F_{k+p}
$$

$$
\begin{aligned}
F_{k+p} &= \ln\left\{\left(\frac{l_{k+p}}{d_{j,k}}\right) + \sqrt{\left[1 + \left(\frac{l_{k+p}}{d_{j,k}}\right)^2\right]}\right\} \\
&\quad - \sqrt{\left[1 + \left(\frac{l_{k+p}}{d_{j,k}}\right)^2\right]} + \frac{d_{j,k}}{l_{k+p}}
\end{aligned}
\tag{7.29}
$$

7.1.4 Conclusions

This chapter has provided the basic formulae required to estimate the parameters of the conductors used to create inductive communication links. It also provides the basis for their simulation.

References

[1] Peter Scholz, Christian Reinhold, Werner John and Ulrich Hilleringmann, "Analysis of Energy Transmission for Inductive Coupled RFID Tags", in Proc. IEEE International Conference on RFID, Gaylord Texan Resort, Grapevine, TX, USA, Mar. 26–28, 2007, pp. 183–190.

[2] Suresh Atluri and Maysam Ghovanloo, "Design of a Wideband Power-Efficient Inductive Wireless Link for Implantable Biomedical Devices Using Multiple Carriers", Proc. of the 2nd International IEEE EMBS Conference on Neural Engineering, pp. 533–537, March 2005.

[3] Noriyuki Miura, Takayasu Sakurai and Tadahiro Kuroda, "Crosstalk Countermeasures for High Density Inductive-Coupling Channel Arrays", IEEE Journal of Solid – State Circuits, Vol. 2, No. 2, Feb. 2007, pp. 410–421.

[4] Youbok Lee, "Antenna Circuit Design for RFID Applications", Mcrochip Technology Inc., 2003.

8

Circuit Models of Near Field Magnetic Induction Communication Links

This chapter presents circuit models of inductive communications. It provides the means to understand the behaviour of the links, how to design and construct the communication links in hardware and simulate them in software.

8.1 Circuit Model of Inductive Links

An inductive communication link is modeled in Figure 8.1 as a transmitter coupling energy to a receiver using mutual inductance principles. The induced current in the receiver is applied to the receiver load. As seen from the figure, the transmitter has an input signal source, a matching network and a mutual inductance voltage.

The current I_1 in the transmitter creates a magnetic flux Φ_{21} which the inductor 2 captures with its area that is exposed to the flux. The mutual inductance and the flux are related by the equation

$$M_{21} = \frac{\Phi_{21}}{I_1} = \frac{1}{I_1} \iint_{\vec{A}_2} \vec{B}_2 d\vec{A}_2 = \frac{\mu_0}{I_1} \iint_{\vec{A}_2} \vec{H}_2 d\vec{A}_2$$

The equation shows that the flux results to flux density $\vec{B}_2$ and magnetic field strength $\vec{H}_2$ respectively. This equation is valid for scalar and constant permeability. In general the mutual inductance, $M = M_{21} = M_{12}$. For example for two circular coils the mutual inductance is

$$M = \frac{\mu \pi N_1 N_2 r_1^2 r_2^2}{2\sqrt{\left(r_1^2 + x^2\right)^3}} \tag{8.1}$$

J. Ihyeh Agbinya (PhD), Principles of Inductive Near Field Communications for Internet of Things, 93–105.

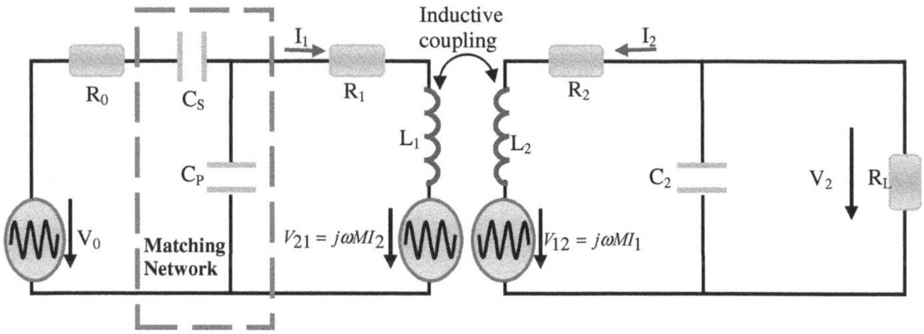

Figure 8.1 Inductive Transceiver with Matching Network

The radii of the transmitter and receiver coils are r_1 and r_2 respectively and the have N_1 and N_2 turns each, are collinear and at distance x apart such that $r_2 < r_1 \ll x$.

For this system the following equation holds

$$V = j\omega LI = j\omega \begin{pmatrix} L_1 & M \\ M & L_2 \end{pmatrix} \begin{pmatrix} I_1 \\ I_2 \end{pmatrix} \tag{8.2}$$

Let us further analyse how power is transferred to the receiver load. This analysis is based on the circuit of Figure 8.3.

The receiver is composed of a matching network formed by its inductor and capacitor L_2 and C_2 respectively. This capacitor and the load resistance R_L form a tank circuit which creates a transformed load given by the expression:

$$Z_{LT0} = j\omega L_2 + R_L \frac{1 - j\omega C_2 R_L}{1 + \omega^2 C_2^2 R_L^2} \tag{8.3}$$

When $1 \ll \omega^2 C_2^2 R_L^2$ the transformed load becomes

$$Z_{LT0} = j\omega L_2 + \frac{1}{\omega^2 C_2^2 R_L} - \frac{j}{\omega C_2} \tag{8.4}$$

At resonance the transformed impedance is purely real (resistive) and hence we have for $\omega = \omega_0$

$$\omega_0 L_2 = \frac{1}{\omega_0 C_2} \Rightarrow \omega_0 = \frac{1}{\sqrt{L_2 C_2}} \tag{8.5}$$

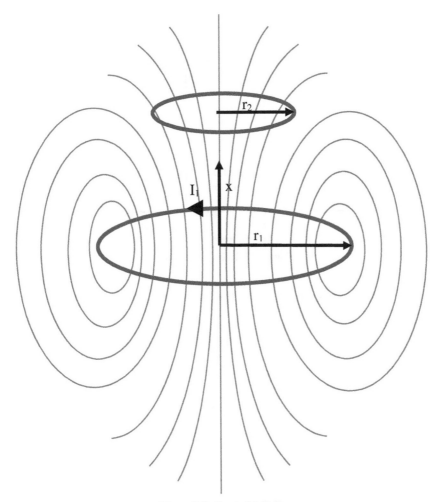

Figure 8.2 Co-Axial Coils

Therefore at resonance the transformed load becomes purely resistive and has value

$$Z_{LT0} = R_{LT} = \frac{1}{\omega_0^2 C_2^2 R_L} = \frac{\omega_0^2 L_2^2}{R_L} \tag{8.6}$$

Therefore the total impedance of the receiver is resistive at resonance and is given by

$$R_R = R_2 + R_{LT} = \omega_0 L_2 \left(\frac{R_2}{\omega_0 L_2} + \frac{\omega_0 L_2}{R_L} \right) = \frac{\omega_0 L_2}{Q_2} \tag{8.7}$$

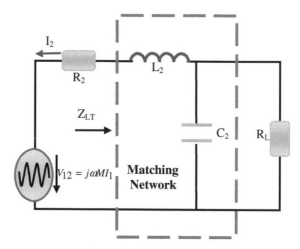

Figure 8.3 Receiver Circuit

The total receiver resistance decreases with the Q of the receiver, where

$$Q_2 = \left(\frac{R_2}{\omega_0 L_2} + \frac{\omega_0 L_2}{R_L} \right)^{-1} = \left(\frac{1}{Q_{22}} + Q_{2L} \right)^{-1} \tag{8.8}$$

Following the analysis method is [1], the impedance of the receiver can be transformed into the network representing the transmitter. This transformation is undertaken by writing the induced current in the mutual inductance in the receiver as follows:

$$I_2 = -\frac{V_{I2}}{R_R} = -\frac{j\omega_0 M I_1}{R_R} \tag{8.9}$$

$$V_{I1} = j\omega_0 M I_2 = \frac{\omega_0^2 M^2 I_1}{R_R} = Z_T I_1 \tag{8.10}$$

Where

$$Z_T = R_T = \frac{\omega_0^2 M^2}{R_R} = \omega k^2 L_1 Q_2 \tag{8.11}$$

This is the transformed receiver impedance at the transmitter circuit (Figure 8.4).

The power delivered by the transmitter to the receiver load at resonance therefore becomes

$$P_R = |I_1|^2 R_R \tag{8.12}$$

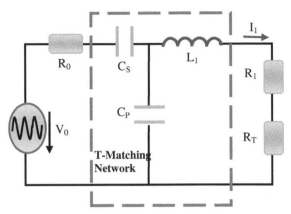

Figure 8.4 Transmitter Circuit

This power may be maximised by either increasing the transmitter current or increasing the receiver load. Increasing the transmitter current can lead to increased crosstalk in the network because higher currents create higher field strengths. Therefore it is preferable to increase the receiver load. Receiver load can be increased by either increasing the coupling coefficient k, the operating resonance frequency or the inductor of the transmitter or Q_2. What is the maximum quality factor of the receiver? This can be obtained by differentiating the expression for its quality factor with respect to L_2 or:

$$\frac{\partial Q_2}{\partial L_2} = \frac{\partial}{\partial L_2}\left[\left(\frac{R_2}{\omega_0 L_2} + \frac{\omega_0 L_2}{R_L}\right)^{-1}\right] = 0 \qquad (8.13)$$

$$L_{2,Q_{2\max}} = \frac{\sqrt{R_2 R_L}}{\omega_0}$$

Hence the choice of the load resistance and the inherent resistance of the coil determine the maximum quality factor of the receiver. This choice of inductive value can be shown to imply that the load resistance be matched to R_2 as demonstrated in [1] or.

$$R_{LT,Q_{\max}} = \frac{\omega_0^2 L_{2,Q_{2\max}}^2}{R_L} = \frac{\omega_0^2}{R_L} \frac{R_L R_2}{\omega_0^2} = R_2 \qquad (8.14)$$

Hence the power delivered to the receiver is split equally between the load resistance and the internal resistance of the coil. It has been shown in [1] that maximum power is delivered to the load for an optimum number of turns of the coil given by the expression

$$N_{2, P_{RL} \text{max}} = \left\lceil \sqrt[3]{\frac{2 R_{20} R_L}{\omega_0^2 L_{20}^2}} \right\rceil \tag{8.15}$$

Where $\lceil \zeta \rceil$ means that the variable ζ is rounded up to the next higher integer.

8.1.1 Coupling and crosstalk

We have shown previously that crosstalk (mutual coupling) exists between transmitters and between receivers of the same system. For example, Figure 8.5 represents a peer-to-peer system of two transceivers in which the transmitter 1 has a voltage source V_{S1} and the other transmitter has source V_{S3}. There exist six coupling coefficients in this peer-to-peer system represented by the k parameters. Power and data transfer using inductive links are normally implemented using a series resonant circuit in the transmitter and a parallel resonant circuit in the receiver as shown in Figure 8.5. Thus the transmitter is loaded with low impedance and is powered with a power amplifier represented in Figure 8.5 as the voltage source V_{S1} and the series resistance R_1. Figure 8.5 is a narrow band model. A broadband model of the system has a series resistance in addition to R_1 in the transmitter. The broadband model is shown in Figure 8.6.

The receivers and transmitters are oriented to ensure that the coils of two transmitters are not directly collinear (not facing each other directly), but rather face the receiver of its peer. Internal coupling inside a transceiver is represented by red lines and the blue lines represent desirable power coupling between one transmitter and the receiver of its peer.

The difference between the narrow band model and the wide band model is in the series resistors. The narrowband model (Figure 8.5) combines the resistors so that $R_1 = R_{SRC} + R_{S1}$ at the transmitter and the resistor $R_2 = R_{p2} R_L / R_{p2} + R_L$ is inserted in series also in the receiver (Figure 8.6).

8.1.2 Narrowband approximation

The load is modelled as a resistor with a peak-to-peak voltage $V_{L,pp}$ and the power across it as $P_L = V_{L,PP}^2 / 8 R_L$. The power transfer function is [3]

$$G_P(s) = \frac{P_L}{P_{SRC}} = \frac{Re(V_L I_{R_l}^*)}{Re(V_S I_1^*)} \tag{8.16}$$

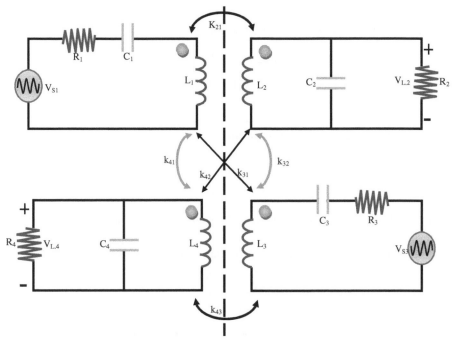

Figure 8.5 Power and Data Transfer Model

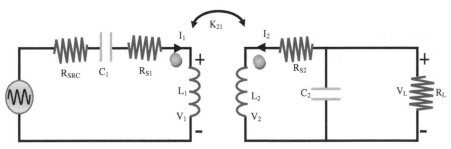

Figure 8.6 Broadband Model

This leads to the power transfer function

$$G_P(\omega) = \frac{\omega^2 k^2 L_1 L_{p2} R_2}{\omega^2 (L_{p2}^2 R_1 + k^2 L_1 L_{p2} R_2) + R_1 R_2^2 (1 - \omega^2 L_{p2} C_2)^2} \cdot \frac{R_{p2}}{R_{p2} + R_L}$$

$$(8.17)$$

The resonant frequency of the tank circuit in the receiver is $\omega_0 = 1/\sqrt{L_{p2}C_2}$ and the maximum efficiency of the receiver is therefore

$$\eta_{\max} = G_P(\omega_0) = \frac{1}{1 + \dfrac{1}{k^2 Q_1' Q_2'}} \cdot \frac{R_{p2}}{R_{p2} + R_L} \tag{8.18}$$

$Q_1' = \omega L_1/R_1$ and $Q_2' = R_2/\omega L_{p2}$.

The optimum value of the capacitor C_1 in the transmitter for maximum power transfer to the receiver load is given by the expression

$$C_{1,opt} = \frac{L_{p2}C_2}{L_1(1 - k^2)} \tag{8.19}$$

Thus when large coupling is desired (k is large), a large capacity is used as well.

The resonant frequency for the non-narrowband approximation is

$$G_P(\omega) = \frac{\sqrt{R_{S2} + R_L}\sqrt[4]{R_{SRC} + R_{S1}}}{\sqrt{C_2 R_L}\sqrt[4]{L_2^2(R_{SRC} + R_{S1}) + k^2 L_1 L_2 R_{S2}}} \tag{8.20}$$

This is a function of the coupling coefficient of the system.

8.1.3 Communications between transmitter and receiver

This explanation follows the analysis in [4] which describes an integrated version of UHF near field RFID system with a fully integrated receiver. Using equation (14) of chapter 4 for a single turn coils in the transmitter and receiver

$$k^2(x) = \frac{\mu_0 A_2^2 \mu_0 r_1^4}{4 L_1 L_2 [(r^2 + x^2)]^3} \tag{8.21}$$

The mutual inductance between the transmitter and receiver respectively for N_1 and N_2 turns in the transmitter and receiver is

$$M = k\sqrt{L_1 L_2} = \frac{N_1 N_2 \mu_0 A_2 r_1^2}{2\sqrt{[(r^2 + x^2)]^3}} \tag{8.22}$$

Given that the area of the circular receiver coils $A_2 = \pi r_2^2$, then the mutual inductance is

$$M = k\sqrt{L_1 L_2} = \frac{\pi \mu_0 N_1 N_2 r_2^2 r_1^2}{2\sqrt{[(r_1^2 + d^2)]^3}} \tag{8.23}$$

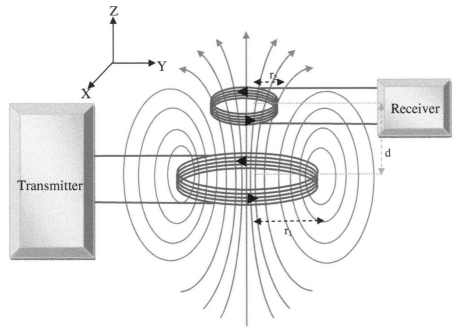

Figure 8.7 Magnetic Induction Communication System

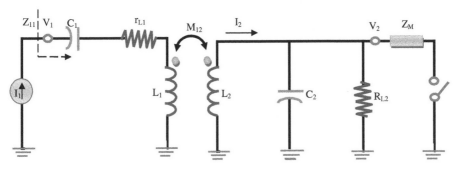

Figure 8.8 Equivalent Circuit of Magnetic Induction Communication System

In this equation **d** is the distance between the inductive transmitter (with coil of radius r_1) and receiver.

The equivalent circuit model of this transmitting system is given by the circuit in Figure 8.8.

The alternating current in the transmitter coil induces a voltage in the tank circuit of the receiver. The receiver responds by modulating the impedance of the transmitter resulting in both amplitude and phase variation there. This phenomenon is referred to as impedance or load modulation.

The communication between the receiver and transmitter is characterised by the trans-impedance which is the ratio of the induced voltage in the receiver coil to the current flowing in the transmitter coil. This impedance is

$$Z_{12} = \frac{V_2}{I_1} = \omega_0 M Q_2 \tag{8.24}$$

Or

$$Z_{12} = \frac{\omega_0 \pi \mu_0 N_1 N_2 \cdot r_2^2 r_1^2 Q_2}{2\sqrt{[(r_1^2 + d^2)]}^{-3}} = N_2 \cdot Q_2 \pi \cdot r_2^2 \left[\frac{(\omega_0 \mu_0 N_1 r_1^2)}{2\sqrt{[(r_1^2 + d^2)]}^{-3}} \right] \tag{8.25}$$

The trans-impedance is a function of the coil parameters, shape, the distance between the coils, the quality factor of the receiver coil, the resonance frequency and the number of turns in the coils. Noting that $Q_2 = R_2/\omega_0 L_2$, the role of the resonance frequency is removed from the above equation.

8.1.4 Receiver to transmitter communications

The reverse communication between the receiver and transmitter is characterised by the variation of the impedance seen at the input to its coil. This impedance is given by the expression

$$Z_{11} = \frac{V_1}{I_1} = r_{L1} + \sqrt{\frac{C_2}{L_2}} \cdot \frac{\omega_0^2 M^2 Q_2}{1 + c \cdot R_{L2}/R_M}$$

$$= \sqrt{\frac{C_2}{L_2}} \cdot \frac{\omega_0^2 Q_2}{1 + c \cdot R_{L2}/R_M} \cdot \left(\frac{\pi \mu_0 N_1 N_2 r_2^2 r_1^2}{2\sqrt{[(r_1^2 + d^2)]}^{-3}} \right)^2 \tag{8.26}$$

Clearly, the roles of the receiver components (capacitor and inductor in the tank circuit), mutual inductance between the transmitter and receiver and the coupling coefficient are prominent in this expression as are the roles of the quality factor of the receiver coil and the resonant frequency. In this equation c represents the state of the modulation switch ($c = 0$ when the switch is open and $c = 1$ when it is closed). The variation of the input impedance is therefore around the two extremes and is given by

$$\Delta Z_{11}(\omega_0) = Z_{11}(\omega_0)|_{k=0} - Z_{11}(\omega_0)|_{k=1} \tag{8.27}$$

$$= \sqrt{\frac{C_2}{L_2}} \cdot \frac{\omega_0^2 M^2 Q_2}{1 + R_M/R_{L2}}$$

The modulation depth of the amplitude modulated signal received by the receiver is defined by the ratio

$$m = \left| \frac{\Delta Z_{11}(\omega_0)}{Z_{11}(\omega_0)} \right| = \frac{\sqrt{\dfrac{C_2}{L_2}} \cdot \dfrac{\omega_0^2 M^2 Q_2}{1 + R_M/R_{L2}}}{r_{L1} + \sqrt{\dfrac{C_2}{L_2}} \cdot \dfrac{\omega_0^2 M^2 Q_2}{1 + c \cdot R_{L2}/R_M}} \tag{8.28}$$

The receiver power determined by the product $|\Delta Z_{11}(\omega_0) I_1|$ should be higher than the receiver sensitivity P_S for communication to be established with the transmitter: $P_S < |\Delta Z_{11}(\omega_0) I_1|$.

8.1.5 General conclusions

1. To integrate the receiver onto a system of body area network with metal traces will reduce the quality factor of the receiver and hence its performance.
2. The coverage range is likely to be limited by the trans-impedance (receiver power up).
3. In general the trans-impedance can be shown to depend strongly on the receiver coil features as, and the constant in this expression relates to the transmitter coil and its shape: $Z_{12} \propto N_2 \cdot Q_2 \pi \cdot r_2^2 \cdot \text{constant}$
4. The optimum radius of the transmitter coil for maximum mutual inductance is given by differentiating the equation for the mutual inductance with respect to r_1. The value is

$$r_1 = \sqrt{2 \cdot d} \tag{8.29}$$

   By knowing the dimensions of the structure on which the transceivers are to be located (eg, a human torso), we can clearly determine where to locate them for optimum mutual coupling of adjacent transceivers.
5. The minimum voltage required for the transmitter is given by the expression

$$I_{TX} = \frac{V_{2,\min}}{Z_{12}(\omega_0)} \tag{8.30}$$

When $d \gg r_1$, we observe that $M \propto d^{-3}$. Therefore, $I_{TX} \propto d^3$. Hence to increase the size of the bubble, it is required to increase the transmitter current by an order of about 8.

8.1.6 Bandwidth of links

The speed at which data is transferred between a source and receiver is dependent on the communication bandwidth. The bandwidth of a link is computed as its 3 dB bandwidth and is obtainable from the transfer function of the link.

References

[1] Peter Scholz, Christian Reinhold, Werner John and Ulrich Hilleringmann, "Analysis of Energy Transmission for Inductive RFID tags", in Proc. IEEE International Conference on RFID, Gaylord Texan Resort, Grapevine, TX, USA, Mar. 26–28, 2007, pp. 183–190.

[2] Noriyuki Miura, Takayasu Sakurai and Tadahiro Kuroda, "Crosstalk Countermeasures for High Density Inductive-Coupling Channel Arrays", IEEE Journal of Solid–State Circuits, Vol. 2, No. 2, Feb. 2007, pp. 410–421.

[3] Kanber Mithat Silay, Catherine Dehollain and Michel Declercq, "Improvement of Power Efficiency of Inductive Links for Implantable Devices", in Proc IEEE, 2008, pp. 229–232.

[4] Amin Shameli, Aminghasem Safarian, Ahmadreza Rofougaran, Maryam Rofougaran, Jesus Castaneda, and Franco De Flaviis, "A UHF Near-Field RFID System with a Fully Integrated Transponder", IEEE Transactions on Microwave Theory and Techniques, Vol. 56, No. 5, May 2008, pp. 1267–1277.

9

Efficiency of Near Field Magnetic Induction Communication Links

The contents of this chapter is a presentation of the analysis of the efficiency of magnetic induction communication links in the near field region. It considers when the transmitters and receiver circuits are compensated and when they are not. Series compensation is used for the transmitters and both series and parallel compensation methods are compared for the receiver. The parallel compensation method always out performs the series receiver compensation in terms of the power delivered to the receiver load.

9.1 Efficiency of Inductive Communication Links

Usually because of the relative dimensions of the transmitter and receiver coils and the comparatively larger distance separating them, the magnetic coupling between them is small, or the so-called loosely coupled system. Thus the flux that is produced by the transmitter and intercepted by the receiver coil is very small. Hence the coupling coefficient is also small and in the order of about 10% or less. This coupling value can be increased by increasing the inductive values of the coils, reducing the separation between them as well as increasing the areas of the coils that intercept the magnetic field. The coupling efficiency is further reduced by the parasitic values of the coils. The approach for correcting for these inherent flaws is called link optimization and uses compensating circuit elements such as capacitors either in parallel or in series with the inductors. Compensating circuits are discussed in this chapter within the concepts of trans-impedance.

9.1.1 Trans-impedance

Following the method in [3], consider the following circuit which describes a voltage source V_G with output impedance Z_G coupled to a load Z_L by

J. Ihyeh Agbinya (PhD), Principles of Inductive Near Field Communications for Internet of Things, 107–124.

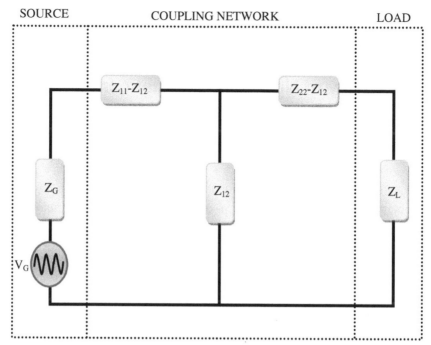

Figure 9.1 Power Coupling from Source to Load Using a Reciprocal Two-Port

a coupling network shown as a T-section in Figure 9.1 [3]. Figure 9.1 is generic and can be used to represent both near field reactive and far field radiative coupling between source and load.

We define the following concepts which are crucial for the analysis of the links. From Figure 9.1, the following impedances are defined. For near field communication without capacitive compensations or additions the following impedances are defined.

Input impedance *at the source* (transmitter) is defined by the expression

$$Z_{11} = R_T + j\omega L_T \tag{9.1}$$

The **input impedance** *at the load* input port (with the other side open) is

$$Z_{22} = R_R + j\omega L_R \tag{9.2}$$

The **transfer impedance** between the source and the receiver is due to the mutual coupling between the coils and is

$$Z_{12} = j\omega M \tag{9.3}$$

With these three impedances, the following concepts are further defined. The *input impedance seen at the source* (transmitter) terminals is

$$Z_T = Z_{11} - \frac{Z_{12}^2}{Z_{22} + Z_L} \tag{9.4}$$

The *input impedance seen at the receiver* terminals is

$$Z_R = Z_{22} - \frac{Z_{12}^2}{Z_{11} + Z_G} \tag{9.5}$$

From these equations, two *reflected impedances* at the transmitter and receiver (the net effect of the mutual coupling between them) are defined as

$$Z_{TREF} = \frac{Z_{12}^2}{Z_{22} + Z_L} \tag{9.6}$$

This impedance is reflected from the load (receiver) back into the transmitter. Similarly the *reflected impedance into the receiver* due to the source is

$$Z_{RREF} = \frac{Z_{12}^2}{Z_{11} + Z_G} \tag{9.7}$$

In general *the intervening medium between the transmitter and receiver may have a non-zero conductivity and hence will introduce finite impedance* of the form $Z_m = R_m + jX_m$. This impedance will modify the input impedances previously defined for the source and the load and also the transfer impedance. For example, if the medium is biological with some conductivity, it will add a non-zero resistive loss and a capacitive reactance coupling the transmitter and receiver.

9.1.2 Uncompensated link

The power transfer efficiency of the magnetic inductive link in the communication system is determined by the flux linkage between the transmitter and the receiver coils. The efficiency of the link is the product of the transmitter and receiver coil efficiencies. The efficiency of each coil is the ratio of the power delivered to the next stage to the power in the coil. The power transfer efficiency and coil power capacity depend on the link efficiency. The link efficiency can be optimized by accounting for the inherent losses or signal leakages in the transmitter and receiver inductors. This is achieved by adding resonant capacitors the so-called coil compensation procedure. This increases the power transfer efficiency of the system. Figure 9.2 shows an uncompensated inductive link and its equivalent circuit in Figure 9.3.

The subscript T and R refer to the transmitter and receiver respectively and M is the mutual inductance.

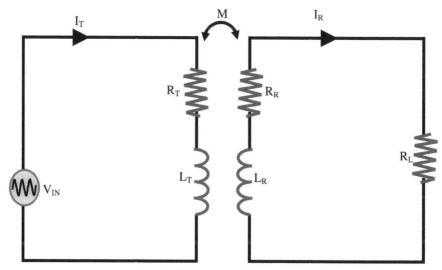

Figure 9.2 Series – Series Topology

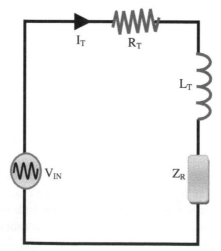

Figure 9.3 Equivalent Circuit of Figure 9.2

The behaviour of an inductive link is normally assessed by analysing the trans-impedance of the link and by inserting the reflected impedance of the receiver circuit in series with the transmitter inductor. This is shown in Figure 9.3. Consider the uncompensated case in Figures 9.2 and 9.3. In the uncompensated case, parasitic capacitors are not included in the circuits.

Consider Figure 9.2 and assume it is driven by an input current source $i_{in} = I_{in}\sin(\omega_0 t)$. For simplicity assume that $Z_G = 0$, hence the following equations are obtained that

$$Z_{in} = Z_G + Z_T = 0 + Z_{11} - \frac{Z_{12}^2}{Z_{22} + Z_L}$$

$$= (j\omega L_T + R_T) - \frac{\omega^2 M^2}{R_L + R_R + j\omega L_R} \tag{9.8}$$

By operating the link at resonance, the reflected impedance is given by the expression

$$Z_{TREF} = \frac{\omega^2 M^2 (R_R + R_L)}{(R_R + R_L)^2 + \omega^2 L_R^2} - j\frac{\omega^3 M^2 L_R}{(R_R + R_L)^2 + \omega^2 L_R^2}$$

$$= R_{TREF} + jX_{TREF} \tag{9.9}$$

The reflected impedance is formed as the sum of the reflected resistance and reflected reactance. The reflected resistance and reactance are due to the mutual coupling, the inductance of the receiver, the resistances in the receiver circuit and load. The link efficiency of Figure 9.2 is defined as the product of the efficiencies of the transmitter section and the efficiency of the receiver which are defined by the expressions:

$$\eta = \eta_T . \eta_R \tag{9.10}$$

Where

$$\eta_T = \frac{Power\ delivered\ to\ the\ receiver\ circuit}{Total\ power\ handled\ by\ the\ transmitter\ circuit} \tag{9.11}$$

and

$$\eta_R = \frac{Power\ delivered\ to\ the\ load}{Total\ power\ handled\ by\ the\ receiver\ circuit} \tag{9.12}$$

Therefore, we have

$$\eta_T = \frac{R_{TREF}}{R_T + R_{TREF}} = \frac{\omega^2 M^2 (R_R + R_L)}{R_T[(R_R + R_L)^2 + \omega^2 L_R^2] + \omega^2 M^2 (R_R + R_L)}$$

$$\eta_R = \frac{R_L}{R_R + R_L}$$

$$\eta = \frac{\omega^2 M^2 R_L}{R_T[(R_R + R_L)^2 + \omega^2 L_R^2] + \omega^2 M^2 (R_R + R_L)}$$

$$= \frac{\omega^2 M^2 R_L}{R_T[a^2 + \omega^2 L_R^2] + a\omega^2 M^2} \tag{9.13}$$

Where $a = (R_R + R_L)$. The link efficiency represents the power delivered to the receiver circuit and to the receiver load compared with the total power handled by the transmitter and receiver.

The link efficiency for an inductor of 2.5 mH at the transmitter and 39 μH at the receiver is shown in Figure 9.4. Hence the power loss in the over all communication system is the difference between the power handled by the transmitter and the link efficiency which is

$$P_{LOSS} = 1 - \frac{\omega^2 M^2 R_L}{R_T[(R_R + R_L)^2 + \omega^2 L_R^2] + \omega^2 M^2 (R_R + R_L)}$$

$$= \frac{R_T[(R_R + R_L)^2 + \omega^2 L_R^2] + \omega^2 M^2 R_R}{R_T[(R_R + R_L)^2 + \omega^2 L_R^2] + \omega^2 M^2 (R_R + R_L)} \tag{9.14}$$

We will henceforth write the power loss as a decibel value with the following type of expressions:

$$P_{LOSS}(dB) = 10 \left[\begin{array}{l} \log_{10}(R_T[(R_R + R_L)^2 + \omega^2 L_R^2] + \omega^2 M^2 R_R) + \\ - \log_{10}(R_T[(R_R + R_L)^2 + \omega^2 L_R^2] \\ + \omega^2 M^2 (R_R + R_L)) \end{array} \right] (dB)$$

$$\tag{9.15}$$

This expression represents all the power losses in the system including in the channel.

9.1.3 Compensated link efficiency

Parasitic capacitances and part and parcel of the coils used in the transmitter and receiver and hence compensation for them is essential. Normally the

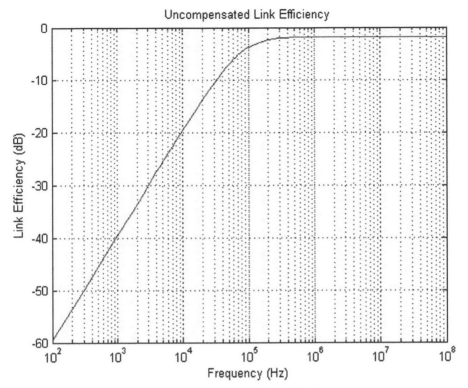

Figure 9.4 Uncompensated Link Efficiency

transmitter is a series tuned circuit and hence it is reasonable to consider a series compensation for it. The receiver is normally a parallel tank circuit. To ensure that we have made the right choices, the receiver will be compensated using both series and parallel compensation which will provide a reasonable guide to which form of compensation is best for the receiver. This choice will be based on which form of compensation delivers optimum power to the receiver load. In Figure 9.5 the parasitic capacitances of the transmitter and receiver are considered in the topologies. This is a high frequency model of the transmitter and receiver respectively.

For some applications, the parasitic capacitances will be very small and are ignored. Figure 9.6 shows the transmitter coil compensation topologies.

The receiver models are also considered as either parallel of series and shown in Figure 9.7.

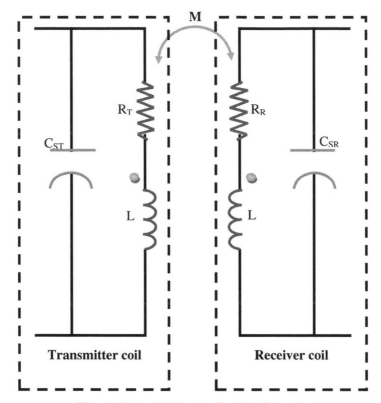

Figure 9.5 Models Showing Parasitic Capacitors

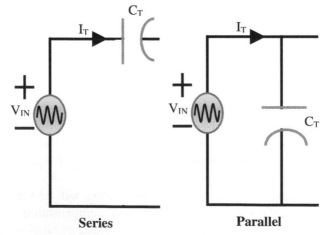

Figure 9.6 Transmitter Coil Compensation Topologies

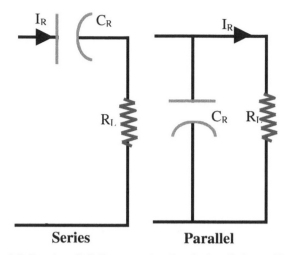

Figure 9.7 Receiver Coil Compensation Topologies: Series and Parallel

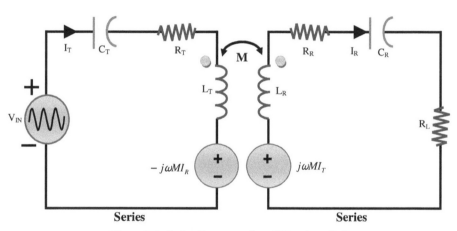

Figure 9.8 Series Compensation of Receiver Coil

In the subsequent analysis, the parasitic capacitances of the transmitter and receiver coils are neglected.

9.1.4 Series compensation of transmitter and receiver coils

The coils in the transmitter and receiver use series compensation topologies.

The reflected impedance of the receiver to the transmitter is given by the expression

$$Z_{REC} = (R_R + R_L) + j\left(\omega L_R - \frac{1}{\omega C_R}\right) \tag{9.16}$$

$$Z_{RREF} = \frac{Z_{12}^2}{Z_{11} + Z_G} = \frac{\omega^2 M^2}{(R_R + R_L) + j\left(\omega L_R - \frac{1}{\omega C_R}\right)} \tag{9.17}$$

Therefore,

$$Z_{TREF} = \frac{\omega^2 M^2 (R_R + R_L)}{(R_R + R_L)^2 + \left(\omega L_R - \frac{1}{\omega C_R}\right)^2}$$

$$- j \frac{\omega^2 M^2 \left(\omega L_R - \frac{1}{\omega C_R}\right)}{(R_R + R_L)^2 + \left(\omega L_R - \frac{1}{\omega C_R}\right)^2} \tag{9.18}$$

The link efficiency is computed as before with the expressions

$$\eta_T = \frac{R_{TREF}}{R_T + R_{TREF}} = \frac{\dfrac{\omega^2 M^2 (R_R + R_L)}{(R_R + R_L)^2 + \left(\omega L_R - \frac{1}{\omega C_R}\right)^2}}{R_T + \dfrac{\omega^2 M^2 (R_R + R_L)}{(R_R + R_L)^2 + \left(\omega L_R - \frac{1}{\omega C_R}\right)^2}}$$

$$= \frac{\omega^2 M^2 (R_R + R_L)}{R_T\left[(R_R + R_L)^2 + \left(\omega L_R - \frac{1}{\omega C_R}\right)^2\right] + \omega^2 M^2 (R_R + R_L)} \tag{9.19}$$

$$\eta_R = \frac{R_L}{R_R + R_L}$$

$$\eta = \eta_T \eta_R$$

$$\eta = \frac{\omega^2 M^2 R_L}{R_T \left[(R_R + R_L)^2 + \left(\omega L_R - \frac{1}{\omega C_R} \right)^2 \right] + a\omega^2 M^2 (R_R + R_L)} \quad (9.20)$$

The effect of the series compensation at the receiver modifies the efficiency of the transmitter through the reactive term in the denominator of the equations. At resonance $\omega_0^2 = 1/L_R C_R$ the efficiency of the peer-to-peer communication link becomes

$$\eta = \frac{\omega^2 M^2 R_L}{R_T \left[(R_R + R_L)^2 + \omega^2 L_R^2 \left(1 - \frac{\omega_0^2}{\omega^2} \right)^2 \right] + a\omega^2 M^2 (R_R + R_L)} \quad (9.21)$$

9.1.5 Series compensation of transmitter & parallel compensation of receiver

In this analysis the receiver has a parallel compensation with the parallel capacitance C_R. The transmitter compensation is series.

The circuit impedance of the receiver is

$$Z_{REC} = j\omega L_R + R_R + \frac{R_L}{1 + j\omega C_R R_L} \quad (9.22)$$

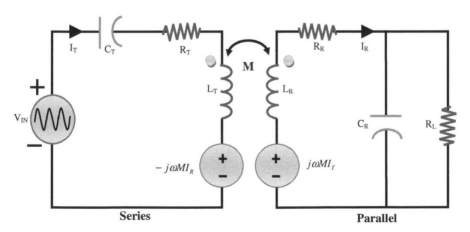

Figure 9.9 Parallel Compensation of Receiver Coil

Or

$$Z_{REC} = \frac{(R_R + R_L) + \omega^2 R_R C_R^2 R_L^2}{1 + \omega^2 C_R^2 R_L^2} + j\left[\frac{\omega L_R - \omega C_R R_L^2 + \omega^3 L_R C_R^2 R_L^2}{1 + \omega^2 C_R^2 R_L^2}\right]$$

(9.23)

As before the reflected impedance from the receiver to the transmitter is

$$Z_{TREF} = \frac{\omega^2 M^2}{j\omega L_R + R_R + \dfrac{R_L}{1 + j\omega C_R R_L}}$$

(9.24)

This simplifies to

$$Z_{TREF} = \frac{\begin{array}{c}\omega^2 M^2((R_R + R_L) + \omega^2 R_R C_R^2 R_L^2) \\ + j[\omega^2 M^2(\omega C_R R_L^2 - \omega^3 L_R C_R^2 R_L^2 - \omega L_R)]\end{array}}{[(R_R + R_L) - \omega^2 L_R R_L C_R]^2 + (\omega L_R + \omega R_R R_L C_R)^2}$$

(9.25)

As in the previous analysis, the transmitter efficiency at resonance is

$$\eta_T = \frac{\omega_0^2 M^2((R_R + R_L) + \omega_0^2 R_R C_R^2 R_L^2)}{\begin{array}{c}\omega_0^2 M^2((R_R + R_L) + \omega_0^2 R_R C_R^2 R_L^2) + R_T([(R_R + R_L) \\ -\omega_0^2 L_R R_L C_R]^2 + (\omega_0 L_R + \omega_0 R_R R_L C_R)^2)\end{array}}$$

(9.26)

Or

$$\eta_T = \frac{\omega_0^2 M^2(a + \omega_0^2 R_R C_R^2 R_L^2)}{\begin{array}{c}\omega_0^2 M^2(a + \omega_0^2 R_R C_R^2 R_L^2) + R_T([a - \omega_0^2 L_R R_L C_R]^2 \\ +(\omega_0 L_R + \omega_0 R_R R_L C_R)^2)\end{array}}$$

(9.27)

The receiver efficiency for this case is given by the expression

$$\eta_R = \frac{Re\left[\dfrac{R_L}{1 + j\omega C_R R_L}\right]}{Re\left[j\omega L_R + R_R + \dfrac{R_L}{1 + j\omega C_R R_L}\right]}$$

(9.28)

This simplifies to

$$\eta_R = \frac{R_L}{(R_L + R_R) + \omega_0^2 R_L R_R C_C^2} = \frac{R_L}{a + \omega_0^2 R_L R_R C_C^2}$$

(9.29)

The efficiency of the link therefore is

$$\eta = \frac{\omega_0^2 M^2 R_L}{R_T[((R_L + R_R) - \omega_0^2 R_R R_L C_R)^2 + \omega_0^2 (L_R + R_R R_L C_R)^2] + \omega_0^2 M^2 ((R_L + R_R) + \omega_0^2 R_R R_L^2 C_R^2)} \quad (9.30)$$

Or by defining $a = (R_L + R_R)$ and $b = R_R R_L C_R$

$$\eta = \frac{\omega_0^2 M^2 R_L}{R_T[(a - \omega_0^2 b)^2 + \omega_0^2 (L_R + b)^2] + \omega_0^2 M^2 (a + \omega_0^2 b R_L C_R)} \quad (9.31)$$

$$\times \eta(\omega' = \omega_0^2) = \frac{\omega' M^2 R_L}{\alpha \omega'^2 + \beta \omega' + \delta}$$

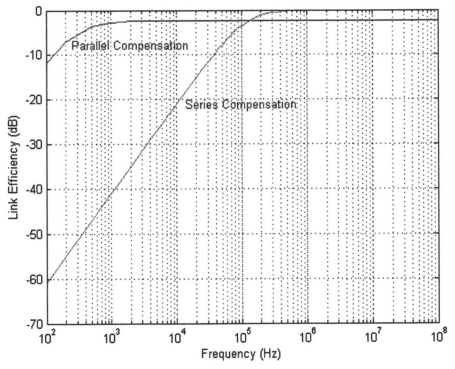

Figure 9.10 Series and Parallel Compensation of Receiver Coil

Where

$$\alpha = b^2 R_T + bM^2 R_L C_R$$
$$\beta = -2abR_T + (L_R + b)^2 R_T + M^2 a \qquad (9.32)$$
$$\delta = R_T a^2$$

Clearly compensation has helped in enhancing the power transfer to the load. The parallel compensation of receiver coil out performs series compensation by boosting the power across the load and the power handled by the receiver circuit more.

9.1.6 Parallel compensation of receiver with parasitic capacitance

With the parasitic capacitance in parallel with the capacitance of the receiver, the equivalent capacitance is $C_{eq} = C_R + C_P$.

For this case the results in the previous section are equally applicable by replacing C_R with C_{eq}. Therefore the link efficiency is

$$\eta = \frac{\omega_0^2 M^2 R_L}{R_T[(a - \omega_0^2 b_{eq})^2 + \omega_0^2(L_R + b_{eq})^2] + \omega_0^2 M^2(a + \omega_0^2 b_{eq} R_L C_{eq})} \qquad (9.33)$$

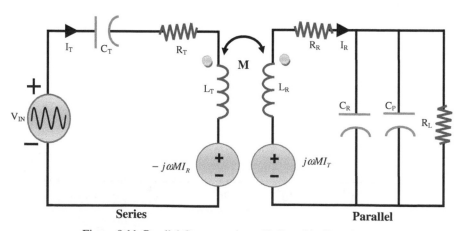

Figure 9.11 Parallel Compensation with Parasitic Capacitance

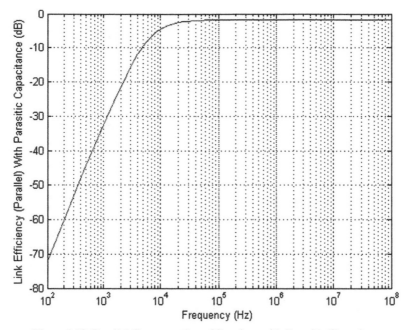

Figure 9.12 Parallel Compensation of Receiver with Parasitic Capacitor

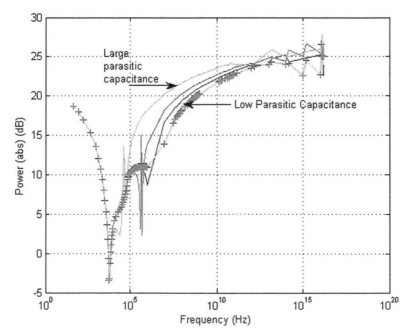

Figure 9.13 Effect of Parasitic Capacitance of Coil on Power Delivery to Load

Where, $b_{eq} = R_R R_L C_{eq}$. Results for the link efficiency are shown in Figure 9.12.

$$\eta(\omega' = \omega_0^2) = \frac{\omega' M^2 R_L}{\alpha \omega'^2 + \beta \omega' + \delta}$$

Where

$$\alpha = b_{eq}^2 R_T + b_{eq} M^2 R_L C_{eq}$$

$$\beta = -2ab_{eq} R_T + (L_R + b_{eq})^2 R_T + M^2 a$$

$$\delta = R_T a^2 \tag{9.34}$$

Consideration of the parasitic capacitance affects only the α coefficient of the denominator quadratic equation of the link. The link efficiency has thus been transformed into a transfer function in ω_0^2. The power delivered to the receiver

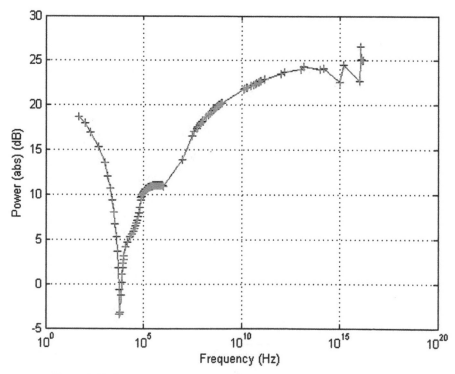

Figure 9.14 Better Power Delivery at Reduced Coil Parasitic Capacitance

load is shown in Figure 9.13. The impact of parasitic capacitors is shown as the spikes in the early part of the graph.

Less parasitic capacitance leads to smoother power delivery both at the low frequency range and also at the very high frequency range. This is shown in the three graphs and the separate graph.

References

[1] Kurt A. Sandquist, Design of a Wireless Power and Data Link for a Cranially–Implanted Neural Prosthesis, MSc thesis, Department of Electrical Engineering, Kansas State University, Manhattan, Kansas, 2004.

[2] Noriyuki Miura, Takayasu Sakurai and Tadahiro Kuroda, "Crosstalk Countermeasures for High Density Inductive-Coupling Channel Arrays", IEEE Journal of Solid–State Circuits, Vol. 2, No. 2, Feb. 2007, pp. 410–421.

[3] Soumyajit Mandal and Rahul Sarpeshkar, "Power-Efficient Impedance-Modulation Wireless Data Links for Biomedical Implants", IEEE Transactions on Biomedical Circuits and Systems, Vol. 2, No. 4, Dec. 2008, pp. 301–315.

[4] Venkat Reddy Gaddam, Remote Power Delivery for Hybrid Integrated Bio-Implantable Electrical Stimulation System, MSc thesis, Department of Electrical and Computer System, Louisiana State University, 2005.

10

Crosstalk in Near Field Magnetic Communication Links

The intention in this chapter is to present analysis of crosstalk in NFMI and how to represent crosstalk in inductive communications. It provides the means to understand the effect of crosstalk in the links and on performance.

10.1 Mutual Inductance

It is a well known thing in Physics that a varying current flowing through a metallic coil creates a magnetic field H and a corresponding magnetic flux (from the field). Any other coil in close proximity that lies in the path of the magnetic flux of the primary coil also experiences this field as its turns cut through the magnetic flux. This section derives the expression of the mutual inductance from classical considerations. In previous chapters we have shown that the ratio of the total flux intercepted by an area A of the secondary coil to the current flowing in it results to self inductance L. Mutual inductance between two coils is defined as:

$$M = \frac{\Psi_S}{I_P} \tag{10.1}$$

The subscripts S and P refer to the secondary and primary sections of the system. Mutual inductance is the ratio of the magnetic flux in the secondary coil to the current in the primary coil. Hence it represents the influence the current in the primary windings have on the secondary coil of the communication system.

For air cored coils, the magnetic flux is given by the integral

$$\Psi_S = \oint_{A_S} B_S(I_P)dA_S = B_S(I_P)N_S A_S = \mu_0 H_S(I_P)N_S A_S \tag{10.2}$$

J. Ihyeh Agbinya (PhD), Principles of Inductive Near Field Communications for Internet of Things, 125–140.

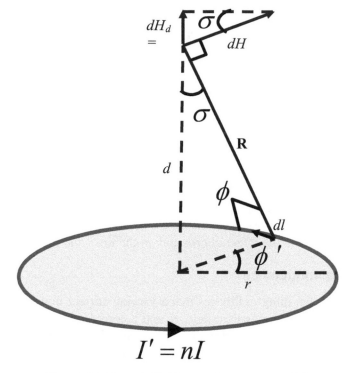

$$I' = nI$$

Figure 10.1 Magnetic Field Setup of a Single Turn Coil

This integration is taken over the area of the secondary coil to obtain the magnetic flux in the secondary coil. H is the magnetic field intensity in the secondary coil. The field intensity is derived. Consider the setup in Figure 10.1.

From Ampere's law the magnetic field intensity is given as

$$dH = \frac{I'dl \sin\phi}{4\pi R^2} = \frac{nIrd\phi'}{4\pi(d^2 + r^2)} \tag{10.3}$$

The magnetic field intensity along the distance d separating the coils is

$$
\begin{aligned}
H_d &= \int dH_d = \int dH \sin\sigma \\
&= \int \frac{rdH}{\sqrt{(d^2 + r^2)}} \\
&= \int_0^{2\pi} \frac{nIr^2 d\phi'}{4\pi(\sqrt{d^2 + r^2})^3} = \frac{nIr^2}{2(\sqrt{d^2 + r^2})^3}
\end{aligned}
\tag{10.4}
$$

We have thus obtained an expression for the total magnetic field at point z in Figure 10.1. This field is seen by the secondary coil and results to the mutual inductance. For near field body area networks the distance d is within the near field region ($d < \lambda/2\pi$) and the radius of the primary coil is far greater than the radius of the secondary coil $r_S \ll r_P$. By combining equations (10.1), (10.2) and (10.4) the resulting expression for the mutual inductance is

$$M = \frac{\mu_0 \pi N_S N_P r_S^2 r_P^2}{2(\sqrt{d^2 + r_P^2})^3} \tag{10.5}$$

The expression for the mutual inductance contains information on the self inductances of the transmitting and receiving coils as well as the coupling coefficient as function of the distance d.

10.1.1 Near field coupling model in SIMO systems

The previous chapters dealt with the mutual coupling between a transmitter and a receiver coil. They referenced and focused on one-to-one systems. We extend this formalism to the case when several receiving coils are involved. Parallel tuned circuits and compensation of receiver coils are used as it provides optimum power transfer to receivers. When several receiver coils are placed within the proximity of a transmitting coil, each one experiences mutual coupling with the transmitter. The impedance of each receiver stage n is

$$Z_R = j\omega L_{Rn} + R_{Rn} + \frac{R_{Ln}}{1 + j\omega C_{Rn} R_{Ln}} \tag{10.6}$$

$$Z_R = \frac{(R_{Rn} + R_{Ln}) + \omega^2 R_{Rn} C_{Rn}^2 R_{Ln}^2}{1 + \omega^2 C_{Rn}^2 R_{Ln}^2}$$

$$+ j \left[\frac{\omega L_{Rn} - \omega C_{Rn} R_{Ln}^2 + \omega^3 L_{Rn} C_{Rn}^2 R_{Ln}^2}{1 + \omega^2 C_{Rn}^2 R_{Ln}^2} \right] \tag{10.7}$$

For each stage we have transfer impedance at the transmitter as

$$Z_T = Z_{11} - \frac{Z_{Tn}^2(n)}{Z_{22}(n) + Z_L} \tag{10.8}$$

$$Z_{Tn}(n) = \frac{V_{2r}}{I_T} = \omega_0 M_{Tn} Q_{rn} \tag{10.9}$$

We assume that the receivers see equal load $R_{L1} = R_{L2} = \cdots = R_{LN} = R_L$

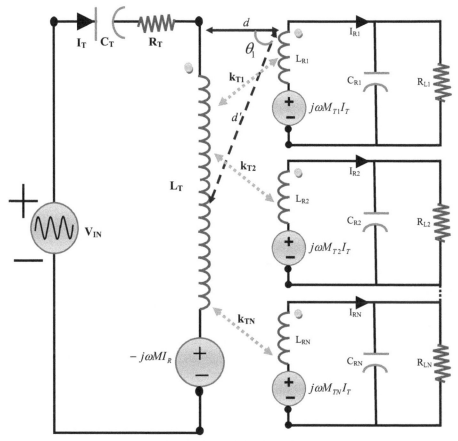

Figure 10.2 SIMO Channels

The reflected impedance for each stage is

$$Z_{TREF} = \frac{\omega^2 M_{Tn}^2}{j\omega L_{Rn} + \left(j\omega C_{Rn} + \dfrac{1}{R_{Ln}}\right)} = \frac{\omega^2 M_{Tn}^2}{j\omega L_{Rn} + R_{Ln}\dfrac{1 - j\omega C_{Rn} R_{Ln}}{1 + \omega^2 C_{Rn}^2 R_{Ln}^2}}$$

(10.10)

At resonance when $1 \ll \omega^2 C_2^2 R_L^2$ the transformed load at the denominator of equation (10.10) becomes purely resistive with the resonance frequency at $\omega_{0n} = 1/\sqrt{L_{Rn} C_{Rn}}$. Each receiver therefore resonates at its own resonant frequency or if they are balanced they resonate at the same frequency with value given as $\omega_{01} = \omega_{02} = \cdots = \omega_{0N} = \omega_0$. Thus several forms of the SIMO system can be formulated as follows.

a) Same resonance SIMO   In this case all the receivers resonate at the same frequency and the following relationships hold as

$$d'_n = d/\cos\theta_n \tag{10.11}$$

$$\omega_{01} = \omega_{02} = \cdots = \omega_{0N} = \omega_0$$

$$k_{Rn}(d'_n) = \frac{M_{TRn}}{\sqrt{L_T L_{Rn}}} = \left(\sqrt{\frac{\sqrt{C_n}}{\sqrt{L_T}}}\right)\omega_0 M_{TRn} \tag{10.12}$$

The power delivered to the receiver n load at resonance is given by the expression

$$I_T = \frac{V_{IN}}{R_T} \quad \text{and} \quad P_T = \frac{0.5 * |V_{IN}|^2}{R_T} \tag{10.13}$$

$$I_{Rn} = \frac{-j\omega_0 M_{TRn} I_T}{R_{Ln}} = \frac{-j\omega_0 M_{TRn} V_{IN}}{R_T R_{Ln}} \tag{10.14}$$

$$P_{Rn} = \frac{\omega_0^2 M_{TRn}^2 |V_{IN}|^2 R_{Ln}}{2(R_T R_{Ln})^2} \tag{10.15}$$

$$\frac{P_{Rn}}{P_T} = \frac{\omega_0^2 M_{TRn}^2 R_T R_{Rn}}{(R_T R_{Rn})^2} = \frac{\omega_0^2 k_{Rn}^2(d'_n) L_T L_{Rn}}{R_T R_{Rn}} = k_{Rn}^2 Q_T Q_{Rn} \tag{10.16}$$

When the transmitter and receiver have efficiencies given as η_T and η_{Rn} respectively the power delivered to the receiver n load is

$$P_{Rn} = \frac{\omega_0^2 P_T \eta_T Q_T}{L_T} C_n Q_{Rn} \eta_{Rn} M_{TRn}^2 \tag{10.17}$$

The total power delivered to the SIMO receiver load is

$$\sum_{n=1}^{N} P_{Rn} = \frac{\omega_0^2 P_T \eta_T Q_T}{L_T} \sum_{n=1}^{N} C_n Q_{Rn} \eta_{Rn} M_{TRn}^2 \tag{10.18}$$

$$= P_T Q_T \eta_T \sum_{n=1}^{N} k_{Rn}^2 \eta_{Rn} Q_{Rn}$$

Thus the total power delivered to the SIMO load is proportional to the quality factor of the transmitter and the product of the quality factor of the receiver and its coupling coefficient. This result is to be expected as we have assumed that the received signals are in phase at the load and at resonance. For equal Q receivers the total power received is

$$\sum_{n=1}^{N} P_{Rn} = P_T Q_T Q_R \eta_T \sum_{n=1}^{N} \eta_{Rn} k_{Rn}^2 \tag{10.19}$$

The efficiency of each link is the next objective.

$$Z_R = \frac{(R_{Rn} + R_{Ln}) + \omega^2 R_{Rn} C_{Rn}^2 R_{Ln}^2}{1 + \omega^2 C_{Rn}^2 R_{Ln}^2}$$
$$+ j \left[\frac{\omega L_{Rn} - \omega C_{Rn} R_{Ln}^2 + \omega^3 L_{Rn} C_{Rn}^2 R_{Ln}^2}{1 + \omega^2 C_{Rn}^2 R_{Ln}^2} \right] \tag{10.20}$$

The efficiency of each transmitter is

$$\eta_T = \frac{\omega_0^2 M^2 (a + \omega_0^2 R_{Rn} C_{Rn}^2 R_{Ln}^2)}{\omega_0^2 M^2 (a + \omega_0^2 R_{Rn} C_{Rn}^2 R_{Ln}^2) + R_T([a - \omega_0^2 L_{Rn} R_{Ln} C_{Rn}]^2} \tag{10.21}$$
$$+ (\omega_0 L_{Rn} + \omega_0 R_{Rn} R_{Ln} C_{Rn})^2)$$

The efficiency of each receiver is

$$\eta_R = \frac{Re\left[\dfrac{R_{Ln}}{1 + j\omega C_{Rn} R_{Ln}} \right]}{Re\left[j\omega L_{Rn} + R_{Rn} + \dfrac{R_{Ln}}{1 + j\omega C_{Rn} R_{Ln}} \right]} \tag{10.22}$$

This simplifies to

$$\eta_R = \frac{R_{Ln}}{(R_{Ln} + R_{Rn}) + \omega_0^2 R_{Ln} R_{Rn} C_{Rn}^2} = \frac{R_{Ln}}{a_n + \omega_0^2 R_{Ln} R_{Rn} C_{Rn}^2} \tag{10.23}$$

By defining $a_n = (R_{Ln} + R_{Rn})$ and $b_n = R_{Rn} R_{Ln} C_{Rn}$

$$\eta_n = \eta_R \eta_T = \frac{\omega_0^2 M_{TRn}^2 R_{Ln}}{R_T[(a_n - \omega_0^2 b_n)^2 + \omega_0^2 (L_{Rn} + b_n)^2]} \tag{10.24}$$
$$+ \omega_0^2 M_{TRn}^2 (a + \omega_0^2 b_n R_{Ln} C_{Rn})$$

There are N links in a SIMO system of N receivers. The efficiency is proportional to and enhanced by the resonance frequency, mutual inductance and the receiver load resistance. It deteriorates with inherent inductor resistance and capacitance in the tuned circuit and the transmitter resistance. A key objective is to test which of the links are best and to determine how to enhance the links to maximize received load power.

*b) **Multiple resonance SIMO*** In this case the receiver stages resonate at different frequencies and hence their performances differ.

$$d'_n = d/\cos\theta_n \tag{10.25}$$

$$\omega_{01} \neq \omega_{02} \neq \cdots \neq \omega_{0N} \neq \omega_0$$

$$k_{Rn}(d'_n) = \frac{M_{TRn}}{\sqrt{L_T L_{Rn}}} = \left(\sqrt{\frac{\sqrt{C_n}}{\sqrt{L_T}}}\right)\omega_{0n}M_{TRn} \tag{10.26}$$

The total power received by a multiple resonant SIMO system is given by the relationship

$$\sum_{n=1}^{N} P_{Rn} = \frac{P_T\eta_T Q_T}{L_T}\sum_{n=1}^{N}\omega_{0n}^2 C_n Q_{Rn}\eta_{Rn}M_{TRn}^2 \tag{10.27}$$

$$= P_T Q_T\eta_T\sum_{n=1}^{N}k_{Rn}^2\eta_{Rn}Q_{Rn}$$

Clearly a single input multiple output system is useful as it aggregates the magnetic fields with several receivers. There is no significant crosstalk between the receivers since none of them is able to induce significant magnetic fields on their neighbours. The efficiency of each link is given by

$$\eta_n = \eta_R\eta_T = \frac{\omega_{0n}^2 M_{TRn}^2 R_{Ln}}{R_T[(a_n - \omega_{0n}^2 b_n)^2 + \omega_{0n}^2(L_{Rn} + b_n)^2]} \tag{10.28}$$
$$+\omega_{0n}^2 M_{TRn}^2(a_n + \omega_{0n}^2 b_n R_{Ln}C_{Rn})$$

The denominator of the efficiency of the link is a quadratic equation in ω_{0n}^2 and we can rewrite it as

$$\eta_n(\omega_n = \omega_{0n}^2) = \eta_R\eta_T = \frac{\omega_n M_{TRn}^2 R_{Ln}}{\alpha\omega_n^2 + \beta\omega_n + \delta} \tag{10.29}$$

Where

$$\alpha = b_n^2 R_T + M_{TRn}^2 b_n R_{Ln}C_{Rn}$$
$$\beta = -2a_n b_n R_T + R_T(L_{Rn} + b_n)^2 + a_n M_{TRn}^2 \tag{10.30}$$
$$\delta = R_T a_n^2$$

10.1.2 Coupling matrix

Coupling matrix for a 2×2 system of transmitters refers to two transmitters with index 1 and 3 and two receivers with index 2 and 4. Thus the coupling coefficients with k_{1n} and k_{3n} refer to coupling coefficients as a result of transmissions from transmitters 1 and 3. Also the coupling coefficients k_{2n} and k_{4n} refer to couplings between receivers and other transmitters and receivers. Thus we have the coupling matrix:

$$C_M = \left\{ \begin{array}{cccc} k_{11} & k_{12} & k_{13} & k_{14} \\ k_{21} & k_{22} & k_{23} & k_{24} \\ k_{31} & k_{32} & k_{33} & k_{34} \\ k_{41} & k_{42} & k_{43} & k_{44} \end{array} \right\} = \left\{ \begin{array}{cccc} 0 & k_{12} & k_{13} & k_{14} \\ k_{21} & 0 & k_{23} & 0 \\ k_{31} & k_{32} & 0 & k_{43} \\ k_{41} & 0 & k_{43} & 0 \end{array} \right\} \qquad (10.31)$$

There is no significant magnetic coupling between receivers. Absence of significant magnetic coupling between receivers is shown in Figure 10.3.

Figure 10.4 depicts a simple 1×2 SIMO transceiver with two receiver coils and one transmitting coil

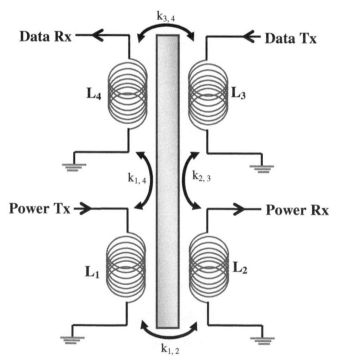

Figure 10.3 Absence of Magnetic Coupling between Receivers

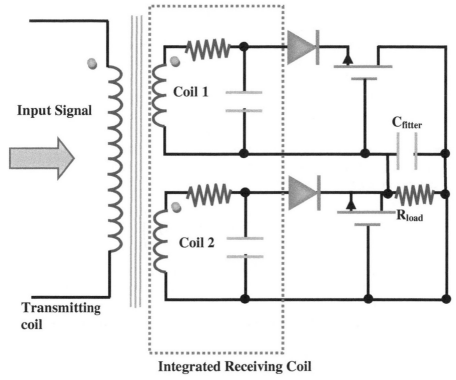

Figure 10.4 A 1 × 2 SIMO Transceiver System

c) Hybrid resonance SIMO In the SIMO setup, if there are no varying currents in the receivers and hence the coupling between neighbouring receivers are individually zero, the total power coupling from transmitter to receivers adds up and can be written as a proportional voltage

$$V_{RL} = Z_d \sum_{\substack{n \neq i}}^{N} k_{Tn} I_n \qquad (10.32)$$

10.1.3 A model of crosstalk

Crosstalk in NFMI communication systems is the interference from leakage signals in neighbouring transmitters and receivers sipping into the signal of the transceiver of interest. The leakage is due to the proximity of the coils and also inadequate isolation of transmitters and receivers. It occurs as cross coupling of magnetic fluxes between neighbouring coils. Crosstalk like normal coupling of magnetic fields is a function of the distance between the coils, their shapes and the mutual inductance between them. Therefore crosstalk is also a function of the inductances of the coils. In this section analysis of crosstalk is undertaken with explanations. The narrow band model of the transceiver is used for the analysis of crosstalk. The effect of crosstalk is to reduce the signal-to-interference ratio in the communication system and hence exacerbate the bit error rate (BER) and reduce the system capacity.

Assuming that a receiver does not cross-feed all its received signal from other receiver sources into its neighbours but only sees the couplings from all the unintended transmitters as interference, then the received signal by the ith receiver is given by the expression

$$V_{Ri} = Z_d \left(k_{ii} I_{Ti} + \sum_{j \neq i}^{N} k_{ij} I_{Tj} \right) \tag{10.33}$$

Where k_{ij} is the coupling coefficient between the ith receiver and the jth transmitter inductor. The first term is the desired signal. The second term is interference. The impedance Z_d is computed as in [2] using trans-impedance concept borrowed from the theory of inductive inter-chip crosstalk. The trans-impedance (the ratio of the induced voltage in the receiver coil to the current in the transmitter coil) of a receiving R, L, C circuit is for ideal coupling (when $k = 1$) is

$$Z_d = \frac{1}{(1 + \omega^2 L_R C_R) + j\omega R_R C_R} \cdot j\omega\sqrt{L_T L_R} \cdot \frac{1}{(1 - \omega^2 L_T C_T) + j\omega R_T C_T} \tag{10.34}$$

Therefore the signal to interference ratio is

$$SIR = \frac{|k_{ii} I_{Ti}|}{\left| \sum_{j \neq i}^{N} k_{ij} I_{Tj} \right|} \tag{10.35}$$

Equation (10.33) decouples the effects of mutual coupling (environment) from the contributions of the physical materials of the transmitter and receiver into the trans-impedance Z_d and the coupling coefficients respectively.

10.1.4 Crosstalk

In implantable devices, it is normal to have embedded inside the human body tissue an inductive transmitter that sends data to an external (outside

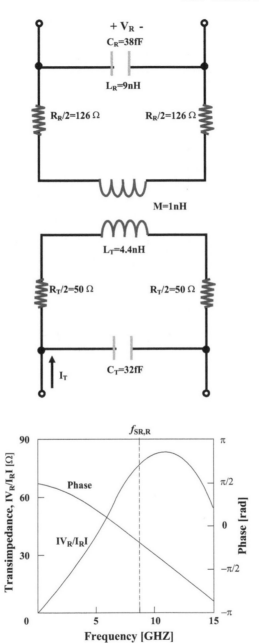

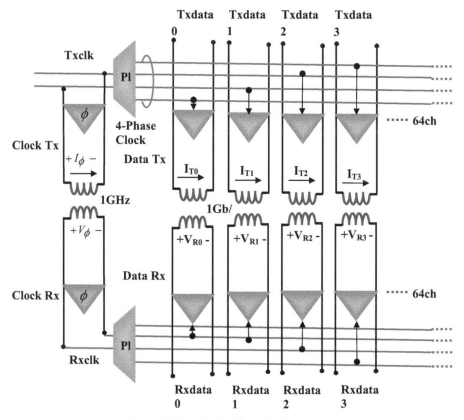

Figure 10.5 Inductive Coupling Transceiver

the body) receiver as in Figure 10.1 [1]. To power the implantable transmitter, an external wireless power transmitter is included and its receiver is buried inside the body of the person close to the data transfer source. Hence two inductive links are intentionally created, first between the data transmitter and its receiver and then second from the power system. Thus in addition to the two intended coupling coefficients k_{12} and k_{34} unintended coupling is created between the buried transmitter (data) k_{23} and receiver (power) and the external power transmitter and data receiver k_{14} as shown in Figure 10.1. The unintended magnetic coupling limits the performance of the system as they appear as interference in the communication system.

While k_{14} affects the low noise amplifier (LNA) of the data receiver, k_{23} causes a shift in the resonant frequency of the data transmitter. To date most medical implantable devices operate within the ISM bands and as such these

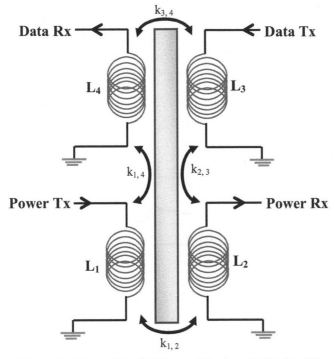

Figure 10.6 Crosstalk in Self-Powered Implantable Devices [1]

coupling interferences do not affect other transmission systems in their neighbourhood and do not require special licenses to operate. Table 10.1 shows the allocated ISM frequencies.

Table 10.1 ISM Frequency Allocations

Frequency (ISM)	Offsets
6.78 MHz	±15 kHz
13.56 MHz	±17 kHz
27.12 MHz	±163 kHz
40.68 MHz	±20 kHz
915 MHz	±13 MHz
2450 MHz	±50 MHz
5800 MHz	±75 MHz
24.125 GHz	±125 MHz
61.25 GHz	±250 MHz
122.5 GHz	±500 MHz
245 GHz	±1.0 GHz

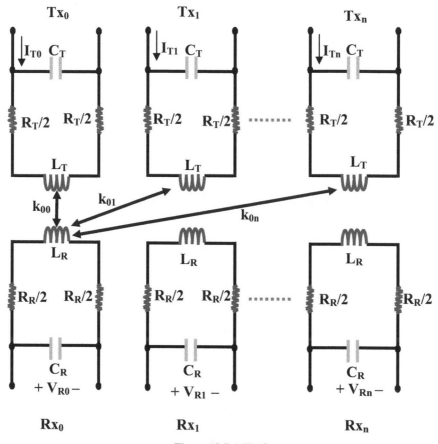

Figure 10.7 MIMO

10.1.5 Circuit model of inductive coupling channel array

Mutual coupling between units of the transmitter (receiver) and receiver (transmitter) result to crosstalk which limits the performance of the system. If there is no crosstalk between transmitters (receivers), such that the only existing coupling is between transmitters and receivers as in Figure 10.7 [2] the system performance is nearly optimum.

10.1.6 Voltage multiplier

This converts RF signal into inductive voltage that is multiplied with the circuit. The output is a DC voltage.

$$V_{DD} = n \cdot (V_{p,RF} - V_{f,D})$$

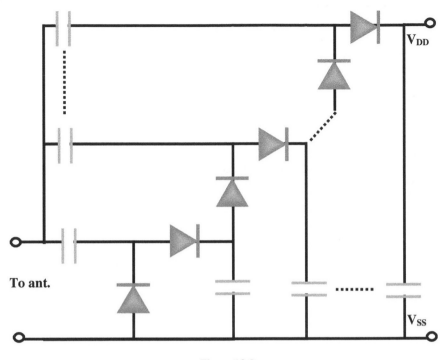

Figure 10.8

Where n is the number of diodes in the circuit and $V_{p,RF}$ and $V_{f,D}$ are the amplitudes of the input RF signal and the forward voltage of the Schottky diodes (approx 200 mV at 7 micro amperes).

References

[1] Kurt A. Sandquist, Design of a Wireless Power and Data Link for a Cranially–Implanted Neural Prosthesis, MSc thesis, Department of Electrical Engineering, Kansas State University, Manhattan, Kansas, 2004.

[2] Noriyuki Miura, Takayasu Sakurai and Tadahiro Kuroda, "Crosstalk Countermeasures for High Density Inductive-Coupling Channel Arrays", IEEE Journal of Solid–State Circuits, Vol. 2, No. 2, Feb. 2007, pp. 410–421.

11

Coding

The previous chapter introduced the basic concept for communication between a near field transmitter and receiver. This chapter presents coding as prelude to modulation.

11.1 Coding of NFMI Communication Signals

A spectrally efficient modulation scheme is a means of using power to save bandwidth while coding is a means of using bandwidth to save power. The objective in coding is to prepare the input bits for modulation to save power. This chapter discusses several coding methods in near filed communications.

11.1.1 Miller coding

Miller code is also referred to as the modified frequency modulation (MFM). Frequency modulation (FM) as described here was introduced for encoding of magnetic fluxes on hard drives and floppy disks. This section is a brief overview of the technique. The section first introduces frequency modulation as used in magnetic recording channels and then we discuss Miller coding.

11.1.2 Frequency modulation

Frequency modulation, completely different from the traditional FM in radio communications is outlined in Table 11.2. The first form of the explanation is in terms of fluxes. The second form will be based on transitions. When the input bit is "0" do not reverse the flux and when the input bit is "1" reverse the flux.

Frequency modulation is implemented by using a flux followed by another data bit. A binary digit "1" is represented by a flux and a "0" by no flux. The scheme helps to neutralize the additive nature of magnetic fields. The coding

J. Ihyeh Agbinya (PhD), *Principles of Inductive Near Field Communications for Internet of Things*, 141–146.

Table 11.1 Frequency Modulation

Input Data Bit	Output Code Sequence	# of 1's	Probability (%)
0	10 (FN)	1	50
1	11 (FF)	2	50
	Weight	1.5	100

F = Flux
N = No flux
1: Reverse the flux
0: Don't reverse the flux

Table 11.2 FM Cell Transitions

Data Bit	Clock Transition	Data Transition
0	T	N
1	T	T

T = Transition
N = No Transition

scheme requires that 4 magnetic areas be used, proving to be a real waste of space.

11.1.2.1 Explanation in terms of transitions

We can also explain what is happening in FM in a second form. This is more attuned to digital signal waveforms. Explained more in magnetic recording parlance, each bit is recorded in what is called a cell that contained two bit information or two potential transitions. The *first part of the cell* is call the clock transition (T) and always contained a transition. If the prior cell contained a north pole, it is reversed to contain a south-pole and vice-versa. The second part of the cell is the section that contains the data transition. This

Table 11.3 Miller Code

Input Data Bit	Output Code Sequence	# of 1's	Probability (%)
0 (preceded by 0)	10	1	25
0 (preceded by 1)	00	0	25
1	01	1	50
	Weight	0.75	100

Table 11.4 FM and MFM Coding

Data Bit	FM Code	MFM Codes
1111	FFFFFFFF	FFFF
0000	FNFNFNFN	NFNFNFN
1010	FFFNFFFN	FNFN
	F = Flux	N = No Flux

part would contain a transition if the data bit is a "1" and no transition (N) if it is a "0".

11.1.2.2 Modified FM

The Miller code or the so-called Modified Frequency Modulation encoding is created from the frequency modulation coding as a means of improving upon the deficiencies of FM.

MFM improved on the FM by adding an extra flux if there exist two consecutive zeros, whereas FM adds one before every piece of data. Thus, a bit sequence shown in Table 11.5 is encoded differently compared to FM.

As shown in Table 11.5, MFM has cut down bandwidth usage of FM by nearly half. However, the input bit pattern determines the amount of data stored on the disk. MFM is still used in floppy disks.

11.1.2.3 Explanation in terms of transitions

Like FM, MFM encoding uses both clock and data transitions. MFM encoding accounts for the fact that in many cases the transition is not necessary, because it is preceded by another transition. Hence MFM reduces the number of transitions required in the code by forcing a clock transition only if one did not previously occur naturally. In a nutshell, a clock transition will therefore only occur for a data bit "0" preceded by a data bit "0".

The modified Miller coding when used for near field magnetic communications can achieve a data rate of about 106 kbps at 100% modulation.

11.1.3 Manchester coding

Manchester coding is specifically specified as the coding technique in near field communication (NFC) standard (ECMA-340 of 2004) [1]. It is a physical layer technique used to create clock edges to facilitate clocking and synchronization of signals. Instead of sending directly the binary bit sequence as generated by the source Manchester encoding replaces bit "1" with 10 and bit "0" with 01. Thus if the encoding scheme is represented as a pulse, it has edges or transition points.

Table 11.5 MFM Transitions

Data Bit	Clock Transition	Data Transition
0 preceded by 0	T	N
0 preceded by 1	N	N
1	N	T

T = Transition
N = No Transition

i) bit "1" is replaced with a "high-to-low" pulse and
ii) bit "0" is also replaced with a "low-to-high" pulse.

Simply perform the mapping:

Table 11.6 Manchester Coding

Input Bit	Output Bit Sequence
1	10
0	01

Represented as a pulse, Figure 11.1 shows the coding in pictorial form.

Each bit has either a positive 90° or negative 90° edge transition, Manchester encoding is also often referred to as a Biphase encoding. The level transitions are extremely helpful in phase locked loop (PLL) demodulation of FSK, BPSQ and QPSK modulated signals. As a penalty for this feature, the encoded signal consumes twice the bandwidth of the original un-encoded input bits.

Example: The following binary sequence was created by an implanted device for transmission: **0 1 1 1 1 0 0 1**

This encodes to: **01 10 10 10 10 01 01 10**

The waveform representation of the coded signal using Figure 11.1 scheme is

The term chip is often used to refer to the levels on either side of a transition or edge. Therefore a Manchester data encoding of a bit requires two chips.

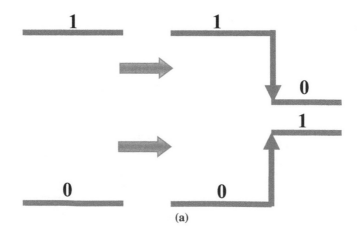

(a)

(b)

Figure 11.1 Manchester Coding

Manchester encoding schemes are used for higher data rate implementations and up to 424 kbps is achieved in the passive mode at a modulation ratio of 10% [2]. In the passive mode, the receiver draws its operating power from the transmitter. In the active mode, both the transmitter and the receiver communicate alternately by generating their own magnetic fields.

References

[1] Near Field Communication Interface and Protocol (NFCIP-1), Standard ECMA-340, 2nd Edition, December 2004, www.ecma-international.org

[2] D. Rinner, H. Witschnig, E. Merlin, "Broadband NFC–A System Analysis for the Uplink", in Proc. CSNDSP08, pp. 292–296.

[3] K. Finkenzeller, *RFID-Handbuch*, 3te Auflage, Carl Hanser Verlag: München Wien, 2002.

12

Modulation

Previously in Chapter 8 the concept of impedance modulation as a means of communication between a near field transmitter and receiver was presented. It considered narrowband implementations. This chapter presents impedance modulation concepts more and also discusses frequency shift keying as two of the most widely modulation techniques in NFMI communications. The modulation methods discussed have been used mostly in biotelemetry but are directly applicable to non-implantable near field short range communication.

12.1 General Principles of NFMI Modulation

Several modulation techniques are used for data and power transmission in implantable devices, the so-called biotelemetry systems. A good summary of the techniques are presented in [1] and [6]. In the downlink (external to internal) several methods are used for the modulation and they include on-off-keying (OOK) or amplitude shift keying (ASK) mostly due to their simplicity and ease of use. Most of the methods are half-duplex in nature in either direction and some of the applications are bi-directional. Asymmetrical modulation techniques are forced by the different data volume requirements and the requirements for implantable conditions to limit power consumption and heating of tissues or the need for different modulation methods for downlink and uplink. The data transmission rate with ASK is usually limited, so also its power transfer capability. Frequency shift keying (FSK) and phase shift keying (PSK) have also been used for the downlink communication [5] and [7]. FSK is most suitable for broadband inductive links which require low quality factors and results to low power transfer efficiencies below 20% [6].

Full duplex communications have been facilitated by using constant amplitude modulation methods such as load shift keying (LSK). LSK also facilitates power transfer. Performance comparison of the various modulation methods is presented in Table 12.1. Some data in this table are derived from [6].

J. Ihyeh Agbinya (PhD), Principles of Inductive Near Field Communications for Internet of Things, 147–174.

Table 12.1 Performance of Implantable Digital Modulation Schemes

Down link Rate	Uplink Rate	Carrier Frequency	Power Consumption	Technology	Ref
		Data Transmission			
Packet detect	Burst of RF energy	2.5 MHz	—	2 μm CMOS	[8]
PWM-ASK (250 kbps)	ASK	1~10 MHz	5 mW	1.2 μm CMOS	[9]
ASK	PWM-ASK (125 kbps)	4 MHz	10–90 mW	3 μm BiCMOS	[10]
ASK(120 k)	BPSK(117–234 k)	10 MHz	4.5 mA	2.5 μm BiC	[11]
OOK(100 kbps)	NO	5 MHz	0.5 mA, 10 V	2 μm CMOS	[12]
OOK	LSK(200 kbps)	6.78 MHz	<120 mW	1.2 μm CMOS	[13]
ASK(170 kbps)	PSK(2.7 Mbps)	50 MHz	6 mW	1.5 μm CMOS	[17]
BPSK 1.12 (Mbps)	LSK	13.56 MHz	414 μW	TSMC 0.18 μm CMOS	[6]
OQPSK	8 Mbps	13.56 MHz	16 μW and 680 μW	TSMC 0.18 μm CMOS	[14]
	FSK(2.5 Mbps)	5 MHz	0.55 mW at 100 kbps	3 μm CMOS	[20]
BPSK(1.69 Mbps)	LSK(100 kbps)	13.56 MHz & 100 kHz	5 mW	0.5 μm CMOS	[21]

We have covered the general principles of impedance reflection from the receiver to the transmitter and from the transmitter to the receiver. We showed that the mutual impedance played a central role as the transfer impedance and couples magnetic flux between the transmitter and receiver. In this section we show the general methods used for modulating the data signal. Several methods are in use. For low data rate transfer transformer action has been used [1]. This will be explained further in a subsequent section.

In figure 12.1, the impedance seen at the transmitter terminal is

$$Z_T = Z_{11} - \frac{Z_{12}^2}{Z_{22} + Z_L} \tag{12.1}$$

The impedance also seen at the receiver terminal is

$$Z_R = Z_{22} - \frac{Z_{12}^2}{Z_{11} + Z_G} \tag{12.2}$$

Thus the transfer impedance acts as an impedance inverter in both the transmitter and receiver. Figure 12.2 is the detailed form of Figure 12.1 in which relevant circuit components are shown. In Figure 12.2

$$Z_{11} = R_1 + j\omega L_1 \tag{12.3}$$
$$Z_{22} = R_2 + j\omega L_2 \tag{12.4}$$

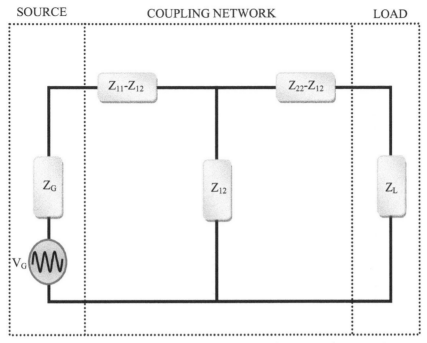

Figure 12.1 Coupling between Source and Load in an Inductive Network

and the transfer impedance is

$$Z_{12} = j\omega M \qquad (12.5)$$

This is the impedance that is used for power coupling between the source and the load. The larger M is the better the power coupling between the two coils.

The next sections discuss impedance modulation or the so called load modulation. Impedance modulation is also referred to as amplitude shift keying in these applications. The ASK is realised by switching the resistive load in the receiver. Passive phase shift keying (PSK) can also be implemented by modulating the capacitive load of the receiver tank circuit. Passive PSK telemetry can share the same wireless inductive link with a powering system which is an advantage.

Both modulation methods affect the reflected impedance at the transmitter which is used to detect a varying voltage that can be demodulated to recover the data.

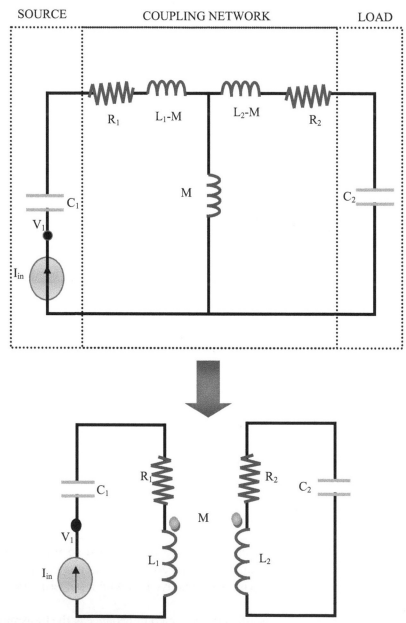

Figure 12.2 A Basic Inductive Communication Link System Circuit Model

12.1.1 Impedance modulation

This section is based on the circuit in Figure 12.2 (a series resonance circuit). Since from our model, Figures 12.1 and 12.2 are identical, we can identity the following circuit parameters:

$$Z_G = 1/j\omega C_1 \quad \text{and} \quad Z_L = 1/j\omega C_2 \tag{12.6}$$

Let the input excitation be a sinusoidal current $i_{in} = I_{in} \sin(\omega_0 t)$. At resonance both the transmitter and receiver operate at the same frequency, hence $\omega_0 = 1/\sqrt{L_1 C_1} = 1/\sqrt{L_2 C_2}$. Define the quality factors of the coils as

$$Q_1 = \frac{\omega_0 L_1}{R_1} = \sqrt{\frac{L_1}{C_1}} / R_1 \tag{12.7}$$

and $Q_2 = \frac{\omega_0 L_2}{R_2} = \sqrt{\frac{L_2}{C_2}} / R_2$. The input impedance at the transmitter is therefore given by

$$Z_{in} = \frac{v_1}{i_{in}} = Z_G + Z_T = Z_G + Z_{11} - \frac{Z_{12}^2}{Z_{22} + Z_L}$$

$$= (R_1 + j\omega L_1) + \frac{1}{j\omega C_1} + \left(\frac{\omega^2 M^2}{R_2 + j\omega L_2 + \frac{1}{j\omega C_2}} \right) \tag{12.8}$$

Z_T is the transfer impedance from the receiver to the transmitter. At resonance the input impedance becomes

$$Z_{in} = \frac{v_1}{i_{in}} = R_1 + \frac{\omega_0^2 M^2}{R_2}$$

$$= R_1 \left[1 + \frac{\omega_0^2 M^2}{R_1 R_2} \right] \tag{12.9}$$

By using the fact that $M = k\sqrt{L_1 L_2}$, this expression becomes

$$Z_{in} = R_1 \left[1 + \frac{k^2 \omega_0^2 L_1 L_2}{R_1 R_2} \right] = R_1 \left[1 + k^2 \frac{\omega_0 L_1}{R_1} \cdot \frac{\omega_0 L_2}{R_2} \right] \tag{12.10}$$

$$= R_1 [1 + k^2 Q_1 \cdot Q_2]$$

$$= R_1 [1 + m]$$

M is the coupling coefficient between the coils and $m = k^2 Q_1 \cdot Q_2$ is the impedance modulation index. Thus the impedance of the transmitter varies in accordance with the modulation index. When $m = 0$, the input voltage is $v_1 = i_{in} R_1$ and bit 0 is transmitted. When $m = k^2 Q_1 \cdot Q_2$, the input voltage $v_1 = i_{in} R_1 [1 + m]$ and bit "1" is transmitted. Thus the impedance modulation causes amplitude shift keying (ASK). For both cases, the input impedance is real, hence there is no phase shift nor phase modulation. The modulation index is a strong function of the coupling coefficient k and since its value reduces very fast as function of distance, adequate power coupling is within a very short distance. Also the input impedance is relatively very small about the order of the dc resistance of the coil (a few Ohms). Therefore the voltage at the input of the transmitter is also small and not big enough for data transfer over a longer range. Hence, impedance modulation over a longer range does not work properly. Impedance modulation is effective for very short communications and suitable mostly in power transfer for implantable devices where the separation between the transmitter and receiver is usually a few centimetres. To improve the performance an alternative is suggested in [1] to convert the series resonance circuit to a parallel resonant circuit.

12.1.1.1 Impedance modulation with parallel resonant circuit
By using parallel resonant circuit for the modulation, the limitations of impedance modulation with series resonant circuit can be improved. The objective is to increase the impedance over which the input voltage v_1 is taken. This modification is demonstrated by [1] using the circuit shown in our Figure 12.3.

The gate of a transistor with transconductance g_m is driven by a voltage $v_{in} \sin(\omega_0 t)$ to create the required input current. The resulting impedance modulation is approximately:

$$Z_{in} = \frac{v_1}{i_{in}} \approx R_1[1 + m] \left(\frac{Q_1}{1 + m} \right)^2 = \frac{R_1 Q_1^2}{1 + m} \tag{12.11}$$

Two assumptions were made to reach this approximation and they are

$$1 + k^2 Q_1 \cdot Q_2 \ll Q_1 \quad \text{and} \quad k \ll 1/Q_2 \tag{12.12}$$

As before $m = k^2 Q_1 \cdot Q_2$. Further insight can be achieved by writing

$$Z_{in} = \frac{v_1}{i_{in}} \approx \frac{(R_1 Q_1)^2}{R_1(1 + m)} = \frac{Z_0^2}{R_1(1 + m)}$$

Thus the use of a parallel circuit has resulted to impedance inversion of the form

$$Z_{in} \Rightarrow \frac{Z_0^2}{Z_{in}} \tag{12.13}$$

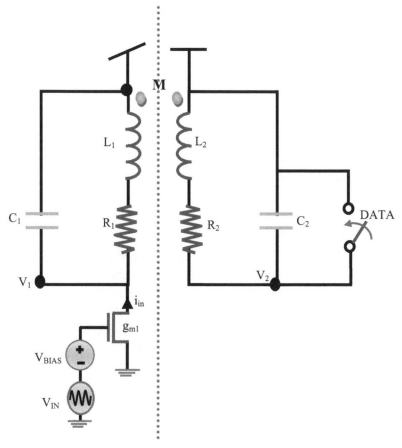

Figure 12.3 Parallel Tuned Circuit in Impedance Modulation [1]

Therefore the effective modulation depth has changed to

$$m_{eff} = \left(1 - \frac{1}{1+m}\right) = \frac{m}{(1+m)} \qquad (12.14)$$

The transmitter resonator voltage switches between v_1 (transmission of bit "1") and $v_1/(1+m)$ (transmission of bit "0"). Define the uplink as the communication from the receiver (implant) to transmitter (external) and downlink as the communication from transmitter (external) to receiver (internal). For implanted devices, during the downlink phase the switch across the internal implant capacitor is kept open. During this period the voltage amplitude is modulated by data and some of the power is coupled by mutual inductance to the internal resonator unit. The relationship between the induced voltage and the input voltage at the transmitter is given by the ratio

$$m_t = \frac{v_2}{v_1} = kQ_2\sqrt{\frac{L_2}{L_1}} \qquad (12.15)$$

$m_t = \sqrt{m}$ when $L_1 = L_2$ and $Q_1 = Q_2$. Therefore $v_2 = m_t v_1$ for the amplitude voltage of the received data from the transmitter to the receiver and $m_{eff} v_1$ in the reverse direction. Observe the weaker dependence of the voltage ratio on the modulation index which has resulted from using a parallel circuit.

12.1.2 Load shift keying

Load shift keying (LSK) or "reflectance modulation" [15] is a digital modulation scheme that has found application in implantable telemetry systems. It is a particular form of ASK. It allows simultaneous transmission of data and power transfer to an implantable device in biotelemetry. The RF power link uses an inductive link formed by two coils to couple magnetic fields to each other. In LSK a change in the receiver load is reflected back to the transmitter as changing impedance (reflected impedance). The so-called "circuit configuration modulator" (CCM) is used to implement the modulation scheme. The CCM uses the acquired data to modulate the secondary load impedance which is reflected to the primary through the inductive mutual coupling.

To realise LSK a parallel tuned RLC circuit is used in the secondary section of the inductive link. Modulation is achieved by modifying the resonant frequency by changing either the capacitive or inductive element of the secondary coil. The shift in resonant frequency is reflected back to the primary coil through the mutual inductance. This appears as changes in the amplitude of the voltage across the inductance of the primary coil.

The encode selects between the binary numbers and also the input load resistance to use for the modulation. The decoder recovers or senses the changing impedance in the transmitter to recover the bit that was sent.

Consider Figure 12.4 in which Zr represents the reflected impedance appearing as part of the impedance of the primary circuit. The reflected impedance is

$$Z_r = \frac{(\omega M)^2}{Z_R} \qquad (12.16)$$

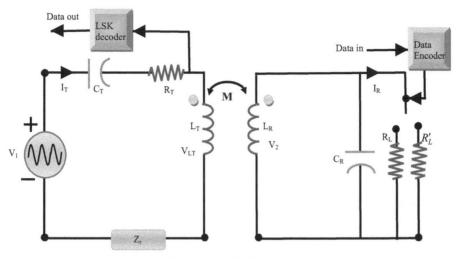

Figure 12.4 LSK Principle

Z_R is the impedance of the receiver or secondary circuit and M is the mutual inductance. Therefore the impedance of the primary circuit becomes

$$Z_T = \frac{V_1}{I_T} = R_T + Z_r + j \left[\omega L_T - \frac{1}{\omega C_T} \right] \tag{12.17}$$

The induced voltage in the receiver circuit is dependent on the current in the transmitter. The total impedance of the receiver circuit therefore is

$$Z_R = \frac{V_2}{I_R} = j\omega L_R + j \left[\frac{1}{R_L} + \omega C_R \right]^{-1}$$

$$= \frac{R_L}{1 + (\omega R_L C_R)^2} + j \left(\omega L_R - \frac{\omega C_R R_L^2}{1 + (\omega R_L C_R)^2} \right) \tag{12.18}$$

The impedances at the transmitter and receiver are purely resistive when they resonate at the same frequency. Hence

$$\omega_0 L_T - \frac{1}{\omega_0 C_T} = 0 \tag{12.19}$$

and

$$\omega_0 L_R - \frac{\omega_0 C_R R_L^2}{1 + (\omega_0 R_L C_R)^2} = 0 \tag{12.20}$$

or

$$\frac{1}{1 + (\omega_0 R_L C_R)^2} = \frac{L_R}{C_R R_L^2} \tag{12.21}$$

We substitute this into the impedance of the receiver at resonance to have

$$Z_R = \frac{R_L}{1 + (\omega_0 R_L C_R)^2} = \frac{L_R}{R_L C_R} \tag{12.22}$$

We can show the influence of the reflected impedance of the receiver on the transmitter by using the equation for the mutual inductance $M = k\sqrt{L_T L_R}$ and the reflected impedance as

$$\begin{aligned} Z_T = R_T + Z_r &= R_T + \frac{(\omega_0 M)^2}{Z_R} \\ &= R_T + \frac{R_L C_R (\omega_0 M)^2}{L_R} \\ &= R_T + \omega_0^2 k^2 R_L C_R L_T \end{aligned} \tag{12.23}$$

This expression shows that at resonance the impedance of the transmitter is being modified by the receiver capacitance and its load.

The modulation depth for the LSK can also be shown through the following analysis. The voltage drop across the transmitter inductance is given as the proportion

$$\begin{aligned} V_{LT} = \frac{(\omega L_T) V_1}{Z_T} &= \left(\frac{1}{\omega C_T Z_T}\right) V_1 \\ &= \frac{V_1}{\omega C_T (R_T + \omega^2 k^2 C_R L_T R_L)} \end{aligned} \tag{12.24}$$

Assume that the secondary load can vary from R_L to R_L', so that

$$V_{LT}' = \frac{V_1}{\omega C_T (R_T + \omega^2 k^2 C_R L_T R_L')} \tag{12.25}$$

The modulation depth becomes [15]

$$m = \left| \frac{V_{LT} - V_{LT}'}{V_{LT} + V_{LT}'} \right| \tag{12.26}$$

Hence the voltage measured across the transmitter inductor changes between two extreme values depending on the depth of modulation. This essentially allows the two extreme voltages to be used to represent bit "0" and bit "1", and is akin to amplitude modulation.

The bit error rate (BER) when the transmitter voltage varies between the above two limits can be estimated as [22, 23]

$$\text{BER} = \frac{1}{2}erfc\left(\frac{|V_{LT} - V'_{LT}|}{4\sqrt{2}\cdot\sigma}\right) \tag{12.27}$$

In terms of the voltage at the input of the receiver, the bit error rate is

$$\text{BER} = \frac{1}{2}erfc\left(\frac{m\cdot|V_0|}{2\sqrt{2}\cdot\sigma}\right) \tag{12.28}$$

where m is the modulation index.

12.1.3 Phase shift keying (PSK)

Phase shift keying was demonstrated in [16] by varying the capacitance of the receiver circuit.

Instead of modulating the real part of the primary impedance, PSK modulates either the imaginary part of the impedance (capacitive or inductive part) [16, 17]. In [16] and [17] the capacitive part of the impedance is modulated by using a transistor switch which switches in and out a pair of series capacitors connected in parallel with the capacitor of the secondary tuned circuit.

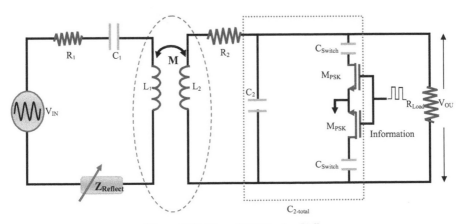

Figure 12.5 Phase Shift Keying [16]

The net effect is a varying capacitive value and hence changing resonant frequency of the circuit. In effect the peer-to-peer system is resonant at multiple frequencies depending on the secondary capacitance.

$$Z_r = \frac{(\omega M)^2}{Z_2} = \frac{(\omega M)^2}{R_2 + 1/(\omega^2 C_{2\_total}^2 R_{load}) + j(\omega L_2 - 1/(\omega C_{2\_total}))}$$

(12.29)

Where the impedance of the transmitter circuit is

$$Z_T = R_1 + Z_r + j\left(\omega L_1 - \frac{1}{\omega C_1}\right)$$

(12.30)

The receiver impedance is

$$Z_2 = R_2 + j\omega L_2 + \left(\frac{1}{R_{load}} + j\omega C_{2\_total}\right)^{-1}$$

$$= R_2 + j\omega L_2 + \frac{R_{load}}{1 + j\omega R_{load} C_{2\_total}}$$

(12.31)

The load quality factor is

$$Q_{2L} = \frac{1}{R_2\sqrt{C_{2\_total}/L_2} + +(1/R_{load})\sqrt{L_2/C_{2\_total}}}$$

(12.32)

The reflected impedance into the transmitter is

$$Z_r = \frac{(\omega M)^2}{Z_2} = \frac{\omega^2 k^2 L_1 L_2}{R_2 + j\omega L_2 + \dfrac{R_{load}}{1 + j\omega R_{load} C_{2\_total}}}$$

(12.33)

or

$$Z_r = \frac{(\omega M)^2}{Z_2} = \frac{R_1 k^2 Q_1 Q_2}{1 + jQ_2 + \dfrac{R_{load}/R_2}{1 + j\omega R_{load} C_{2\_total}}}$$

(12.34)

The value of the reflected impedance is modified in amplitude and phase by varying the total capacitance in the receiver. The value of the total capacitor varies as

$$C_{2\_total} = \begin{cases} C_2 \\ C_2 + \dfrac{C_{switch}}{2} \end{cases}$$

(12.35)

and the modulation has two phase frequencies

$$\begin{cases} \omega_{01} = 1/\sqrt{L_2 C_2} \\ \omega_{02} = 1/\sqrt{L_2 \left(C_2 + \frac{C_{switch}}{2} \right)} \end{cases} \qquad (12.36)$$

By extending this method recursively, several resonant frequencies can be used for communication with each one switched in when required. The resonance value is determined by the capacitance in the receiver.

12.1.4 Binary phase shift keying (BPSK)

Binary phase shift keying was introduced to improve the data transmission rates and power efficiency [6]. BPSK doubles the bandwidth efficiency of PSK without loss of performance. Hence it is typically used in the downlink communication to downlink data to implants. BPSK is often combined with LSK in the uplink. In BPSK data is used to change the phase of the received waveform to either zero ($0°$) or $180°$. This is equivalent to multiplying the carrier with a bit stream of "1" (high) and "-1" (low) states respectively. Among methods for demodulating BPSK the COSTAS loop is most popularly used. Its ease of implementation lends itself for wide applications. A COSTAS loop consists of two parallel phase locked loops (PLL) as in Figure 12.6 [6] and the analysis in this section is heavily based on the contents in [6].

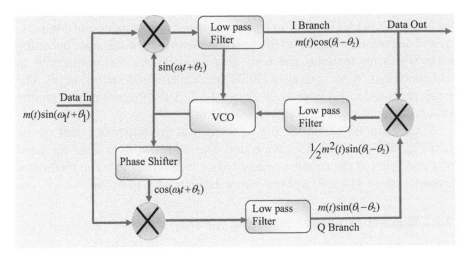

Figure 12.6 COSTAS Loop for Demodulating BPSK [6]

The phase errors of the PLLs are multiplied to control the frequency of the oscillator. Consider the input signal

$$V_{in}(t) = m(t) \sin(\omega_1 t + \theta_1) \qquad (12.37)$$

$m(t)$ is the data stream ("1" or "−1"). The carrier is $\sin(\omega_1 t + \theta_2)$. As shown in Figure 12.6, the input data is pre-mixed by two signals out of phase by 90°. The outputs when the signals pass through the in phase upper and quadrature lower low pass filters are

$$V_{OI}(t) = m(t) \cos(\theta_1 - \theta_2) \qquad (12.38)$$

and

$$V_{OQ}(t) = m(t) \sin(\theta_1 - \theta_2) \qquad (12.39)$$

These two signals are used to form the control signal of the voltage controlled oscillator (VCO) by mixing them to have

$$V_M(t) = \frac{1}{2} m^2(t) \sin 2(\theta_1 - \theta_2) \qquad (12.40)$$

Thus the control signal is proportional to the phase difference of the signals in the I and Q arms of the PLL. In phase lock the phase differences are equal ($\theta_1 = \theta_2$) and the recovered signal becomes

$$V_{OI}(t) = m(t) \cos(\theta_1 - \theta_2) = m(t) \cos 0^0 = m(t) \qquad (12.41)$$

Traditional COSTAS loop suffer from high power consumption and complexity, and demodulators also suffer from the sensitivities of the input operating points which for implants and body area networks are disadvantageous in field operations. A more modern BPSK demodulator is described in [6]. The circuit proposed in [6] is given in Figure 12.7. A comparator is used to first convert the input signal into square waveform.

A hard limited COSTAS loop is formed at the quadrature half of the circuit by adding a second comparator. The objective is to relax the need for a multiplier at the centre as explained by the authors [6]. The modulator consumes about 414 μW and provides a data rate of 1.12 Mbps.

12.1.5 Quadrature phase shift keying (QPSK)

BPSK traditionally uses one carrier whose phase is modified by the binary inputs to have two polarities depending on when a "1" or a "−1" is input.

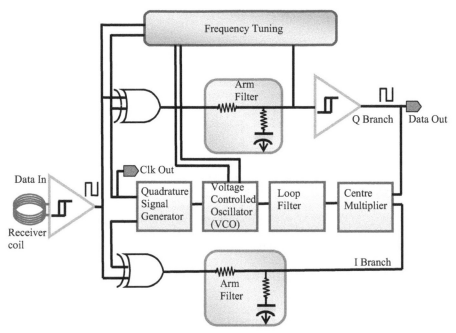

Figure 12.7 Improved BPSK Demodulator

QPSK on the other hand uses two carriers. One carrier is used in the in phase (I) signal component and the second in the quadrature (Q) arm of the circuit [18]. The two carriers are thus out of phase by 90°. The polarities of the carriers are switched by the binary inputs "1" and "−1". Hence with two arms, four different phases can be handled. This leads to a doubling of the data rate compared with the traditional BPSK. For example, consider the input signal in Figure 12.8.

The first signal is the data input and this signal is mapped into the I and Q arms of the network such that the in phase signal has all the even bits and the quadrature component all the odd bits. This QPSK mapping doubles the bit duration that goes into the I and Q arms or the symbol duration is twice the bit duration. Therefore if the input bit stream is

$$B(t) = b_0 \ b_1 \ b_2 \ b_3 \ b_4 \ b_5 \ \ldots \ b_N \tag{12.42}$$

Then the in phase component is

$$I(t) = b_0 \ b_2 \ b_4 \ b_6 \ b_8 \ b_{10} \ \ldots \ b_{2n} \tag{12.43}$$

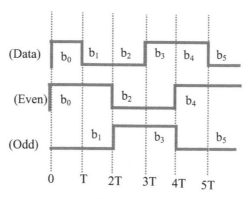

Figure 12.8 QPSK Signals

The quadrature component also is

$$Q(t) = b_1 \ b_3 \ b_5 \ b_7 \ b_9 \ b_{11} \ \dots \ b_{2n+1} \tag{12.44}$$

Thus the transmission rate is $1/2T$ bits per second and the signals are aligned synchronously with coinciding transitions. The in phase component $I(t)$ is multiplied by a carrier $\cos(2\pi f_c t + \theta_1)$ which is orthogonal to the quadrature carrier $\sin(2\pi f_c t + \theta_1)$ that is used to multiply $Q(t)$ to form the Q-channel. Thus both arms of QPSK are individually BPSK signals with symbol duration 2T. The two signals are summed to form the QPSK signal. Therefore the QPSK signal is given by the expression

$$s(t) = I(t) \cdot \cos(2\pi f_c t + \theta_1) + Q(t) \cdot \sin(2\pi f_c t + \theta_1) \tag{12.45}$$

Due to the orthogonal nature of the two components of QPSK, each one can be demodulated clearly and independently of the other. Using the well known sum of cosines, the QPSK signal can also be written as

$$s(t) = \cos(2\pi f_c t + \theta) \tag{12.46}$$

where the angle $\theta = 0° \pm 90°$ or $\theta = 180°$ depending on the four combinations of $I(t)$ and $Q(t)$ {$(-1, -1)$; $(-1, 1)$; $(1, -1)$; and $(1,1)$}. In QPSK, due to the alignment of the signals (Figure 12.8) the modulated carrier fades away temporarily whenever a data signal changes polarity and the demodulator also temporarily loses lock or loses the signal it tries to lock on to. This problem is solved by intentionally mis-aligning the $I(t)$ and $Q(t)$ signals by one time period T [19]. This offset leads to the so-called Offset QPSK or (OQPSK).

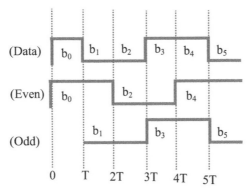

Figure 12.9 OQPSK Signals

12.1.6 Offset quadrature phase shift keying (OQPSK)

Offset QPSK is similar to QPSK except that the signals which are used as inputs to the I and Q arms of the system are offset from each other (Figure 12.9) by one bit duration (T). OQPSK has the same data streams as QPSK, possess the same spectral efficiency, same power spectral density and error performance but is able to lock on to the signal being demodulated of the sign of the signal changes.

The expression for the signal is

$$s(t) = I(t) \cdot \cos(2\pi f_c t + \theta_1) + Q(t - T) \cdot \sin(2\pi f_c t + \theta_1) \qquad (12.47)$$

Figure 12.10 [14] is the diagram for OQPSK except for the dotted line.

12.1.6.1 Demodulation of QPSK

There are three popular demodulation schemes including COSTAS loop, re-modulation loop and the multiply-filter-divide (squaring) loop methods. Among these, COSTAS loop is most popular. The re-modulation scheme is equivalent to the hard-limited COSTAS loop. The squaring loop is used for high-data rate burst mode systems. The analysis in this section is based on the COSTAS loop as presented in [14]. The demodulator consists of two phase locked loops (the I-channel and Q-channel). Two orthogonal carriers $\cos(2\pi f_c t + \theta_2)$ and $\sin(2\pi f_c t + \theta_2)$ are generated by the VCO.

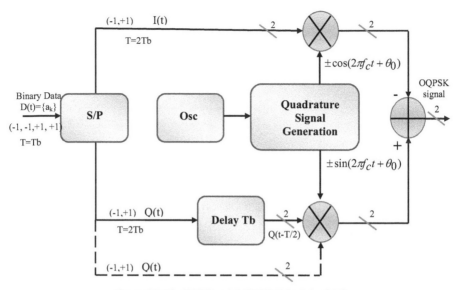

Figure 12.10 QPSK and OQPSK Principles [14]

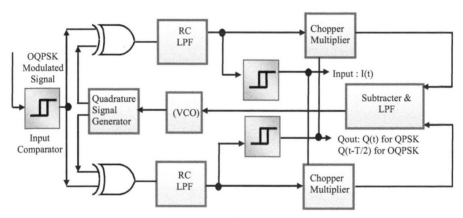

Figure 12.11 OQPSK Demodulator

These are used to pre-multiply the received input OQPSK signal. The signal in the I-channel is recovered through the expression

$$s_I(t) = I(t) \cdot [\cos(\theta_1 - \theta_2) + \cos(4\pi f_c t + \theta_1 + \theta_2)] + Q(t - T)$$
$$\times [\sin(\theta_1 - \theta_2) + \sin(4\pi f_c t + \theta_1 + \theta_2)] \qquad (12.48)$$

The low pass filter blocks the terms in $4\pi f_c t$ and produces the output signal

$$s_I(t) = I(t) \cos(\theta_1 - \theta_2) + Q(t - T) \sin(\theta_1 - \theta_2) \qquad (12.49)$$

When the PLL is locked to the signal, the phase terms cancel out ($\theta_1 - \theta_2 = 0$) and hence, $s_I(t) = I(t)$. The output of the Q-channel is similar to the one from the I-channel and is given by the expression

$$s_Q(t) = -I(t) \sin(\theta_1 - \theta_2) + Q(t - T) \cos(\theta_1 - \theta_2) \qquad (12.50)$$

Thus at phase lock as well, the output of the channel is $s_Q(t) = Q(t - T)$. This output is delayed with respect to the I-channel output by a small time equal to T. Note that in the traditional QPSK since $I(t)$ and $Q(t)$ signals coincide in time, their carrier phases shifts of $\theta = 0° \pm 90°$ or $\theta = 180°$ occur every $2T$ seconds. In OQPSK the small offset leads to staggering of the signals and eliminates the phase term at $\theta = 180°$ and the phase changes are thereby limited to $\theta = 0°$ *and* $\pm 90°$ every T seconds.

OQPSK possess some distinct inherent advantages. It avoids out of band interference. In addition, the offset time T between the $I(t)$ and the $Q(t)$ signals leads to immunity to phase jitter in the presence of additive white Gaussian noise. These advantages accrue to OQPSK due to avoiding the large phase change of 180° that is associated with QPSK.

12.1.7 Frequency shift keying

ASK has limitation in the amount of data that can be sent and the effects of the modulation index limits its range as well. Some implant applications such as the ones that interface with the central nervous system (and visual implants) need to transfer a large amount of data in real time and ASK are clearly not suited to that. Hence, consideration is given to frequency shift keying. In FSK binary data is sent at different frequencies, with bit "0" being sent at frequency f_0 and bit "1" at frequency f_1. An FSK signal may be considered as a summation of two ASK carriers $f_0(t)$ and $f_1(t)$ of equal amplitude V_m.

$$f(t) = f_0(t) \sin(2\pi f_0 t + \phi) + f_1(t) \sin(2\pi f_1 t + \phi)$$

Observe that FSK uses two carrier frequencies while QPSK and OQPSK use one carrier in phase quadrature. The mode of FSK transmission for implants is distinctly different from the traditional FSK in communications. In the traditional FSK, the data rate is hundreds of times slower than either f_0 or f_1. This makes it difficult if one needs data rates in excess of 1 Mbps. In wireless biomedical implants, the carrier frequencies (and hence the intended baud rates) are chosen to be in the tens of MHz to limit tissue damage which results from high frequency applications. A new approach was devised for the FSK used in biomedical implants to allow data transfers that are close to f_1($f_0 = 2f_1$). In the method devised, bit "1" is transmitted with one cycle of the carrier f_1 and bit "0" is transmitted by two cycles of f_1 (Figure 12.12(a)).

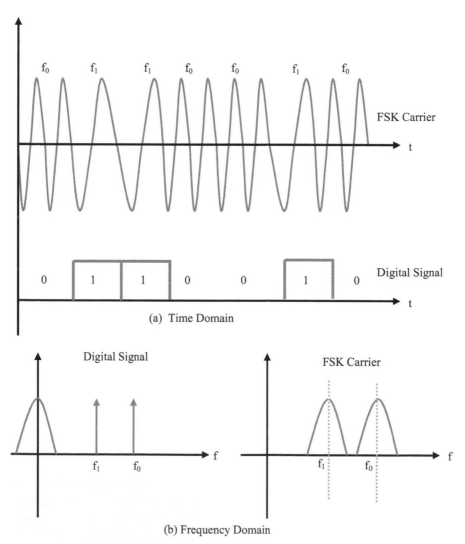

(b) Frequency Domain

Figure 12.12 Frequency Shift Keying in Time and Frequency Domains

This leads to a steady data transfer rate of f_1 bits per second and the transmitter frequency switches at a small fraction of a cycle and only at zero crossings [20].

FSK has excellent characteristics in that its amplitude or power is maintained constant which is suitable for implantable devices and for short range NFMI communications as well. This is irrespective of the frequency, modulation index and data content. Hence, since

$$|f_0(t)| = |f_1(t)| = V_m$$

It also means the power generated is also constant as

$$P_{rms}(FSK) = \frac{1}{2}V_m^2 = \text{constant}$$

Thus a load sees a constant power at each frequency location. NFMI communication is often across a line of sight and over short distances and hence the effect of the medium between the transmitter and receiver is common and minimal. Furthermore, the great power means a longer relative distance is obtainable compared with ASK modulation. It also does not produce the type of power fluctuations inherent when ASK is used with its varying modulation index. When all other system parameters are constant, the received induced voltage reduces as an inverse third power of the distance the transmitter and receiver coils $(1/d^3)$. The induced signal in ASK is highly sensitive to the patient's motion (in implantable devices) and to vibrations in RFID tags and smart cards. High quality factor coils are required in ASK to ensure significant modulation index to enable clear separation (detection) of signal amplitudes at the receiver. In FSK however, low Q coils can be used because the data bits (mark and space) are carried on entirely different frequencies and amplitude variations can be accommodated during detection.

In the ASK data transmission, the tank circuit which forms the receiver should have a frequency response with high quality factor (Q) and centred at the carrier frequency to facilitate high enough amplitude variation for data detection. On the other hand in FSK data transmission, the passband is centred between the two frequencies $f_0(t)$ and $f_1(t)$ with a low Q so that enough power is passed at both carrier frequencies. For applications in biomedical implants this is an advantage because the quality factor of the receiver coil is generally low particularly when the receiver coil is integrated with high resistivity.

Despite these apparent advantages that FSK possess over ASK, synchronisation of ASK signals is easier comparatively. For example for implantable

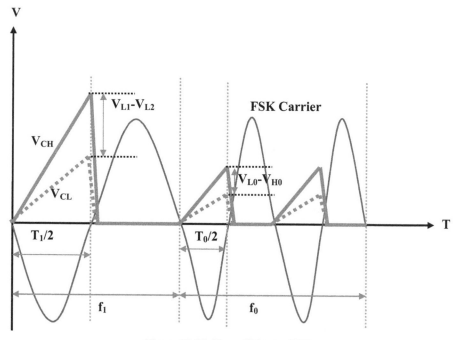

Figure 12.13 Phase Coherent FSK

devices, the internal clocks can be derived from the carrier frequency commonly used by the receiver and transmitter. Several versions of FSK synchronisation exist for magnetic communications. They include in phase-coherent and non-coherent versions. In phase coherent FSK at the onset of each bit, the carrier frequencies $f_0(t)$ and $f_1(t)$ have a fixed phase. An internal frequency is derived from the combination of the two carrier frequencies based on the FSK protocol. In the non-coherent case, the carrier phase is random and the data bits are encoded by using self-clocking schemes such as Manchester and Miller encoding.

In [5] data rate-to-carrier frequency ratio is maximized by using the phase coherent FSK. The carrier frequencies are related as $f_0 = 2f_1$ and using a single cycle of f_1 to encode "1" and two cycles of f_0 to encode a "0".

The carrier frequency is switched at a fraction of a cycle either at the leading (positive-going) or trailing (negative-going) edge zero crossings. Therefore the bit length is consistently equal to $1/f_1$ and a constant data rate of f_1. Thus for average carrier frequency $f_{avr} = (f_0 + f_1/2)$, the data-rate-to-carrier frequency ratio can be as high as 67% [5]. Also any odd number of consecutive cycles of f_0 is an indication of data transmission error in the system.

12.1.7.1 FSK demodulation

Common FSK demodulation techniques such as quadrature detector circuits, phase locked loops and FM discriminations need analog filtering which for integrated circuits consume chip space. This makes them unsuitable for implantable devices as they tend to increase the size of the devices. In [5] the FSK signal was considered as a baseband signal in an attempt to avoid the use of common demodulation methods. In doing so, detection is converted to measuring the periods of the FSK signal cycles. If the period is higher than a set threshold, bit "1" is said to have been detected and if it is less bit "0" is received. This approach speeds up the demodulation technique as it has thus been reduced to finding just a time interval.

As shown in Figure 12.14 the detector is a clock recovery system connected to the receiver tank circuit. Bearing in mind that the period of one of the carriers is twice the other in this system, even if there is some spreading in the signal duration, it is fairly possible to recover the two signal periods more uniquely. The data detector serves to discriminate between the short and long periods in the received data and hence a clear discrimination of the data is possible. In the subsequent sub-sections, time measurement systems for analog and digital FSK detection techniques are described.

12.1.7.2 Analog FSK detection

Charging a capacitor at constant current and monitoring its voltage is a well known technique in circuit theory for time measurements. In FSK demodulation, charging and discharging of the capacitor is synchronized with the FSK carrier frequency. Detection of a bit is therefore set as a voltage measurement process across the capacitor. If the voltage is above a certain value, a bit "1"

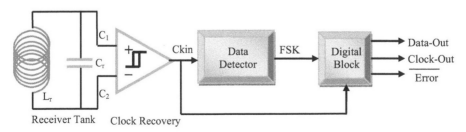

Figure 12.14 FSK Detection System

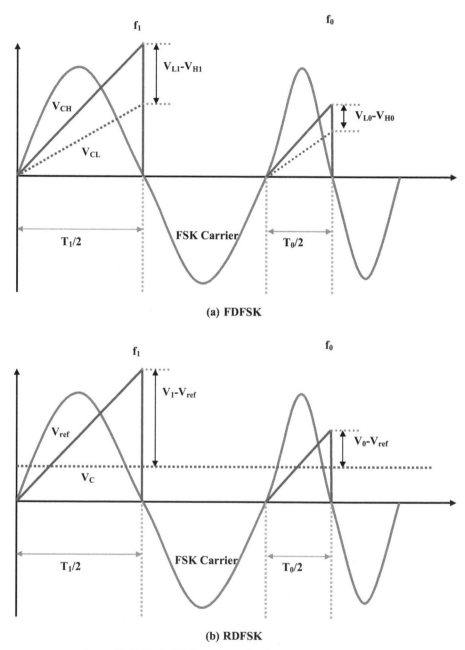

(a) FDFSK

(b) RDFSK

Figure 12.15 Fully Differential and Referenced Differential FSK

is detected and if below it is a bit "0". There are two techniques for doing this based on capacitive charging. They are the fully differential FSK (FDFSK) demodulation and referenced differential FSK (RDFSK) demodulation.

In a fully differential FSK demodulation, two unequal capacitors are charged with two different currents and their charged voltages are compared using a hysteresis comparator.

A referenced differential FSK demodulator on the other hand uses a single capacitor and its charged voltage is compared with a reference to decide if bit "1" or a "0" is detected.

12.1.7.3 Digital FSK detection

The duration of the received carrier cycles are measured in digital demodulation of FSK using a constant-frequency clock time-base f_{tb} [5] at a rate that is much higher than the anticipated carriers f_0 and f_1. During the positive envelope of the carrier, an n-bit counter is made to run. When the envelope goes negative, the counter is stopped. A digital comparator is used to decide if a long or short cycle is received, by comparing the count with a reference count held by the comparator. The counter is then re-set to initiate the next count. The following conditions must hold:

1) The period of the time base $1/f_{tb}$ must be smaller than the time difference between the half cycles of f_0 and f_1 for accurate discrimination between them for the measurement to be accurate. Hence

$$f_{tb} > \frac{2f_0 f_1}{f_0 - f_1} \qquad (12.51)$$

2) The minimum width of the counter (n) should also be

$$2^n > \frac{f_{tb}}{f_0} \qquad (12.52)$$

3) To simplify the complexity of the demodulator and reduce dynamic power consumption, the comparator and counter are combined by choosing the following design parameters

$$f_0 > \frac{f_{tb}}{2^n} > f_1 \qquad (12.53)$$

Hence the most significant bit (MSB) of the counter determines whether a long or short cycle is received and the constant reference number is

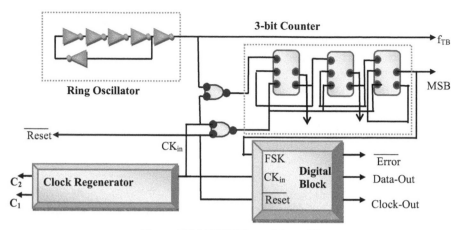

Figure 12.16 DFSK Demodulator [5]

then 2^n. For example if as in [5] the following choices are made:

$$f_0 = 8 \text{ MHz}$$
$$f_1 = 4 \text{ MHz}$$

Then

$$f_{tb} > 16 \text{ MHz}$$

The following range of values results if an $n = 3$ (3-bit) counter is used

$$64 \text{ MHz} > f_{tb} > 32 \text{ MHz} \tag{12.54}$$

Clearly a time base more than 32 MHz and less than 64 MHz would be adequate. The choice of such a time base ensures that phase noise and temperature-dependent frequency variations of the ring oscillator will not affect the performance of the demodulator. A 5-stage ring oscillator was used to generate the time base of 49 MHz in [5].

References

[1] Soumyajit Mandal and Rahul Sarpeshkar, "Power-Efficient Impedance-Modulation Wireless Data Links for Biomedical Implants", IEEE Transactions on Biomedical Circuits and Systems, Vol. 2, No. 4, Dec. 2008, pp. 301–315.

[2] D. Rinner, H. Witschnig and E. Merlin "Broadband NFC–A System Analysis for the Uplink", in Proc. IEEE 6th international conference on Communication Systems, Networks and Digital Signal Processing, July 25, 2008, CSNDSP08, pp. 292–296.

[3] Rahul Sarpeshkar, Woradorn Wattanapanitch, Scott K. Arfin, Benjamin I. Rapoport, Soumyajit Mandal, Michael W. Baker, Michale S. Fee, Sam Musallam and Richard A. Andersen, "Low-Power Circuits for Brain–Machine Interfaces", IEEE Transactions on Biomedical Circuits and Systems, Vol. 2, No. 3, Sept. 2008, pp. 173–183.

[4] Zoubir Hamici, Roland Itti and Jacques Champier, "A High-Efficiency Biotelemetry System for Implanted Electronic Device", in Proc. 1995 IEEE-EMBC and CMBEC, pp. 1649–1650.

[5] Maysam Ghovanloo and Khalil Najafi, "A Wideband Frequency-Shift Keying Wireless Link for Inductively Powered Biomedical Implants", IEEE Trans. On Circuits and Systems-1: Regular Papers, Vol. 51, No. 12, Dec. 2004, pp. 2374–2383.

[6] Mohamad Sawan, Yamu Hu and Jonathan Coulombe, "Wireless Smart Implants Dedicated to Multichannel Monitoring and Microstimulation", IEEE Circuits and Systems Magazine, First Quarter 2005, pp. 21–39.

[7] J. Parramon, et al., "ASIC-based battery less implantable telemetry microsystem for recording purposes," Engineering in Medicine and Biology Society, in *Proc. of the 19th Annual International Conference*, Vol. 5, 1997, pp. 2225–2228.

[8] G. J. Suaning and N. H. Lovell, "CMOS neurostimulation ASIC with 100 channels, scaleable output, and bidirectional radio-frequency telemetry," *IEEE Trans. on Biomedical Engineering*, vol. 48, Feb. 2001. pp. 248–260.

[9] W. Liu, et al., "A neuro-stimulus chip with telemetry unit for retinal prosthetic device," *IEEE J. Solid-State Circuits*, vol. 35, Oct. 2000, pp. 1487–1497.

[10] T. Akin, K. Najafi and R. M. Bradley, "A wireless implantable multichannel digital neural recording system for a micromachined sieve electrode," *IEEE J. Solid-State Circuits*, vol. 33, Jan. 1998, pp. 109–118.

[11] J. Parramon, et al., "ASIC-based battery less implantable telemetry microsystem for recording purposes," Engineering in Medicine and Biology Society, in *Proc. of the 19th Annual International Conference*, vol. 5, 1997, pp. 2225–2228.

[12] G. Gunnar, E. Bruun, and H. Morten, "A Chip for an Implantable Neural Stimulator," *Analog Integrated Circuits and Signal Processing*, vol. 22, 1999, pp. 81–89.

[13] B. Smith, Z. Tang et al., "An externally powered, multichannel, implantable stimulator-telemeter for control of paralyzed muscle," *IEEE Transactions on Biomedical Engineering*, vol. 45, Apr. 1998, pp. 463–475.

[14] Zhijun Lu, and Mohamad Sawan, "An 8 Mbps Data Rate Transmission by Inductive Link Dedicated to Implantable Devices", 2008, pp. 3057–3060.

[15] Zhengnian Tang, Brian Smith, John H. Schild, and P. Hunter Peckham, "Data Transmission from an Implantable Biotelemeter by Load-Shift Keying Using Circuit Configuration Modulator", IEEE Transactions On Biomedical Engineering, Vol. 42. No. 5. May 1995, pp. 524–528.

[16] Nattapon Chaimanonart, Michael A. Suster, Wen H. Ko, and Darrin J. Young, "Two-Channel Data Telemetry with Remote RF Powering for High-Performance Wireless MEMS Strain Sensing Applications", 2005, pp. 285–288.

[17] Michael Suster, Jun Guo, Nattapon Chaimanonart, Wen H. Ko, and Darrin J. Young, A Wireless Strain Sensing Microsystem with External RF Power Source and Two-Channel Data Telemetry Capability", in Proc. IEEE International Solid–State Conference, Feb. 14, 2007, pp. 380–382.

[18] Shihong Deng, Yamu Hu and Mohamad Sawan, "A High Data Rate QPSK Demodulator for Inductively Powered Electronics Implants", in Proc. IEEE ISCAS, 2006, pp. 2577–2580.

[19] Subbarayan Pasupathy, "Minimum Shift Keying: A Spectrally Efficient Modulation", IEEE Communications Magazine, vol. 17, No. 4, Jul. 1979, pp. 14–22.

[20] Maysam Ghovanloo and Khalif Najafi, "A High Data Transfer Rate Frequency Shift Keying Demodulator Chip for the Wireless Biomedical Implants", in Proc. The 2002 45th Midwest Symposium on Circuits and Systems, vol. 3, Aug. 4–7 2002, pp. III-433–III-436.

[21] Sameer Sonkusale and Zhenying Luo, "A Complete Data and Power Telemetry System Utilizing BPSK and LSK Signaling for Biomedical Implants", in Proc. 30th Annual International IEEE EMBS Conference, Vancouver, British Columbia, Canada, Aug. 20–24, 2008, pp. 3216–3219.

[22] Franco Fuschini, Carmine Piersanti, Francesco Paolazzi, and Gabriele Falciasecca, "On the Efficiency of Load Modulation in RFID Systems Operating in Real Environment", IEEE Antennas and Wireless Propagation Letter, vol. 7, 2008, pp. 243–246.

[23] G. De Vita and G. Iannaccone, "Design criteria for the RF section of UHF and microwave passive RFID transponders," *IEEE Trans. Microwave Theory Tech.*, vol. 53, no. 9, 2005, pp. 2978–2990.

13

Broadband Near Field Magnetic Communications

The previous chapter introduced the concept of impedance modulation as a means of communication between a near field transmitter and receiver. It considered narrowband implementations. This chapter presents concepts for broadband inductive communications.

13.1 Circuit Model of Broadband Inductive Links

For broadband applications, data transfer requires bandwidth that are best obtained at higher frequencies. High Q results to narrow bandwidth around resonance but power efficiency is maximized at resonance. Losses in biological tissues are high at broadband, hence for implants, narrowband circuits are desirable. However when large amounts of data need to be transferred quickly broadband is required. A broadband inductive communication link is modeled as a transmitter coupling energy to a receiver using mutual inductance principles.

Manchester encoding schemes are used for higher data rate implementations and up to 424 kbps is achieved in the passive mode at a modulation ratio of 10% [2]. In the passive mode, the receiver draws its operating power from the transmitter. In the active mode, both the transmitter and the receiver communicate alternately by generating their own magnetic fields.

13.1.1 Spectral masks

In [2] PSK is proposed as the preferred modulation scheme for enhancing the data rate. One of the early standards for broadband NFC is the ISO 14443 standard. The European standard [3] specifies spectral masks. It defines the maximum field strength at 10 m from the transmitting antenna to be 60 dB$\mu A/m$ for the 13.56 MHz carrier at a bandwidth of ± 900 kHz as in Figure 13.4.

J. Ihyeh Agbinya (PhD), Principles of Inductive Near Field Communications for Internet of Things, 175–180.

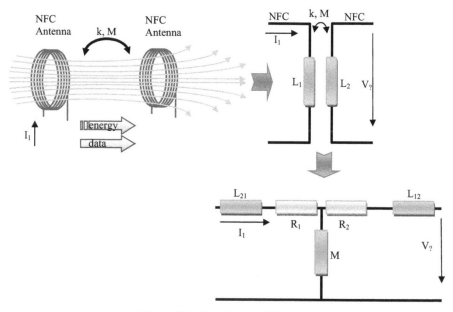

Figure 13.1 Broadband NFC System

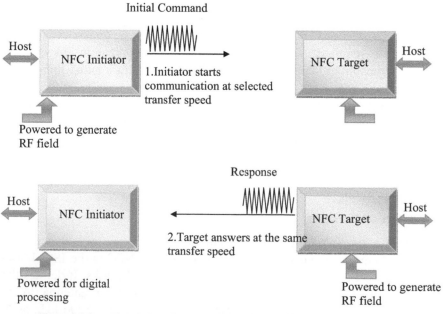

Figure 13.2 Active-Active Communication Mode Using Miller Encoding [2]

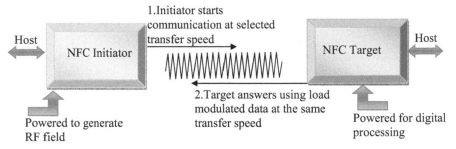

Figure 13.3 Active-Passive Communication Mode Using Miller Coding [2]

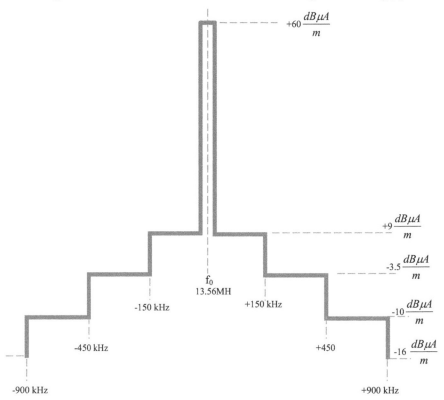

Figure 13.4 Spectral Mask

The spectral mask therefore sets limits on the achievable data rates which can be inferred because the standard limits the symbol duration to $16/f_c$.

Higher order PSK (N-PSK) can be used to increase the data rates as shown in Table 13.1 [2]. The first four rows are specified in the standards for Europe and Asia. The last two rows are proposed in [2]. This is based on

Table 13.1 Achievable Data Rates

Symbol duration	180°	90°	45°	22.5°	11.25°	5.625°	2.81°	1.4°
$128/f_c$	106	212	318	424	530	636	742	**848**
$64/f_c$	212	424	636	848	1059	1271	1483	**1695**
$32/f_c$	424	848	1271	1695	2119	2543	2966	**3390**
$16/f_c$	848	1695	2543	3390	4238	5085	5933	**6780**
$8/f_c$	1695	3390	5085	6780	8475	10170	11865	**13560**
$4/f_c$	**3390**	**6780**	**10170**	**13560**	**16950**	**20340**	**23730**	**27120**

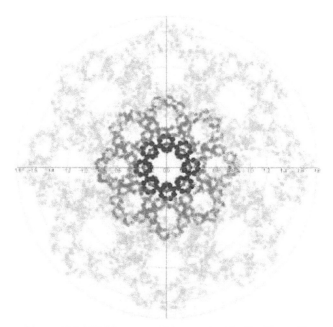

Figure 13.5 I/Q Diagram with Increasing Quality Factor [2]

reducing the symbol duration (column 1 of Table 13.1) essentially increasing the bandwidth of the modulated and transmitted symbols. An error free communication is possible up to a Q-factor of 10 after which the received symbols interfere with each other limiting symbol detection. Essentially for constant Q-factors below 10 and constant coupling factors, the received voltage, power, constant bandwidth and scattering but with reduced symbol rate from $16/f_c$ to $8/f_c$ data rate can be increased.

In the quest for achieving higher bit rates by increasing the phase states within the original standard single phase state, we also increase the probability

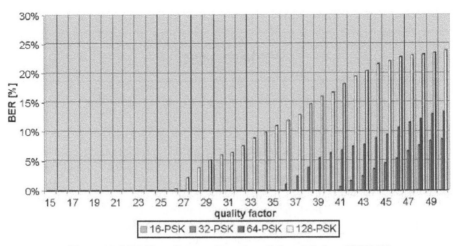

Figure 13.6 Bit Error Rates as Function of Q and Order of PSK [2]

of symbols mixing. For example when we have 8 or 16 phase states instead of one, we increase the mixing of detected symbols [2, 4].

The higher the modulation rate, the lower the quality factor required to limit inter-symbol interference.

Bit error rates increase with the order of modulation (in this case PSK). The higher the modulation order, the lower the quality factor required for required bit error rate. With a 32-PSK, error free transmission is possible up to $Q = 41$. A 64-PSK can be transferred error free at $Q = 35$. A 128 PSK requires a $Q = 25$ for error free transfer. With nearly no significant extra data processing, data rates up to 4.24 Mbps was achieved in lab tests by the authors in [2].

References

[1] Soumyajit Mandal and Rahul Sarpeshkar, "Power-Efficient Impedance-Modulation Wireless Data Links for Biomedical Implants", IEEE Transactions on Biomedical Circuits and Systems, Vol. 2, No. 4, Dec. 2008, pp. 301–315.

[2] D. Rinner, H. Witschnig and E. Merlin "Broadband NFC – A System Analysis for the Uplink", in Proc. IEEE 6th international conference on Communication Systems, Networks and Digital Signal Processing, July 25, 2008, CSNDSP08, pp. 292–296.

[3] K. Finkenzeller, *RFID-Handbuch*, 3te Auflage, Carl Hanser Verlag: München Wien, 2002.

[4] D. Rinner, *NFC-Systemrealisierungen für hohe Datenraten*, Diplomarbeit, Institut für Breitbandkommunikation der TU Graz: Graz 2007.

14

Effects of Trans-Impedance on Crosstalk

Chapter thirteen introduced the basic concepts for communication between a near field transmitter and receiver. This chapter presents their applications in wireless power transfer, body area networks, near field contact communication, short range sensing, implantable devices and personal area networks.

14.1 System Topologies and Applications

When and why does one have to use series (transmitter) and parallel (receiver) RLC circuits or vice versa? "The series resonance circuit in the reader loads the transmitter with a low impedance, whereas, the parallel tank in the implant improves the driving performance for non–linear loads, in this case, the rectifier [1]". Four configurations are possible [2].

14.1.1 Topologies

Four different topologies are possible depending on whether the transmitter and receiver are series or parallel tuned RLC circuits. The input source may also be either a voltage or a current source as shown in Figures 14.1 to 14.4.

14.1.1.1 Series–parallel topology

The transmitter is a series tuned RLC circuit and the receiver is a parallel RLC tuned circuit as in Figure 14.1. The series transmitter delivers maximum current (as produced by the source) to induce maximum possible magnetic flux to the receiver. The receiver is better a parallel tuned circuit so that the total induced voltage in the receiver coil is applied to the receiver load. This provides maximum power to the load.

The parasitic resistors of the transmitter and receiver coils are shown in Figure 14.1 as R_T and R_R respectively. The Kirchoff's voltage law equation

J. Ihyeh Agbinya (PhD), *Principles of Inductive Near Field Communications for Internet of Things*, 181–200.
© 2011 *River Publishers. All rights reserved.*

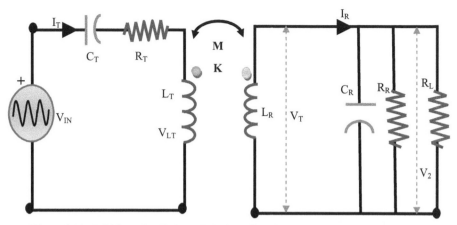

Figure 14.1 T-Voltage in (Series), R-Voltage Out (Parallel)-Popular in Implantable

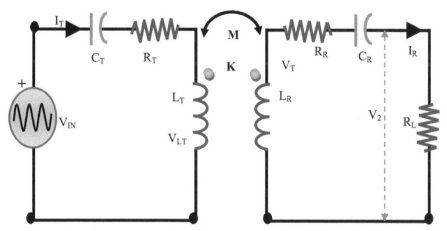

Figure 14.2 T-Voltage in (Series), R-Current Out (Series)

for the transmitter and receiver circuits

$$\begin{pmatrix} V_{IN} \\ 0 \end{pmatrix} = \begin{pmatrix} Z_T & -j\omega M \\ -j\omega M & Z_R \end{pmatrix} \begin{pmatrix} I_T \\ I_R \end{pmatrix} \tag{14.1}$$

The transmitter and receiver impedances are:

$$Z_T = R_T + j\omega L_T + \frac{1}{j\omega C_T} = R_T + j\omega \left(\omega L_T - \frac{1}{\omega C_T} \right) \tag{14.2}$$

$$Z_R = R_R + j\omega L_R + \frac{R_L}{1 + j\omega R_L C_R} \tag{14.3}$$

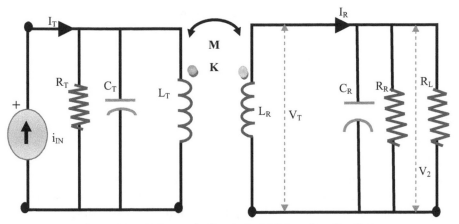

Figure 14.3 *T*-Current in (Parallel), *R*-Voltage Out (Parallel)

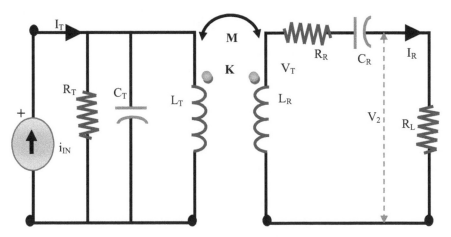

Figure 14.4 *T*-Current in (Parallel), *R*-Current Out (Series)

The resonance frequency of the receiver is

$$\omega_0^2 = \frac{1}{L_R C_R} - \left(\frac{1}{C_R R_L}\right)^2 \tag{14.4}$$

The resonance frequency of the receiver is a function of the load resistance R_L, the receiver inductance L_R and receiver capacitance C_R. A specified range of load is required for resonance and when the load resistance falls outside that range, the receiver can shift out of resonance. Hence the receiver capacitor is used for tuning it through the expression

$$C_R = \frac{R_L \pm \sqrt{R_L^2 - 4L_R^2\omega_0^2}}{2L_R R_L \omega_0^2} \tag{14.5}$$

At this value of resonance frequency maximum power is delivered to the load and the capacitor retains a value based on the choice of resonance frequency provided that $0 < \sqrt{R_L^2 - 4L_R^2\omega_0^2}$ or $2L_R\omega_0 < R_L$. What is the value of the coupling coefficient at which maximum power is delivered to the load? The transmitter will deliver maximum power to the load when the reflected impedance of the receiver $(\omega M)^2/Z_R$ is equal to the impedance of the transmitter at resonance. This point is given by the expression

$$k_c = \sqrt{R_T C_T \left(\frac{1}{C_R R_L} + \frac{R_R}{L_R}\right)} \tag{14.6}$$

The critical mutual inductance therefore is

$$M_c = k_c\sqrt{L_R L_T} = \sqrt{L_R L_T} * \sqrt{R_T C_T \left(\frac{1}{C_R R_L} + \frac{R_R}{L_R}\right)} \tag{14.7}$$

This requires coils have high Q and small parasitic resistances. This will increase the range of coverage. This however means that in a surveillance situation, the transmitter is vulnerable to detection because of the large coverage range.

High system efficiency can be obtained when the ratio of power delivered to the load is high compared to the power created at the transmitter. This efficiency is defined by the expression

$$\eta = \frac{P_{R_L}}{P_T} = \frac{|I_{R_L}|^2 R_L}{Re(I_T)V_{IN}} \tag{14.8}$$

The receiver capacitance which results to the highest efficiency is obtained by differentiating the expression for the efficiency and is given as $\frac{\partial \eta}{\partial C_R} = 0$.

$$C_r = \frac{L_R R_T}{R_T R_R^2 + \omega^2 \left(k^2 L_R L_T R_R + R_T L_R^2\right)} \tag{14.9}$$

The load is a critical part of the tuned value of the capacitor in the receiver circuit.

14.1.1.2 Series–series topology

Both the transmitter and receiver use series tuned RLC circuits. Transmitter delivers an input voltage. The current through the receiver load is proportional to the impedance of the receiver.

As in the previous analysis for the parallel tuned receiver case, the Kirchoff's voltage equations result to

$$\begin{pmatrix} V_{IN} \\ 0 \end{pmatrix} = \begin{pmatrix} Z_T & -j\omega M \\ -j\omega M & Z_R \end{pmatrix} \begin{pmatrix} I_T \\ I_R \end{pmatrix} \tag{14.10}$$

For this case the impedance of the transmitter is as before and that for the receiver is

$$Z_R = R_R + R_L + j\omega L_R + \frac{1}{j\omega C_R} \tag{14.11}$$

The resonance frequency occurs with the impedance is all real or

$$j\omega L_R + \frac{1}{j\omega C_R} = 0 \tag{14.12}$$

$$\omega_0 L_R = \frac{1}{\omega_0 C_R} \Rightarrow \omega_0^2 = \frac{1}{L_R C_R} \tag{14.13}$$

The resonance frequency does not depend on the load as in the parallel case. Hence the circuit can be tuned always to $C_R = \frac{1}{\omega_0^2 L_R}$ without the load resistance interfering. The maximum power delivered to the load at resonance is

$$P_R = \frac{k^2 \omega_0^2 L_R L_T R_L V_{IN}^2}{\left[(R_L + R_R) R_T + k^2 \omega_0^2 L_R L_T \right]^2} \tag{14.14}$$

The system efficiency is

$$\eta = \frac{k^2 \omega_0^2 L_R L_T R_L}{(R_L + R_R) \left[(R_L + R_R) R_T + k^2 \omega_0^2 L_R L_T \right]^2} \tag{14.15}$$

The value of the receiver inductance which maximizes the system received power is

$$L_R = \frac{(R_L + R_R) R_T}{k^2 \omega_0^2 L_T} \tag{14.16}$$

The critical coupling coefficient for maximum power transfer is given by the expression

$$k_c = \sqrt{\frac{1 + \dfrac{R_L}{R_R}}{Q_T Q_R}} \tag{14.17}$$

14.1.1.3 Parallel–parallel topology

Both the transmitter and receiver use parallel tuned RLC circuits for different reasons. A parallel series transmitter is desirable if small currents must be delivered by the transmitter to the coupled receiver. For example in low power, tissue protective application provided there is enough power to establish efficient communication link.

The input current is divided into three components and only a portion passes through the transmitter inductor. The current flowing through the transmitter is only a portion of the input current i_{in}, which is smaller than the total input current

$$I_{Lt} = \frac{Z_L i_{in}}{Z_{TP}} \tag{14.18}$$

This current induces a smaller voltage in the receiver compared with when the total input current from the source is used.

14.1.1.4 Parallel–series topology

In this topology, the transmitter is a parallel tuned RLC circuit and the receiver is a series tuned RLC circuit.

As in the parallel-parallel topology, minimum current is delivered to the transmitter coil and also the load current is minimized through the series voltage division or reduced current receiver.

14.1.2 Effects of trans-impedance on induction voltage in receivers

Assuming that a receiver does not cross-feed all its received signal from other receiver sources into its neighbours but only experiences the couplings from all the unintended transmitters as interference, then the received signal by the ith receiver is given by the expression

$$V_{Ri} = Z_d \left(k_{ii} I_{Ti} + \sum_{j \neq i}^{N} k_{ij} I_{Tj} \right) \tag{14.19}$$

Where k_{ij} is the coupling coefficient between the ith receiver and the jth transmitter inductor. The first term is the desired signal. The second term is interference. Therefore the signal to interference ratio is

$$SIR = \frac{|k_{ii} I_{Ti}|}{\left| \displaystyle\sum_{j \neq i}^{N} k_{ij} I_{Tj} \right|} \tag{14.20}$$

Two types of receiver circuits can be used and the choice depends on application. A tuned parallel RLC circuit is able to deliver most of the induced voltage *e* at the terminals of the receiver inductor to the load. Hence maximum voltage across the receiver load is the objective. For the series RLC receiver circuit, the objective is to deliver the total current at the terminals of the receiver inductor through the load. Hence the power across the load is dependent on high current rather than high voltage. This section analyses the performances of both types of receiver circuits and compares them using their trans-impedance.

14.1.3 Parallel tuned receiver circuit

Tran-impedance is the ratio of the voltage across the receiver load to the current flowing in the transmitter inductor. Therefore, it is a good measure of how efficient the transmitter is in transferring power to the receiver load. In general:

$$Z_d = \frac{V_L}{i_T} \tag{14.21}$$

Here V_L is the voltage across the receiver load and i_T is the current that causes this voltage to be induced and flows through the transmitter coil. The impedance Z_d is computed as in [2] using trans-impedance concept borrowed from the theory of inductive inter-chip crosstalk. The trans-impedance (the ratio of the induced voltage in the receiver coil to the current in the transmitter coil) of a receiving R, L, C circuit for ideal coupling (when $k = 1$) is

$$Z_d = \frac{1}{(1 + \omega^2 L_R C_R) + j\omega R_R C_R} \cdot j\omega \sqrt{L_T L_R} \cdot \frac{1}{(1 - \omega^2 L_T C_T) + j\omega R_T C_T} \tag{14.22}$$

Noting that the basic principles for power and data transfer through an inductive link is the same as in transformers except for the weak coupling we can then borrow the methods used in transformer theory. The weakly coupled transformer can be replaced with the expression

$$n' = k\sqrt{\frac{L_2}{L_1}} \tag{14.23}$$

$$i_2 = \frac{i_1}{n'} \tag{14.24}$$

With low coupling the impedance seen by this current is approximately equal to $k^2 L_2$. Therefore the voltage induced by this current is

$$V_T = j\omega L_2 k^2 \left(\frac{i_1}{n'}\right) = j\omega L_2 k^2 \left(i_1 \frac{1}{k}\sqrt{\frac{L_1}{L_2}}\right) \tag{14.25}$$

$$= j\omega k \sqrt{L_1 L_2} i_1$$

V_T is the voltage across the inductor L_2 induced by the current i_1. In general the voltage across the receiver load is given by a voltage divider expression

$$V_2 = \frac{Z_L * V_T}{Z_L + Z_{rem}}$$

$$= \frac{\text{Receiver Load Impedance} * \text{Voltage across receiver inductor}}{\text{Receiver Load Impedance} + \text{Remaining receiver Impedance}} \tag{14.26}$$

The voltage across the load is given by the expression

$$V_2 = \frac{\left(\frac{R_L}{1 + j\omega C_2 R_L}\right) V_T}{(R_R + j\omega L_2) + \frac{R_L}{1 + j\omega C_2 R_L}} = \frac{V_T}{1 + (R_R + j\omega L_2)\left(\frac{1}{R_L} + j\omega C_2\right)} \tag{14.27}$$

The trans-impedance version of this equation is obtained by substituting for the mutual inductance

$$\frac{V_2}{i_1} = \frac{\omega k \sqrt{L_1 L_2}}{\left(1 - \omega^2 L_2 C_2 + \frac{R_R}{R_L}\right) + j\left(\frac{\omega L_2}{R_L} + \omega R_R C_2\right)} \tag{14.28}$$

Table 14.1 System Parameters

Parameter	L_1	L_2	C_2	R_R	R_L	ω
value	43.5 μH	3.7 μH	330 pF	1 Ω	1 $k\Omega$	$2\pi \cdot f(MHz)$

Table 14.2 Induction Factor Equations

Frequency (MHz)	$\lvert Z_d \rvert$	$V_{RL} = Z'_1\left(k_{ii}I_{Ti} + \sum_{j \neq i}^{N} k_{ij}I_{Tj}\right)$
4	$1279.2 * k(x)$	$1279.2\left(k_{ii}I_{Ti} + \sum_{j \neq i}^{N} k_{ij}I_{Tj}\right)$
6.78	$356.2 * k(x)$	$356.2\left(k_{ii}I_{Ti} + \sum_{j \neq i}^{N} k_{ij}I_{Tj}\right)$
13.56	$137.3 * k(x)$	$137.3\left(k_{ii}I_{Ti} + \sum_{j \neq i}^{N} k_{ij}I_{Tj}\right)$

By substituting the expression for the voltage at the terminals of the inductor $L2$, this expression can further be simplified to

$$\frac{V_2}{i_1} = \lvert Z_d \rvert = \frac{\omega k \sqrt{L_1 L_2}}{\sqrt{\left(\frac{\omega L_2}{R_L} + \omega R_R C_2\right)^2 + \left(1 - \omega^2 L_2 C_2 + \frac{R_R}{R_L}\right)^2}} \qquad (14.29)$$

This expression establishes the relationships between the network elements and the system trans-impedance which forms a part of the crosstalk coupling between multiple transceiver elements in a body area network. The value of the trans-impedance is given as a function of k as follows:

At 4 MHz we have the function

$$\lvert Z_d \rvert = 1279.2 k(x) \qquad (14.30)$$

or

$$V_2 = 1279.2 * k(x) \cdot i_1 \qquad (14.31)$$

Figure 14.2 shows the induction factor for the parameters shown in Table 14.1. We define the voltage induction factor as the ratio of the trans-impedance to the coupling coefficient $\lvert Z_d \rvert / k$. The value of maximum voltage induction factor is an interesting design parameter not often available to the designer at the start of design. Hence the availability of the induction factor is significant. Figure 14.6 reveals that the best induction point is at resonance.

This should not be a surprise! The peak induction factor in Figure 14.6 is at 5 MHz and it is 1660 ohms. The value can be inferred by differentiating the equation for the induction factor with respect to frequency and setting it to zero $\left(\frac{\partial Z'}{\partial \omega} = 0\right)$ and

$$Z' = \frac{|Z_d|}{k} = \frac{\omega\sqrt{L_1 L_2}}{\sqrt{\left(\frac{\omega L_2}{R_L} + \omega R_R C_2\right)^2 + \left(1 - \omega^2 L_2 C_2 + \frac{R_R}{R_L}\right)^2}} \qquad (14.32)$$

This occurs at the resonance frequency. These results indicate that we can estimate or predict properly the level of induced voltage in the receiver components. Apart from at resonance which has just the one maximum induction factor for the receiver, there are always two frequencies with identical induction factors (IF).

14.1.3.1 Effects of changing receiver inductor
The results in Figure 14.6 motivate Figures 14.7 and 14.8 to check how specific system components affect the induction factor or trans-impedance. In Figure 14.5 all other system parameters are kept constant. Changes are only in the receiver inductor. In Figure 14.7 only the receiver inductor is changed. Low frequency resonance is established with very high values of the inductor. As the inductor reduces, the resonance frequency moves to higher frequencies.

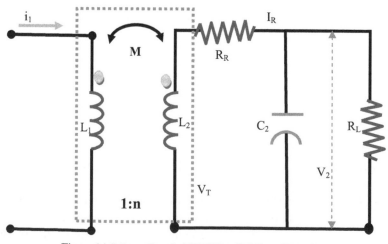

Figure 14.5 Low Coupled RLC Parallel Tuned Receiver

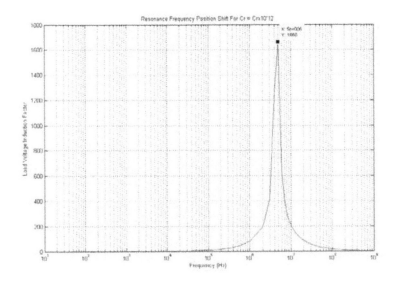

Figure 14.6 Voltage Induction Factors Due to Trans-Impedance for Receiver

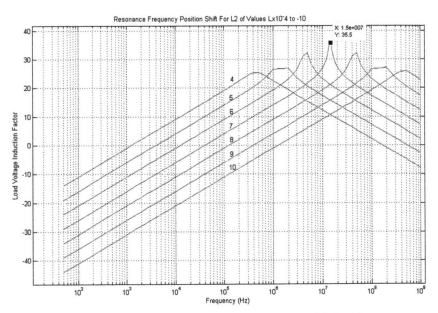

Figure 14.7 Induction Factor Graphs for Changing $L2$ in Receiver

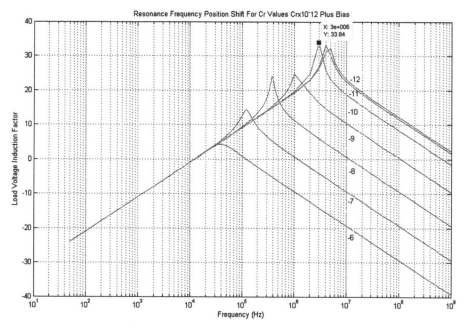

Figure 14.8 Induction Factor Graphs for Changing C Receiver

14.1.3.2 Effect of changing receiver capacitor

There are also preferred choices of parameters for maximum voltage induction factor with respect to the capacitor in the receiver and Figure 14.6 shows this.

In Figure 14.8, we change only the capacitor in the receiver circuit by connecting capacitive biases in parallel with it. These graphs show that it is possible to estimate the worst crosstalk signals from a resonating coil. The maximum voltage induction factor should be used to assess the worst possible crosstalk that coil can produce. Figures 14.7 and 14.8 show that parasitic inductance and capacitors have the effects of shifting the resonance frequencies. If the parasitic capacitance is in parallel with the system capacitance, the shift is to lower frequencies. For reduced or series capacitance, the shift is to higher frequencies. Inductors have the opposite effect as capacitors.

14.1.3.3 Effects of changing receiver resistances

There are two resistors in this receiver circuit. R_L is the load resistance and R_R is the self resistance of the receiver inductor.

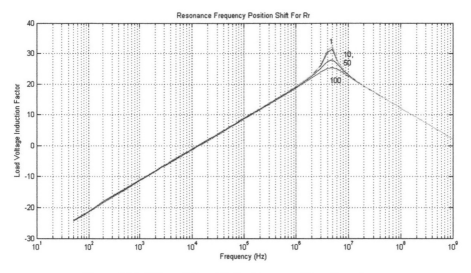

Figure 14.9 Effect of Parasitic Resistance on Trans-Impedance

This figure shows that the effect of increasing the receiver load resistance is an increase in the peak value of the induction factor. The peak position is not altered.

14.1.3.4 Effects of changing transmitter inductor
The variation of trans-impedance with L_t is shown in Figures 14.7 and 14.8.

The figure shows that by making the value of the transmitter inductor bigger, we also raise the value of the induction factor. If nothing else is altered at the receiver, the resonance point remains fixed.

14.1.4 Series receiver circuit

In the series RLC receiver circuit, the voltage at the terminal of the receiver inductor is shared by the load, the resistive losses of the inductor and capacitor.

For that case the expression for the trans-impedance is modified slightly to be

$$V_2 = \frac{R_L V_T}{\left(R_R + R_L + j\omega L_2 + \dfrac{1}{j\omega C_2}\right)} = \frac{j\omega C_2 R_L V_T}{\left(1 - \omega^2 L_2 C_2\right) + j\omega C_2 (R_R + R_L)}$$

(14.33)

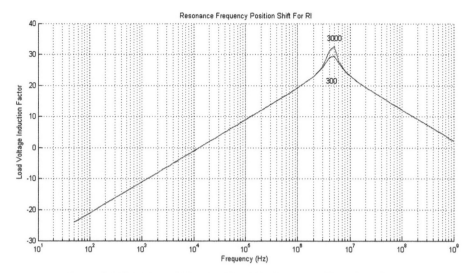

Figure 14.10 Effects of Receiver Load Resistance on Trans-Impedance

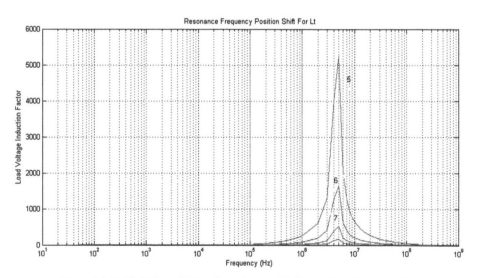

Figure 14.11 Variation of Trans-Impedance with Transmitter Inductor (L_1)

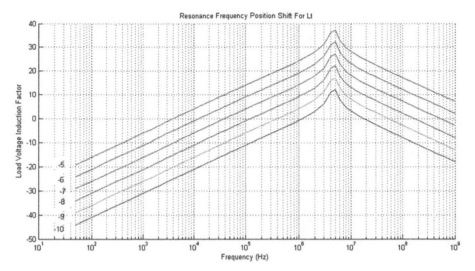

Figure 14.12   Variation of Trans-Impedance with Transmitter Inductor (L_1)

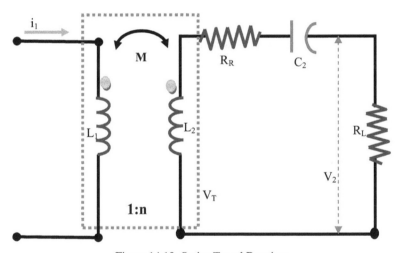

Figure 14.13   Series Tuned Receiver

As in equation (14.25)

$$V_T = j\omega k \sqrt{L_1 L_2} i_1$$

and

$$\frac{V_2}{i_1} = |Z_d| = \frac{k\omega^2 C_2 R_L \sqrt{L_1 L_2}}{\sqrt{\left(1 - \omega^2 L_2 C_2\right)^2 + \omega^2 C_2^2 (R_R + R_L)^2}} \tag{14.34}$$

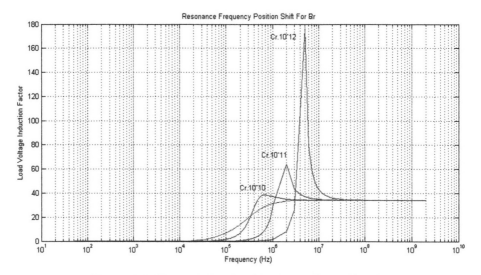

Figure 14.14 Effects of Receiver Capacitor on Power Transfer

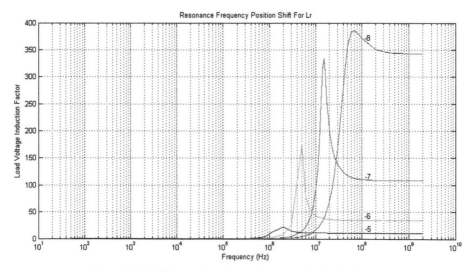

Figure 14.15 Effects of Receiver Inductor on Induction Factor

The induction factor in this case is

$$\gamma = \frac{\omega^2 C_2 R_L \sqrt{L_1 L_2}}{\sqrt{(1 - \omega^2 L_2 C_2)^2 + \omega^2 C_2^2 (R_R + R_L)^2}} \tag{14.35}$$

14.1.4.1 Effects of receiver capacitor

To compare the behaviour of the series to the parallel tuned receiver, the same parameters in Table 14.1 are used. The characteristics of this induction factor are shown in the next sets of figures. Figure 14.14 shows that the series tuned receiver circuit require very small capacitors for pairs of frequencies to share identical induction factors. With the sets of values in Table 14.1, pairs of frequencies can have equal induction factors when capacitors in the range of nano Farads and smaller are used. Induction factors are significantly higher for low receiver resistor set to 10 Ohms. In general weaker coupling results compared to the parallel tuned circuit with increasing load resistance. The load resistances were kept below 100 Ohms in all the Figures that follow, as that helps to achieve two frequencies with identical and significant induction factor. At higher resistive values, this double frequency effect is sacrificed.

14.1.4.2 Effects of changing receiver inductor

Changing the receiver inductor from a high value of 37 μH to a small value of 37 nH (nano Henry) has the effects of shifting the 'peak' induction factor frequency to higher frequencies and increasing the induction factor of the circuit.

Induction factors above 400 are obtainable for only single frequencies as no pairs of frequencies have equal induction factors. The effects of reducing the inductor value are similar to that of reducing the capacitor value.

14.1.4.3 Effects of changing receiver load resistance

The receiver load resistance has a significant impact on the magnitude of the induction factor. At low load resistances comparable to the parasitic resistance, the induction factor is small, typically around 35. Increasing the load resistance boosts the induction factor to about 410 and still provides two frequencies with equal induction factors.

Above load resistance of 130 Ohms no two frequencies have equal induction factors and are significantly higher as shown in Figure 14.13.

14.1.4.4 Effects of changing receiver parasitic resistance

Figure 14.18 shows that low parasitic resistance is desirable for high induction factor. This leads to better power transfer to the load at resonance.

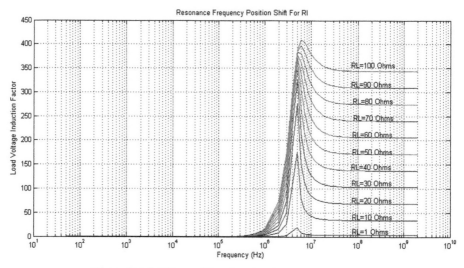

Figure 14.16 Effects of Receiver Load on Induction Factor

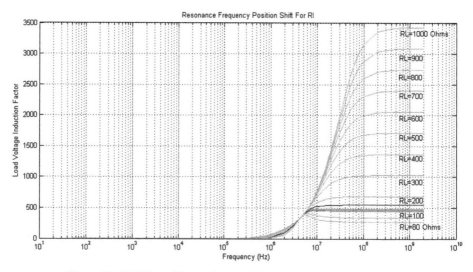

Figure 14.17 Effect of Increasing Load Resistance on Induction Factor

Parasitic resistance above 30 Ohms cause flat responses in terms of trans-impedance. Instead of power being directed to the load, it is consumed by the parasitic resistance.

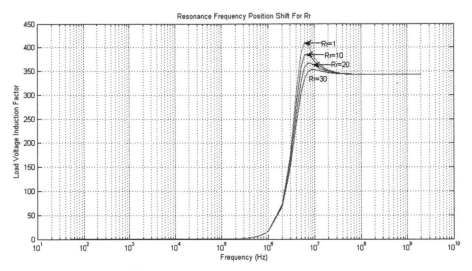

Figure 14.18 Effects of Receiver Parasitic Resistance on Induction Factor

14.1.5 Trans-impedance and wireless power transfer

Induction factor is an essential parameter for understanding how energy is transferred from the transmitter to the receiver and how much of it is available as useful energy and interference. The product of the induction factor and the coupling coefficient typify the trans-impedance and the product of the trans-impedance with the current in the transmitter gives the amount of desirable signal and the interference from neighbouring undesirable sources.

References

[1] Christian Sauer, Milutin Stanacevic, Gert Cauwenberghs, and Nitish Thakor, "Power Harvesting and Telemetry in CMOS for Implanted Devices", IEEE Trans. On Circuits and Systems–I: Regular Papers, vol. 52, No. 12, Dec. 2005, pp. 2605–2613.

[2] Douglas C. Galbraith, Mani Soma and Robert L. White, "A Wide-Band Efficient Inductive Transdermal Power and Data Link With Coupling Insensitive Gain", IEEE Transactions on Biomedical Engineering, vol. BME-34, No. 4, Apr. 1987, pp. 265–275.

15

Magneto-Induction Link Budgets

In any communication system, one of the fundamental concepts is determination of link budget, which refers to the difference between transmitted signal power and signal power at receiver [1]. It specifies how power at transmitter is allocated along the communication chain to receiver. The reason for having a link budget is that signal loses some of its power as it travels through a wireless channel (path loss) due to undesirable environmental and signal propagation effects such as fading, diffraction, reflection and scattering [1]. Therefore it is crucial to determine the signal strength at receiver, subsequently communication range for determination of other factors affecting network planning and hardware design.

15.1 Introduction

Figure 15.1 illustrates how the transmitted signal loses its power in the communication channel. To be able to extract information form the received signal, receiver requires the received signal power to be greater or at least equal to the sensitivity of the receiver.

The concept of link budget has been widely explored for traditional terrestrial radio frequency (RF)-based electromagnetic (EM) wave communications and there are a number of channel models available for planning the network in different environmental types, such as urban, suburban and rural areas, flat or hilly terrains and dense or scattered vegetation areas, because transmitted signal experiences different path losses in different environments.

RF-based EM communications channel models have taken into account different environmental characteristics and basically are the function of transmitter and receiver antenna gain, transmission power and all the possible losses. RF-based communication channel models have been studied from different perspectives, for example when direct line-of-sight (LoS) exists or not, or if detailed and precisely gathered data is required (deterministic models) or

J. Ihyeh Agbinya (PhD), Principles of Inductive Near Field Communications for Internet of Things, 201–220.
© 2011 *River Publishers. All rights reserved.*

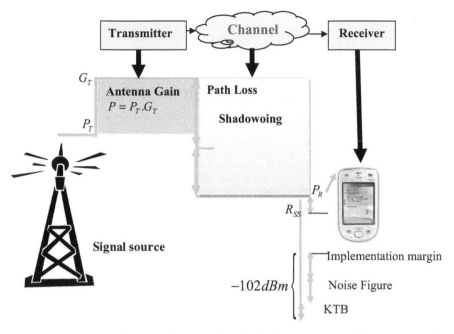

Figure 15.1 Received Signal Power Level After Passing through the Wireless Channel

modelling can rely on measured mean path losses for different environmental types (empirical models) [1].

Some of the channel models concerning with RF-communication link budget are the Free Space model, suitable when direct LoS exists between transmitter and receiver and uses Friis equation. Two-ray model concerns with communications where there are a direct LoS and a reflected path between transmitter and receiver [1]. Hata-Okamura model is valid in 150 to 1500 MHz frequency range and is suitable for distances more than 1Km; however, modified form of this model known as COST 231 has extended the frequency range to 2 GHz [1]. In COST 231 modelling, similar to SUI and Erceg model, the environment is categorised into three environment types with specific characteristics [1]. SUI and Erceg Model, which are particularly suitable for WiMAX, are available at frequency band 3.5 GHz and 1.9 GHz respectively [1, 4]. Walficsh-Ikegami is another model, recommended by WiMAX Forum and proper for channel modelling at 800 MHz to 200 MHz frequency range. This model can distinguish between the LoS and NLoS (non-line-of-sight) for WiMAX situations [1].

In spite that there are a wide range of channel models available for EM RF-based communication systems, neither of those models deal with magnetic induction (MI) communication system; because environmental effects on the magnetic waves are different from EM waves. Therefore, in this chapter three link budget models, proposed by the authors are discussed. Although the first two models known as Agbinya-Masihpour (AM) and Agbinya-Masihpour1 (AM1) are applied to a peer-to-peer (one section) communication, the third link budget model called Agbinya-Masihpour2 (AM2) pertains to a multiple section MI communication link budget. These link budget models determine the communication range based on different system parameters.

15.1.1 Agbinya-Masihpour link budget model

AM model approaches the MI link budget from two perspectives. Firstly when the coils used at the transmitting and receiving antennas are air-cored and the second approach is when antenna coils are wound around ferrite material. It is discussed later that the ferrite-cored coils enhance the achievable communication range. However, both approaches are based on LoS situation. In other word, direct line-of-sight between transmitting and receiving antennas exist in AM model.

15.1.1.1 Received signal power estimation in an NFMIC system

The link budget of a typical communication system is basically how the transmitted signal is 'spent' as it travels through the channel and received by the receiver. It is thus a balance sheet of energy gain and consumption in the communication chain. A link budget is derived from the power relations in the chain of communication along the path from the transmitter to the receiver. For an MI system the parameters of interest in the link budget relate to the magnetic antenna and include the mutual coupling, coupling coefficients and the role of Q-factors and coil efficiencies as well as permeability of the core material, in delivering fluxes from transmitter to receiver.

The block diagram of a magnetically inductive communication system is shown in Figure 15.2. Transmitter and the receiver are composed of two coils each of radius r_T and r_R and separated from each other by a distance d. The communication link between them is an inductive coupling k. The coupling

Figure 15.2 Inductively Coupled Near Field System

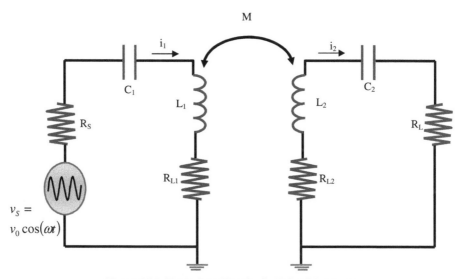

Figure 15.3 Equivalent Circuit of a Pair of Antennas

between the two magnetic antennas can be estimated using equation (15.1)

$$k = \frac{M}{\sqrt{L_T L_R}} \tag{15.1}$$

In this equation, M is the mutual coupling between two inductive circuits and L_T and L_R are inductive values of transmitter and receiver respectively. The lump circuit model of Figure 15.1 is shown in Figure 15.3.

The efficiencies of coils are given to describe how effective they are in transmitting power and by definition they are:

$$\eta_T = \frac{R_S}{R_S + R_{LT}}; \quad \eta_R = \frac{R_L}{R_L + R_{LR}} \tag{15.2}$$

Also quality factors are given by definition to be:

$$Q_T = \frac{\omega_0 L_T}{R_{LT} + R_S}; \quad Q_R = \frac{\omega_0 L_R}{R_{LR} + R_L} \tag{15.3}$$

In [3] it is shown that the power delivered to the receiver load is given by equation (15.4):

$$\frac{P_L(\omega)}{P_S} = \frac{V_C}{V_D} \eta_T \eta_R Q_T Q_R = \eta_T \eta_R Q_T Q_R k^2(d) \tag{15.4}$$

$$k^2(d) = \frac{V_C}{V_D} \tag{15.5}$$

Where

$$P_L(\omega) = \frac{|i_R|^2 R_L}{2} = \frac{P_S \eta_T \eta_R Q_T Q_R k^2}{\left(1 + Q_T^2 \dfrac{(2\Delta\omega)^2}{\omega_0^2}\right)\left(1 + Q_R^2 \dfrac{(2\Delta\omega)^2}{\omega_0^2}\right)} \tag{15.6}$$

At resonance this expression simplifies to

$$P_L(\omega) = P_S \eta_T \eta_R Q_T Q_R k^2(d) \tag{15.7}$$

For the air-cored coils, reactive power at load is computed by:

$$k^2(d) = \frac{V_C}{V_D} = \frac{\mu_0 A_R^2 \mu_0 r_T^4}{4 L_T L_R \left[(r_T^2 + d^2)\right]^3} \tag{15.8a}$$

However, if coils are wound around the ferrite material with relative permeability μ_T and μ_R at transmitting and receiving antennas respectively, the magnetic flux will be enhanced. Therefore the coupling coefficient increases and is:

$$k^2(d) = \frac{V_C}{V_D} = \frac{\mu_T \mu_0 A_R^2 \mu_R \mu_0 r_T^4}{4 L_T L_R \left[(r_T^2 + d^2)\right]^3} \tag{15.8b}$$

On the other hand, the self inductance of coils can be expressed as:

$$L = \frac{\mu_0 \pi r^2 N^2}{l + 0.9r} = \frac{\mu_0 \pi r^2}{l + 0.9r}; \quad N = 1 \tag{15.9}$$

We also assume that the length of the wire used at the coils are far larger than the radius of them: $r \ll l$ and $l = 2\pi r$. Hence:

$$L_T \approx \frac{\mu_0 r_T}{2}; L_R \approx \frac{\mu_0 r_R}{2}; \quad A_R = \pi \cdot r_R^2; N = 1 \tag{15.10}$$

Therefore we can express the coupling coefficient as a function of the coil radii and separation distance between them (equation (15.11)).

$$k^2(d) = \frac{r_T^3 r_R^3 \pi^2}{\left(r_T^2 + d^2\right)^3} \tag{15.11}$$

Where $r_R \ll r_T$.

From the equivalent circuit model of Figure 15.2, the following relationship (equation (15.12)) at the resonance frequency $\omega_0 = \frac{1}{\sqrt{L_T C_T}} = \frac{1}{\sqrt{L_R C_R}}$ between the antennas is maintained. Hence the two coils are chosen so that they resonate at the same frequency and hence permit optimum mutual communication to be established between them.

Finally power at the receiver at resonant frequency becomes:

$$P_R(\omega) = P_T Q_T Q_R \eta_T \eta_R \frac{r_T^3 \mu_0 \mu_T r_R^3 \mu_0 \mu_R \pi^2}{\left(r_T^2 + d^2\right)^3} \tag{15.12}$$

15.1.1.2 Air-cored coils—line of sight

An MI peer-to-peer communication system and equivalent circuit model are shown in Figure 15.2. In this figure, there are magnetic transmitter (T) and receiver (R), which are located at boresight of each other. The antennas are separated by a distance d. as mentioned earlier in section 1.1, received signal power can be computed by equation (15.4), when the coils are aired-cored. We express the equation (15.4) as:

$$(d^2 + r_T^2)^3 = \frac{P_T Q_T Q_R \eta_T \eta_R r_T^3 r_R^3 \pi^2}{P_R} \tag{15.13}$$

Let

$$Q = (Q_T + Q_R)(dB) \tag{15.14}$$
$$\eta = (\eta_T + \eta_R)(dB) \tag{15.15}$$
$$P = (P_T - P_R)(dBm) \tag{15.16}$$

Therefore:

$$10 \cdot \log\left(d^6 \left(1 + \frac{r_T^2}{d^2}\right)^3\right) = P + Q + \eta + 30 \cdot \log(r_T r_R) + 20 \cdot \log \pi \tag{15.17}$$

or

$$\log d = \frac{P + Q + \eta + 30 \cdot \log(r_T r_R) + 20 \cdot \log \pi - 30 \cdot \log\left(1 + \frac{r_T^2}{d^2}\right)}{60}$$

(15.18)

The factor of 60 in equation (15.18) comes from the fact that in MI communication, inductive power decays to the sixth power of distance. Therefore the link budget equation becomes:

$$d = 10^{\frac{P + Q + \eta + 30 \cdot \log(r_T r_R) + 20 \cdot \log \pi - 30 \log\left(1 + \frac{r_T^2}{d^2}\right)}{60}}$$

(15.19)

There are terms in equation (15.18) which contribute significantly to the determination of distance and those that have very little effect on d. In the next equation, those factors are separated. Therefore, let

$$d = 10^{\frac{P + Q + \eta + 20 \cdot \log \pi}{60}} \cdot 10^{-\frac{1}{2}\left[\log(r_T r_R) - \log\left(1 + \frac{r_T^2}{d^2}\right)\right]}$$

(15.20)

Alternatively this expression can be written as:

$$d = d' \cdot \Delta d$$

(15.21)

Where:

$$d' = 10^{\frac{P + Q + \eta + 20 \cdot \log \pi + 30 \log(r_T r_R)}{60}}$$

(15.22)

and the correction term is:

$$\Delta d = 10^{-\frac{1}{2}\left[\log\left(1 + \frac{r_T^2}{d'^2}\right)\right]}$$

(15.23)

Since Δd is almost 1, it can be neglected to simplify calculations for network planning purposes. Thus the link budget equation can be stated as:

$$d = d' = 10^{\frac{P + Q + \eta + 20 \cdot \log \pi + 30 \log(r_T r_R)}{60}}$$

(15.24)

By simulation results provided in chapter 18, we have shown that Δd is too small to affect network planning in a significant manner. However this equation has not yet taken into account the relative permeability of magnetic former that is used to enhance magnetic field.

15.1.1.3 Ferrite-cored coil—line-of-sight

Magnetic flux density, B at the point d in a near field situation is enhanced by the relative permeability of core material on which antenna coil is wound. This field is:

$$B = \frac{\mu_r \mu_0 I \cdot N \cdot r_T^2}{2d^3} \qquad (15.25)$$

Where μ_r is the relative permeability of ferrite core, I is loop current, N is the number of turn of coil. Assume that receiver also uses a ferrite core former of relative permeability μ_R. Hence both transmitted and received fluxes are amplified by factors equal to the relative permeability of each core. Thus inductive antennas have gains equal to their relative permeability of ferrite cores. Using ferrite core ensures that the coil of smaller radii can be used for antennas. Hence the achievable range of communication, based on the power equation (equation (15.12)) is:

$$d = 10^{\dfrac{P+Q+\eta+\mu+30\cdot\log(r_T r_R)+20\cdot\log\pi-30\log\left(1+\frac{r_T^2}{d^2}\right)}{60}} \qquad (15.26)$$

Where:

$$\mu = (\mu_T + \mu_R)(dB) \qquad (15.27)$$

This equation includes two new factors for the gains of antennas. Therefore transmission range will be extended by about $10^{\mu_T(dB)+\mu_R(dB)/60}$ times, if ferrite core material at transmitter and receiver have permeability μ_T and μ_R respectively.

15.1.2 Agbinya-Masihpour1 link budget model

Figure 15.3 illustrates MI transceivers within a peer-to-peer communication system. α is the angle between the receiver coil and the x axis; in other word the angle between the axes of transceivers is $(90 - \alpha)$, however in this study α is assumed to be $\pi/2$. Figure 15.3 also shows circuit model for this transmission system. Transmitter circuit consists of a voltage source U_S which creates a current in the circuit ($I = I_0 \cdot e^{-j\omega t}$). It also has an inductor with inductive value L_T and self resistance of R_T. Inductor in transmitter creates magnetic field around itself which induces a current in the receiver circuit through coupling of magnetic fields with mutual inductance M. However the receiver circuit has a coil with inductive value of L_R and self resistance R_R as well as a load with impedance Z_L. Relationships between these components

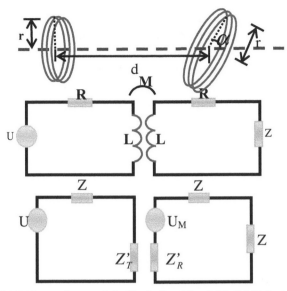

Figure 15.4 MI Transceivers and the Equivalent Circuit Model (Adapted from [5])

are shown in equation (15.28) [5, 6]:

$$Z_T = R_T + j\omega L_T; \quad Z_T' = \frac{\omega^2 M^2}{R_R + j\omega L_R + Z_L};$$

$$Z_R = R_R + j\omega L_R; \quad Z_R' = \frac{\omega^2 M^2}{R_T + j\omega L_T};$$

$$U_M = -j\omega M \frac{U_S}{R_T + j\omega L_T} \tag{15.28}$$

Since the aim of this section is to develop a link budget equation for the MI channel model cited in [5, 6] by *Sun and Akyildiz*, and compare it with *Agbinya-Masihpour* link budget, firstly we need to formulate *Sun and Akyildiz* model to AM model. Therefore we establish a common formalism for presenting these two models. Common formalisms are for the efficiencies of coil:

$$\eta_T = \frac{R_S}{R_S + R_T} \cong 1$$

$$\eta_R = \frac{R_L}{R_L + R_R} \cong 1 \tag{15.29}$$

Where R_S is source resistance and assumed to be far larger than coil resistance therefore the efficiency of transmitter is almost equal to 1. Similarly, load resistance is quite larger than receiver coil self-resistance therefore receiver efficiency is also assumed to be 1. However the self inductance of coils is a function of the number of turns, relative permeability of core material of coil, radius of coil, the length of wire used at coils as well as the area of a single turn A:

$$L = \frac{\mu A \cdot N^2}{l + 0.9r} \tag{15.30}$$

On the other hand the quality factor of coils are

$$Q_T = \frac{\omega L_T}{R_T}; \quad Q_R = \frac{\omega L_R}{R_R} \tag{15.31}$$

Where,

$$R_R = 2\pi \cdot r_R N_R R_0; \quad R_T = 2\pi \cdot r_T N_T R_0 \tag{15.32}$$

R_0 is the resistance of a unit length of the loop [6].

Based on the assumption that $l = 2\pi \cdot r \cdot N \gg 0.9r$ and $N_T = N_R = N$:

$$Q_T = Q_R = Q \tag{15.33}$$

Hence the power equation derived earlier (equation (15.4)) reduces to:

$$P_R = P_T Q^2 k^2(d) \tag{15.34}$$

In the power equation, $k(d)$, could be viewed as the gain of NFMIC system in distance d. Coupling coefficient based on this assumption is:

$$k^2(d) = \frac{\mu^2 A_R^2 \cdot r_T^4}{4 \cdot L_T L_R \left(d^2 + r_T^2\right)^3} \tag{15.35}$$

We also make a further assumption that the radius of transmitting coil is far smaller than communication range therefore path loss becomes:

$$k^2(d) = \frac{\pi^2 r_T^3 r_R^3}{N_T N_R d^6} \tag{15.36}$$

The above formulation is used to develop a link budget for the NFMIC system shown in Figure 15.3.

Agbinya-Masihpour 1 channel model is derived from [5, 6] based on the power equation cited in [6] (equation (15.37)):

$$P_R = \frac{P_T \omega \mu N_R r_T^3 r_R^3 \sin^2 \alpha}{16 \cdot R_0 d^6} \tag{15.37}$$

Using the same formalism as above, equation (15.37) cast as a link budget equation becomes:

$$d_{AM\,1} = 10^{\beta_{(AM1)}} \tag{15.38}$$

$$\beta_{(AM\,1)} = \frac{\begin{array}{c} P + 10 \cdot \log \mu + 10 \cdot \log \omega + 30 \log(r_T r_R) \\ -10 \cdot \log 16 - 10 \cdot \log R_0 + 10 \cdot \log N_R \end{array}}{60}$$

To be able to compare AM and AM1 link budget models, the two models are required to be stated based on the same formulism.

We have previously introduced AM channel model in section 1. Based on this model (AM) and the assumption used to formulate AM1 model, signal power received at receiver for AM model becomes:

$$P_R = \frac{\omega^2 \mu^2 \cdot r_T^3 r_R^3 P_T}{16 N_R N_T R_0^2 d^6} \tag{15.39}$$

Therefore, the AM model as a function of the same system parameters as AM1 model becomes:

$$d_{AM} = 10^{\beta_{(AM)}}$$

Where,

$$\beta_{(AM)} = \frac{\begin{array}{c} P + 20 \cdot \log \mu + 20 \cdot \log \omega + 30 \log(r_T r_R) - 10 \cdot \log 16 \\ -20 \cdot \log R_0 - 10 \cdot \log N \end{array}}{60}$$

$$P = (P_T - P_R) dBm \tag{15.40}$$

$$N = N_T \cdot N_R$$

By comparing these two models (equations (15.38) and (15.40)), it is obtained that received signal power for AM model is higher than received power in AM1 model by a factor equal to:

$$\frac{P_{R(AM)}}{P_{R(AM1)}} = \frac{\omega \mu}{N_R^2 N_T R_0} \tag{15.41}$$

Comparing these two models suggests that the achievable communication range is higher by loading the inductors with capacitive elements. Therefore AM model provides longer communication range than the AM1 link budget model. However the achievable distance difference between the two models is more significant at higher frequencies.

15.1.3 Agbinya-masihpour2: A link budget model for multiple section NFMIC

Even though NFMIC has a number of benefits over the EM RF-based communication systems, such as higher reliability, security, power efficiency and link robustness, very short communication range is a major limitation in such systems. To extend the communication range, a cooperative relaying method for NFMIC has been proposed by researchers in the field known as magneto inductive waveguide model. It mainly focuses on inducing received flux on neighbouring coils to achieve larger range within the system. A few studies have been done on waveguide model [5, 6, 7]. *Solymar* and *Shamonina* have studied magneto-inductive waveguide model from power transmission point of view and have discussed termination of the waveguide chain to decrease power loss due to power reflection [5]. *Sun* and *Akyildiz* examined the application of magneto-inductive waveguide for underground communications. They suggest that magneto-inductive waveguide can be implemented for data transmission in underground environment where channel condition is unstable and EM waves fail to operate properly [6]. In this section, firstly basic waveguide model is discussed and then a link budget model for an $n + 1$ section waveguide is introduced.

Waveguide model consists of a number of cooperative relaying nodes between transmitter and receiver. They relay transmitted data until it reaches the target receiver. Mutual coupling occurs between neighbouring coils. An n-relay-node waveguide system and equivalent circuit model are illustrated in Figure 15.5. Where current in circuit n is induced due to mutual inductive coupling with the adjacent coils $n + 1$ and $n - 1$, relationship between them is [5]:

$$ZI_n + X(I_{n+1} + I_{n-1}) = 0 \tag{15.42}$$

In equation (15.42), $Z = j(\omega L - 1/\omega c)$ is loop impedance, $X = j\omega M$ is impedance of coupling and $I_n = I \cdot e^{-jkxn}$ shows the current flowing in *nth* circuit. However k is propagation constant and x is separation distance between adjacent coils.

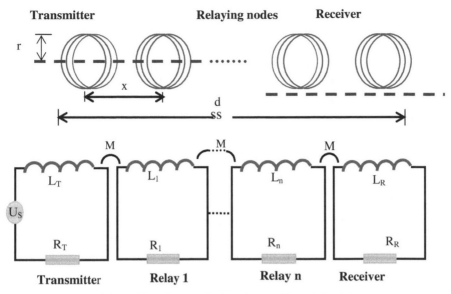

Figure 15.5 Magnetic Waveguide and Circuit Model

Sun and Akyildiz in [5, 6] proposed a complex path loss model which relates the transmitted power and received power to system parameters such as permeability of medium, per unit resistance of wire used to create transmitting and receiving coils, number of turns of the coils, their radii, operating frequency and the communication range. Based on their model, for a peer-to-peer communication, the received power to that of transmitted power is shown in equation (15.43).

$$\frac{P_R}{P_T} \approx \frac{\omega^2 \mu^2 N_T N_R r_T^3 r_R^3 \sin^2 \alpha}{8d^6} \cdot \frac{1}{4R_0 \left(2R_0 + \frac{1}{2} j\omega\mu N_T\right)} \quad (15.43)$$

By assuming a low resistance loop, and when high signal frequency and large number of turns are employed in transmitter and receiver circuits ($R_0 \ll \omega\mu N_T$), power ratio reduces to [6]:

$$\frac{P_R}{P_T} \approx \frac{\omega\mu N_R r_T^3 r_R^3 \sin^2 \alpha}{16R_0 d^6} \quad (15.44)$$

However, Recently Sun and Akyildiz [5, 6] extended the work on magneto inductive waveguide, on the basis of the mentioned peer-to-peer communication model and applied it to underground wireless where EM RF-based

systems perform poorly [5]. Since channel condition in underground communications is dynamic and material within the channel (soil, rock and water)
contribute to a high path loss and large bit error rate, the magneto inductive
waveguide has been seen as a promising method to address range limitation
associated with the use of EM as physical layer. In this model, intermediate
nodes operate without the need for individual source of power. However, this
study has greatly simplified this model.

In this section we build upon the work of [5, 6, 7] a link budget for a magneto inductive waveguide communication system has been developed using
an array of n antenna within the channel between transmitter and receiver.
The pioneering work of *Syms, Shamonina* and *Solymar* [7], have established
some of the theories for magneto inductive waveguide used in this section.
Sun and Akyildiz proposed a complex path loss. In this section this model is
appreciably simplified. In their work, each intermediate coil is loaded with a
capacitor C to neutralize self-induction and C has value [6]:

$$C = \frac{2}{\omega^2 N^2 \mu \cdot \pi \cdot r} \tag{15.45}$$

Based on this component, the path loss expression is given as the value:

$$\frac{P_R}{P_T} = \frac{\omega^2 N^2 \mu^2 r_T^3 r_R^3}{4x^6 2R_0 \left(4R_0 + \frac{\omega^2 N^2 \mu^2 r_T^3 r_R^3}{4x^6 2R_0}\right)} \left[\frac{j}{\frac{4R_0}{\omega\mu N}\left(\frac{x}{r_R}\right)^3 + \frac{\omega\mu N}{4R_0}\left(\frac{r_R}{x}\right)^3}\right]^{2n} \tag{15.46}$$

In this expression, variable x is the separation distance between any two
sections of waveguide (distance separation between two neighbouring coils).
Thus communication range d is divided into $n + 1$ sections of length x, or
$d/x = (n + 1)$. Path loss is a real number and for odd values of n has a
negative value. In this section the path loss expression is simplified for analysis and modelling into so-called *AM2* link budget equation and thus provides
insight into the achievable range based on selected system variables. By using
the expressions for Q-factors, the path loss equation can be expressed as
$\frac{P_R}{P_T} = P_1 P_{2n}$.

Where:

$$P_1 = \frac{\omega^2 N^2 \mu^2 r_T^3 r_R^3}{4x^6 2R_0 \left(4R_0 + \frac{\omega^2 N^2 \mu^2 r_T^3 r_R^3}{4x^6 2R_0}\right)} \tag{15.47}$$

By substituting $Q = \frac{\omega\mu}{4\pi \cdot R_0}$ at equation (15.47), it becomes:

$$P_1 = \frac{Q_T \cdot Q_R N^2 \pi^2 k^2(x)}{\left(2 + Q_T \cdot Q_R \cdot N^2 \pi^2 k^2(x)\right)} \tag{15.48}$$

By defining $k^2(x) \approx \frac{r_T^3 r_R^3}{x^6}$ (P_1 is similar to neglecting Δd in the *AM* model described in section 1). The second term is:

$$P_{2n} = \left[\frac{j}{\frac{4R_0}{\omega\mu N} \left(\frac{x}{a}\right)^3 + \frac{\omega\mu N}{4R_0} \left(\frac{a}{x}\right)^3} \right]^{2n} \tag{15.49}$$

Therefore

$$P_{2n} = \left(Q_T \cdot Q_R N^2 \pi^2 k^2(x)\right)^n \left[\frac{j}{\left(1 + Q_T \cdot Q_R N^2 \pi^2 k^2(x)\right)} \right]^{2n} \tag{15.50}$$

Hence the path loss equation becomes

$$\frac{P_R}{P_T} = \frac{Q_T \cdot Q_R N^2 \pi^2 k^2(x)}{\left(2 + Q_T \cdot Q_R \cdot N^2 \pi^2 k^2(x)\right)} \left(Q_T \cdot Q_R N^2 \pi^2 k^2(x)\right)^n$$

$$\times \left[\frac{j}{\left(1 + Q_T \cdot Q_R N^2 \pi^2 k^2(x)\right)} \right]^{2n} \tag{15.51}$$

Define $\beta = Q_T Q_R N^2 \pi^2 k^2(x)$ as section path loss. Waveguide path loss is therefore given equation (15.52):

$$\frac{P_R}{P_T} = \frac{\beta \cdot \beta^n}{(2 + \beta)} \left[\frac{j}{(1 + \beta)} \right]^{2n} = \left(\frac{\beta^{n+1}}{(2 + \beta)} \right) \left[\frac{j}{(1 + \beta)} \right]^{2n} \tag{15.52}$$

Since $j = e^{\frac{j\pi}{2}}$ we therefore have

$$\frac{P_R}{P_T} = \frac{\beta \cdot \beta^n}{(2 + \beta)} \left[\frac{j}{(1 + \beta)} \right]^{2n} = \left(\frac{\beta^{n+1}}{(2 + \beta)} \right) \left(\frac{1}{(1 + \beta)} \right)^{2n} e^{j\pi n} \tag{15.53}$$

When n is odd the path loss has negative value. For $n = 0$ we have the simple peer-to-peer MI system path loss equation. If design is such that $r_T = r_R = x$, then $k(x) = 1$. Coupling coefficient can not be more than unity or $r_T \cdot r_R$

can not be greater than a section length, therefore, in practice $r_T \cdot r_R \leq x^2$. $r_T \cdot r_R > x^2$ is an invalid situation as it is impractical for the receiver to collect more flux than all flux created by transmitter. For $n = 0$ and β is unity, received power is a third of transmitted power and we have the same result as in equation (15.43) by *Sun and Akyildiz* [5, 6]:

$$\frac{P_R}{P_T} = \frac{1}{3}\left(\frac{1}{2}\right)^{2n} \tag{15.54}$$

This means that $\beta = Q_T Q_R N^2 \pi^2 k^2(x)$, the design equation required for determining optimum power transfer between sections is a function of the number of turns used, coupling coefficients and coils quality factors. In equation (15.43) we have per turn quality factor of coils. It is possible from this to literally design the value of received power based on selected system parameters from these equations.

15.1.3.1 One section peer-to-peer link budget: line of sight
The case when $n = 0$ is particularly interesting for study because it is one transmitter and one receiver case with one path loss equation:

$$\frac{P_R}{P_T} = \frac{Q_T \cdot Q_R N^2 \pi^2 k^2(x)}{\left(2 + Q_T \cdot Q_R \cdot N^2 \pi^2 k^2(x)\right)} \tag{15.55}$$

i) In this equation, when $Q_T \cdot Q_R \cdot N^2 \pi^2 k^2(x) \gg 2$ the power equation approaches the value $\frac{P_R}{P_T} \approx 1$.
This is an invalid case. The received powers in most communication systems (inductive and radiative) are usually not equal due to losses in the channel and the transceiver circuitry.

ii) When $Q_T \cdot Q_R \cdot N^2 \pi^2 k^2(x) = 2$, $\frac{P_R}{P_T} = \frac{1}{2}$.
Although this approximation promises that half the transmitted power can be received, engineering the scenario could prove very difficult. Hence a better approximation is required.

iii) When $Q_T \cdot Q_R \cdot N^2 \pi^2 k^2(x) \ll 2$, which means that $k^2(x)$ is small, then a more practical situation is:

$$\frac{P_R}{P_T} \approx \frac{Q_T \cdot Q_R N^2 \pi^2 k^2(x)}{2} \tag{15.56}$$

Thus by using different assumptions, we arrive at a familiar expression. The reduced equation and AM models are equivalent (to within a constant multiplier). This proves the fact that *Agbinya-Masihpour* model and the *Sun* and *Akyildiz* model [6] are identical and have correctly modelled MI communications.

15.1.3.2 Multiple section waveguide link budget

1) Using the approximation $Q_T \cdot Q_R \cdot N^2 \pi^2 k^2(x) \gg 2$ is used in the expression for a multiple-section waveguide, it simplifies to the equation:

$$\frac{P_R}{P_T} \approx \left(Q_T \cdot Q_R N^2 \pi^2 k^2(x) \right)^n \left[\frac{j}{\left(Q_T \cdot Q_R N^2 \pi^2 k^2(x) \right)} \right]^{2n} \quad (15.57)$$

Therefore the power loss equation can be written in terms of 10^{α_1}.
Where

$$\alpha 1 = \frac{\begin{array}{l} P_T(dBm) - P_R(dBm) - 10 \cdot n[\log Q_T + \log Q_R + 10 \\ +2 \cdot \log N + 3 \cdot \log(r_R r_T)] \end{array}}{60 \cdot n}$$

$$(15.58)$$

This is an unrealistic situation and hence not practical

2) However, when $Q_T \cdot Q_R \cdot N^2 \pi^2 k^2(x) = 2$, the power difference between transmitted signal and received signal is 3 dBm. In this case the communication distance approaches a threshold in which waveguide model performs in a reverse manner. This means that if transmitted power is larger than received power for more than 3 dBm, by increasing the number of relaying nodes the achievable range decreases. However this distance threshold is a function of coil characteristics. For example by increasing the radius of transmitting and receiving coil or coil Q-factors, the distance threshold increases.

Equation (15.59) shows the relationship between the distance threshold and coil attributes. When $\beta = 2$ then $P_T - P_R = 3 \ dBm$ therefore the distance threshold is:

$$d_{\text{threshold}} = 10^t$$

$$t = \frac{\begin{array}{l} 3 + 10 \cdot \log Q_T + 10 \cdot \log Q_R + 20 \cdot \log N \\ +20 \cdot \log \pi + 30 \cdot \log(r_T \cdot r_R) \end{array}}{60} \quad (15.59)$$

Based on this assumption the communication distance $d = n \cdot x$, for an $n+1$ section magneto inductive waveguide system becomes identical for any number of relaying nodes. However when $Q_T \cdot Q_R \cdot N^2 \pi^2 k^2(x) = 2$, the approximation gives:

$$\frac{P_{R(n+1)}}{P_{T(n)}} = \frac{2^n}{4} \left[\frac{j}{3}\right]^{2n} \tag{15.60}$$

Equation (15.60) shows the power ratio for any adjacent nodes.

3) The third assumption and the most practical situation is when $Q_T \cdot Q_R \cdot N^2 \pi^2 k^2(x) \ll 2$. Therefore the power equation simplifies to

$$\frac{P_R}{P_T} \approx \frac{1}{2} \left(Q_T \cdot Q_R N^2 \pi^2 k^2(x)\right)^{n+1} e^{j\pi \cdot n} \tag{15.61}$$

This expression truly shows an $n + 1$ section power terms and if the term in bracket is less than one, received power is always less than transmitted power and depends on the number of sections in magneto inductive waveguide. Therefore the power loss equation for this case can be written in terms of $10^{\alpha 2}$.

Where

$$\alpha 2 = \frac{P_T(dBm) - P_R(dBm) - 3 + 10 \cdot (n + 1)}{\left[\log Q_T + \log Q_R + 10 + 2 \cdot \log N + 3 \cdot \log(r_R r_T)\right]}{60 \cdot (n + 1)} \tag{15.62}$$

15.1.4 Summary

In this chapter, three link budget models are discussed:

- AM and AM1 channel models are applied to a peer-to-peer NFMIC system when LoS exist. The two models describe how system parameters such as transmitting and receiving coil radii, quality factor, relative permeability of the core material and transmission and received power impact the achievable communication range.
- AM2 link budget model is applicable to a multiple section magneto inductive model. It studies the magneto inductive waveguide for

practicability of different situations. AM2 introduces the section path loss $\beta = Q_T Q_R N^2 \pi^2 k^2(x)$ and suggest that the only practical situation is when $\beta \ll 2$. However, in this chapter we have shown that waveguide model operates inversely after waveguide distance threshold.

"MATLAB Codes available for download: http://riverpublishers.com/river_publisher/book_details.php?book_id=90"

References

[1] J. I Agbinya and M. Masihpour; "Part III: WiMAX nd LTE Link Budget"; in *"Planning and Optimisation of 3G and 4G Wireless Networks"* (ed. Agbinya JI), ISBN: 978-87-92329-24-0 © 2009 River Publishers, pp. 105–136.

[2] J. Milanovic, S. Rimac-Drlje, K. Bejuk, "Comparison of Propagation Models Accuracy for WiMAX on 3.5 GHz." *IEEE International Conference*, pp. 111–114, 2007.

[3] J. I. Agbinya, N. Selvaraj, A. Ollett, S. Ibos, Y. Ooi-sanchez, M. Brennan, and Z. Chaczko, "Size and characteristics of the 'Cone of Silence' in Near Field Magnetic Induction Communications", *IMILCIS2009*, Canberra, 2009.

[4] J. Milanovic, S. Rimac-Drlje, K. Bejuk, "Comparison of Propagation Models Accuracy for WiMAX on 3.5 GHz." *IEEE International Conference*, 2007, pp. 111–114.

[5] Ian F. Akyildiz, Zhi Sun and Mehmet C. Vura, "Signal propagation techniques for wireless underground communication networks", in (Elsevier) Physical Communication 2 (2009) pp. 167–183.

[6] Zhi Sun, Ian F. Akyildiz, Underground wireless communication using magnetic induction, in: Proc. IEEE ICC 2009, Dresden, Germany, June 2009.

[7] R.R.A Syms, L. Solymar and E. Shamonina, "Absorbing terminations for magneto-inductive waveguide": Proc. IEE 2005, Vol. 152.

16

Magneto – Inductive Waveguide Devices

In the previous chapters we have introduced the concepts of magneto-inductive waves and showed that unless they are terminated properly, there will be signal reflections in the waveguide. In this chapter, we show how the presence of the power reflections is used to advantage in creating MI devices.

16.1 Introduction

Several magneto-inductive devices are discussed and are explained in this chapter. First, two-port devices are presented, then three-port devices. The chapter concludes by discussing the effects of resistive losses in the loop impedance.

16.1.1 Two-port devices

This section describes two-port devices based on MI mirrors and their extension and applications including Fabry-Perot, Bragg grating, tapers and couplers.

16.1.1.1 Magneto-inductive mirror

One of the most important applications of optics is the creation of mirrors. In optics, a mirror results to an image of an object due to reflection. A mirror effect is caused by a discontinuity in the medium through which light propagates causing the light to bend backwards and be received at a source. This creates an illusion of seeing the object in a mirror at a distance. A similar effect exists in magnetic communications. A magneto-inductive mirror is caused by an electrical discontinuity or abrupt change in impedance. This causes currents to be reflected back from the discontinuity into the circuit.

An MI mirror is formed by connecting two sections of an MI waveguide. Each section supports a forward wave. The loop impedance and coupling

---

Co-authored by Mehrnoush Masihpour & Agbinya JI, University of Technology, Sydney

J. Ihyeh Agbinya (PhD), Principles of Inductive Near Field Communications for Internet of Things, 221–249.

are as in the traditional waveguide. However a section of the waveguide is coupled by $Z_{m1} = j\omega M_1$, where M_1 is the modified mutual inductance. The modified mutual inductance is achieved by varying the loop separation, locating the first coil from the source by a distance other than d used for the second section (Figure 16.1.a). The currents in the coils are taken as I_n where n is the loop number. Due to the modified coupling, reflected waves are induced in the section with the modification.

In the analysis here, we focus on the application of nearest neighbor interactions and recursively extend the concept to the overall structure. The nearest neighbour equation at loop n is

$$ZI_n + Z_m (I_{n-1} + I_{n+1}) = 0 \tag{16.1}$$

where $Z = R + j\omega L + 1/j\omega C$. An abruptly terminated MI waveguide could have zero reflection if the last element were terminated with a characteristic impedance [2] such that

$$(R + j\omega L + 1/j\omega C + Z_0) I_n + Z_m I_{n-1} = 0 \tag{16.2}$$

By comparing equations (16.1) and (16.2) the conditions lead to the equation

$$Z_0 I_n = Z_m I_{n-1} = j\omega M I_{n-1} \tag{16.3}$$

Assuming a current wave $I_n = I_0 \cdot \exp(-jkna)$, we have that

$$Z_0 = j\omega M \cdot \exp(-jka) \tag{16.4}$$

At the junction the following equations hold and need to be solved

$$ZI_{-1} + Z_{m1}I_0 + Z_m I_{-2} = 0$$
$$ZI_0 + Z_m I_1 + Z_{m1}I_{-1} = 0 \tag{16.5}$$

Due to the discontinuity in the input line (a change in the mutual inductance impedance from Z_m to Z_{m1}), a reflected wave is created. We will refer to the section of the waveguide with incident (I) and reflected waves (R) as the input line and the one with only the transmitted wave (T) as the output. The MI model should naturally account for conservation of energy. The input energy is distributed into three components: (i) the power absorbed by the elements due to Ohmic losses (ii) the power taken by the terminating impedance when used and (iii) the radiated power. The essence in a high loss model is to minimize the radiated power in other to contain and conserve the

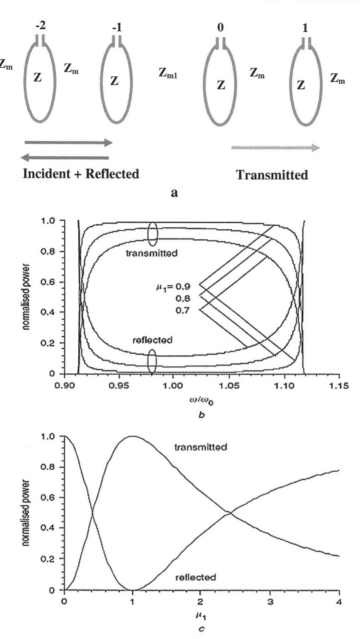

Figure 16.1 Two-Port Devices with Simple Mirror
(a) Magneto-inductive waveguide mirror
(b) variation of $|\Re|^2$ and $|T|^2$ with ω/ω_0, for $\kappa = 0.2$ and different values of μ_1
(c) variation of $|\Re|^2$ and $|T|^2$ with μ_1, for $\omega = \omega_0$

required communication power within a defined 'bubble' of interest. From a theoretical point of view, since the characteristic impedance of an MI wave is $Z_0 = j\omega M \cdot \exp(-jka)$ [3], the power in the wave can be computed.

The voltage across the characteristic impedance is $V_0 = I \cdot Z_0$. The power carried by the wave is then given by the expression:

$$P = \frac{1}{2}\mathrm{Re}\,al\left[V_0 I^*\right] = \frac{1}{2}\mathrm{Re}\,al\left[Z_0 I \times I^*\right]$$

$$= \frac{\mathrm{Re}\,al\left[j\omega M \cdot I^2 \exp(-jka)\right]}{2} = \frac{\mathrm{Re}\,al\left[\omega M \cdot I^2 \exp\left(-j\left(ka - \frac{\pi}{2}\right)\right)\right]}{2}$$

$$= \frac{\mathrm{Re}\,al\left[\omega M \cdot I^2\left(\cos\left(ka - \frac{\pi}{2}\right) - j\sin\left(ka - \frac{\pi}{2}\right)\right)\right]}{2}$$

$$= \frac{\omega M \cdot I^2 \cos\left(ka - \frac{\pi}{2}\right)}{2}$$

$$= \frac{\omega M \cdot I^2 \sin(ka)}{2} \tag{16.6}$$

This power is proportional to the mutual inductance between two coils, the current carried by the transmitting coil and the phase shift per section 'ka'. The value of k is traditionally complex to account for Ohmic losses and as such this power too is a function of the Ohmic losses.

16.1.1.2 Reflection and transmission coefficients

In other to analyse the performance of the MI waveguides, let the current in loop n be

$$I_n = I\exp(-jkna) + \mathrm{Re}\,xp(jkna) \quad n \le -1$$
$$I_n = T\exp(-jkna) \quad n \ge 0 \tag{16.7}$$

For all loops with negative index, the current is a sum an incident and reflected wave and for positive index n, only a transmitted wave exists. The governing equation away from the junction is satisfied by the equations which we solve together to obtain

$$I + R = \mu_1 T$$
$$\mu_1[I\exp(jka) + \mathrm{Re}\,xp(-jka)] = T\exp(jka) \tag{16.8}$$

Where, $\mu_1 = Z_{m1}/Z_m$. By defining the reflection and transmission coefficients as $\Re = R/I$ and $\Im = T/I$ we have

$$\Re = (\mu_1^2 - 1)\exp(jka)/\left[\exp(jka) - \mu_1^2\exp(-jka)\right] \tag{16.9}$$

$$\Im = \mu_1[\exp(jka) - \exp(-jka)]/\left[\exp(jka) - \mu_1^2\exp(-jka)\right] \tag{16.10}$$

The reflected ($|\Re|^2$) and transmitted power ($|\Im|^2$) coefficients are

$$|\Re|^2 = (\mu_1^2 - 1)^2/\left[(\mu_1^4 + 1) - 2\mu_1^2\cos(2ka)\right] \tag{16.11}$$

$$|\Im|^2 = \mu_1^2[2 - 2\cos(2ka)]/\left[(\mu_1^4 + 1) - 2\mu_1^2\cos(2ka)\right] \tag{16.12}$$

In general:

$$|\Re|^2 + |\Im|^2 = 1 \tag{16.13}$$

Thus if $\mu_1 \neq 1$, a reflected wave will occur in the system. There will be no transmitted wave when $\cos(2ka) = 1$ or $2ka = 0$ or 2π. At that point $a = \pi/k$, meaning the presence of a discontinuity causes power to be totally reflected. Syms et al [1] have shown the variations of the reflected and transmitted powers as functions of ω/ω_0 and μ_1 in Figures 16.1b) and c) for $\kappa = 0.2$ respectively. Near the centre of the band the reflected and transmitted coefficients are approximately

$$\Re \approx (\mu_1^2 - 1)/(\mu_1^2 + 1) \tag{16.14}$$

$$\Im \approx 2\mu_1/(\mu_1^2 + 1) \tag{16.15}$$

16.1.2 Fabry–Perot resonator filters

In optics, a Fabry-Perot is formed as a mirror with two reflecting surfaces so that an incident light reflects from both surfaces and also allows transmissions to take place from them. The device is named after its inventors Fabry and Perot. Similarly, in MI Fabry-Perot resonators two MI waveguides are used in which the first MI waveguide with loop numbers $n \leq -1$ is coupled to a loop with an impedance $Z_{m1} = j\omega M_1$ and a second waveguide with loop numbers $n \geq 1$ is also coupled to the loop with an impedance $Z_{m2} = j\omega M_2$. Each waveguide behaves as an MI mirror. Because there are now two discontinuities one each on either side of the excitation loop, reflections occur in both waveguides and behaves like a Fabry-Perot. When the reflection coefficient is approximately unity, the mirror (Fabry-Perot) behaves like a narrow band pass filter.

The equations near the junction are:

$$ZI_{-1} + Z_{m1}I_0 + Z_mI_{-2} = 0$$
$$ZI_0 + Z_{m2}I_1 + Z_{m1}I_{-1} = 0$$
$$ZI_1 + Z_mI_2 + Z_{m2}I_0 = 0 \qquad (16.16)$$

The solutions away from the junction are of the form

$$I_n = I \exp(-jkna) + R_{FP} \exp(jkna) \quad n \le -1$$
$$I_n = T_{FP} \exp(-jkna) \quad n \ge 0 \qquad (16.17)$$

The current I_0 is unknown and need to be determined. The solutions to these two equations are:

$$\Re_{FP} = \frac{R_{FP}}{I} = \frac{(1 - \mu_1^2) \exp(jka) + (1 - \mu_2^2) \exp(-jka)}{(\mu_1^2 + \mu_2^2 - 1) \exp(-jka) - \exp(jka)} \qquad (16.18)$$

$$\Im_{FP} = \frac{T_{FP}}{I} = \frac{\mu_1\mu_2(\exp(-jka) - \exp(jka))}{(\mu_1^2 + \mu_2^2 - 1) \exp(-jka) - \exp(jka)} \qquad (16.19)$$

$$\Gamma_0 = \frac{I_0}{I} = \frac{\mu_1(\exp(-jka) - \exp(jka))}{(\mu_1^2 + \mu_2^2 - 1) \exp(-jka) - \exp(jka)} \qquad (16.20)$$

Where, $\mu_2 = Z_{m2}/Z_m$. The power reflection and transmission coefficients are given by the expressions

$$|\Re_{FP}|^2 = \frac{\left[(\mu_1^2 - 1)^2 + (\mu_2^2 - 1)^2\right] - 2(1 - \mu_1^2)(\mu_2^2 - 1)\cos(2ka)}{\left[1 + (\mu_1^2 + \mu_2^2 - 1)^2\right] - 2(\mu_1^2 + \mu_2^2 - 1)\cos(2ka)}$$
$$(16.21)$$

$$|\Im_{FP}|^2 = \frac{2\mu_1^2\mu_2^2\{1 - \cos(2ka)\}}{\left[1 + (\mu_1^2 + \mu_2^2 - 1)^2\right] - 2(\mu_1^2 + \mu_2^2 - 1)\cos(2ka)} \qquad (16.22)$$

Assuming that $\mu_1 = \mu_2$ and for two different values of μ_1, Figure 16.2b) shows the variation of $|\Im_{FP}|^2$ as a function the ratio ω/ω_0 for $\kappa = 0.2$.

The magnetic wave is transmitted totally when $ka = \pi/2$ and $\omega/\omega_0 = 1$. It falls to zero on both sides. The frequency response of the power reflection coefficient is complimentary. The device is therefore a band pass filter similar to a Fabry – Perot cavity. As shown in figure (16.2.b) the bandwidth increases with increasing $\mu_1(= \mu_2)$ and a Fabry-Perot type response is obtained only in the range $\mu_1 < 0.6$. In the range $\mu_1 > 0.6$ the response is very broad [1].

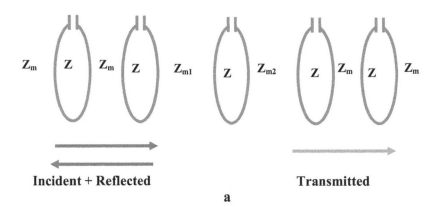

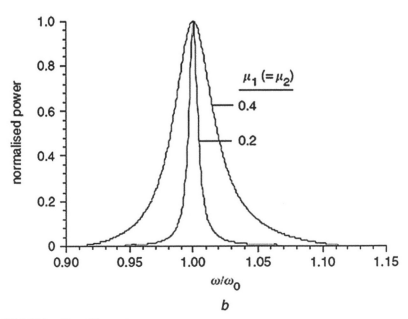

Figure 16.2 Fabry-Perot Resonators
(a) Magneto-inductive waveguide Fabry-Perot interferometer
(b) Variation of $|T_{FP}|^2$ with ω/ω_0, for $\kappa = 0.2$ and different values of $\mu_1(=\mu_2)$

16.1.2.1 The general algorithm

The algorithm for computing and analyzing the currents in these types of MI devices was given in [1]. A larger set of simultaneous equations need to be solved if there are more discontinuities in the device system.

Step 1

The general governing Kirchoff's voltage law equation for these types of devices is given by the relationship

$$Z_{m,n-1}I_{n-1} + ZI_n + Z_{m,n+1}I_{n+1} = 0 \qquad (16.23)$$

Where the device structure is revealed by the differences in the indexes of the impedances Z_m and $Z_{m,n}$.

The recursive equation for the propagating current is

$$I_{n+1} = -\frac{\left(Z_{m,n-1}I_{n-1} + ZI_n\right)}{Z_{m,n+1}} \qquad (16.24)$$

Step 2

To excite the device structure for wave propagation, the starting position may be at the middle where $n = 0$ or 1 and the impedance $Z_{m,n}$ is Z_m. The imposed current in the starting loop is set to: $I_0 = 0$ and $I_1 = \exp(-jka)$. The recursive equation is then used for computing other currents i_n in the device. If there are N perturbed mutual inductances $Z_{m,n}$, then $N + 2$ equations are required to determine I_{N+2} currents.

Step 3

To satisfy the boundary condition for pure wave propagation, a backward going wave has to be launched by assuming that $I_0 = 0$ and $I_1 = \exp(+jka)$. Again the recursive equation is used for computing the induced currents j_n in the loops.

The full solution is a combination of the two solutions as

$$I = i_n + R_G j_n \qquad (16.25)$$

R_G is the reflection coefficient.

Step 4

This step is used to satisfy the output boundary conditions to ensure that the currents I_{N+2} and I_{N+1} are related by a phase change of a traveling wave. This means that

$$I_{N+2} = I_{N+1} \exp(-jka) \qquad (16.26)$$

The reflection coefficient for the system is

$$R_G = -\frac{(i_{N+2} - i_{N+1}\exp(-jka))}{(j_{N+2} - j_{N+1}\exp(-jka))} \tag{16.27}$$

The transmission coefficient at loop $N+1$ then is

$$T_G = i_{N+1} + R_G j_{N+1} \tag{16.28}$$

16.1.2.2 Multiply-resonant Fabry–Perot cavities

Multiply-resonant MI cavities can also be created based on the above algorithm. Figure 16.3 is the response of a two resonant loop cavity structure. The figure is for the case when $\kappa = 0.2$ and $\mu_1(=\mu_2) = 0.2)$.

(Variation of $|\Im_{FP}|^2$ with ω/ω_0 for two-loop cavity with $\kappa = 0.2$ and

$$\mu_1(=\mu_2) = 0.2 \tag{16.29}$$

The two transmission peaks lie within the magneto-inductive band. They are located at the positions $ka = 2\pi/3$ and $ka = \pi/3$. By introducing more resonant loops into the cavity, more transmission peaks can be created within the MI band. If there are M resonant loops in the cavity, the locations of the transmission peaks are at. $ka = v\pi/(M+1)$, where $v = 1, 2, \ldots, M$. Resonances occur when the accumulated phase change in the round trip in the cavity is a multiple of 2π or $2(M+1)ka = v2\pi$.

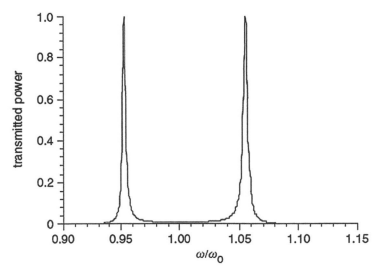

Figure 16.3 Doubly Resonant Fabry-Perot [1]

16.1.3 Bragg gratings

In optic fiber a Bragg grating is a short length of optical fiber that filters out a particular wavelength. This may be achieved by using periodically spaced zones in the fiber core which are altered to have different refractive indexes slightly higher than the core index. The structure selectively reflects a very narrow range of wavelengths and transmits others. Fiber Bragg gratings are used in laser technology to stabilize their outputs and to filter out wavelengths in a wavelength division multiplex (WDM) system. This section explains how to achieve the same wavelength processing devices using MI technology.

Bragg gratings are obtained when weak reflectors (weak mirrors) are arranged in series and aid each other in a manner that reflections add up coherently. This is unlike in Fabry–Perot devices which rely for their performance on strong reflectors. MI Bragg gratings are obtained by alternating the mutual impedances (discontinuities) Z_{m1} and Z_{m2} (Figure 16.4). Hence the mutual inductance ratio also varies from $\mu_1 = Z_{m1}/Z_m$ to $\mu_2 = Z_{m2}/Z_m$ over the period $\ell = 2a$. For example if the variation in impedance ratio is $\mu_1 = 1.1$ and $\mu_2 = 0.9$, the reflectivity alternates between $+0.1$ and -0.1 respectively [1].

Constructive reflections from adjacent periods take place when $2k\ell = 2\pi$, corresponding to when $ka = \pi/2$ and corresponds to high reflections at the band centre.

The overall reflection and transmission coefficients vary as

$$R_B = \tanh(2MR) \tag{16.30}$$

$$\Im_B = 1/\cosh(2MR) \tag{16.31}$$

Where R is the reflectivity of a single elementary reflector and M is the number of periods. It was shown in [1] that $M = 15$ periods results to 99% power reflection.

The reflective and transmitted power variations $|\Re|^2$ and $|\Im|^2$ as a function of the periods for $\omega = \omega_0$, $\kappa = 0.2$, $\mu_1 = 1.1$ and $\mu_2 = 0.9$ is shown in Figure 16.4.

The frequency response of a Bragg grating has a bandpass characteristic as shown in Figure 16.5.

The reflective power frequency response ($|\Re|^2$) is for a 20 period grating superimposed on the response of a raised cosine (dashed line) taper as a function of ω/ω_0. A raised cosine taper is used to reduce the sidelobes in the response.

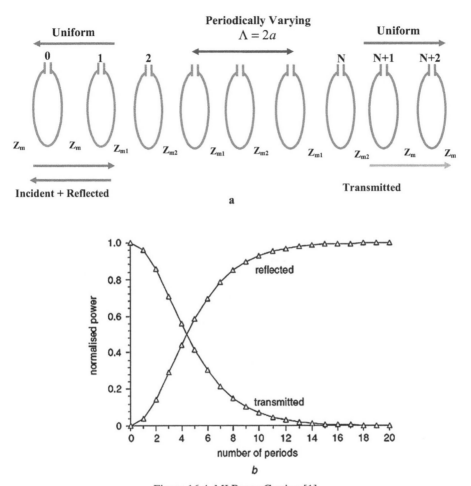

Figure 16.4 MI Bragg Grating [1]

16.1.4 Tapers

In the previous section a raised cosine taper was used to remove or limit the sidelobes from the Bragg grating. In tapers the properties of an MI waveguide change monotonically from a starting value to an end value. MI tapers are obtained by varying the mutual inductance from Z_m to Z_{m1} and the loop impedance is kept as Z.

Two modifications are made to obtain a taper. First, the input and output waveguide are different. This is achieved by using two mutual inductances. Second, a modification is made to the propagation constant is used as in the

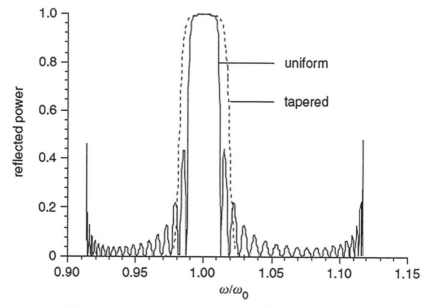

Figure 16.5 Frequency Response of an MI Bragg Grating [1]

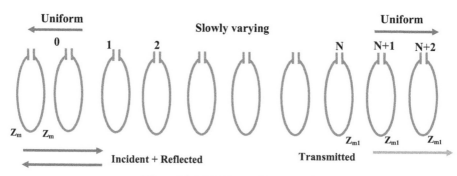

Figure 16.6 MI Waveguide Taper [1]

equation:

$$\cos(k_1 a) = \left\{ (\omega/\omega_0)^2 - 1 \right\} / \kappa_1 \qquad (16.32)$$

Where $\kappa_1 = \mu_1 \kappa$ and $\mu_1 = Z_{m1}/Z_m$. When $|\mu_1| < 1$, then $k_1 a$ is real and there is always a propagating transmitted wave at the output (wave-guide) of the taper. $k_1 a$ may be imaginary at some frequencies if and when $|\mu_1| > 1$ making the taper to reflect the waves perfectly and thus cutting off the transmitted wave.

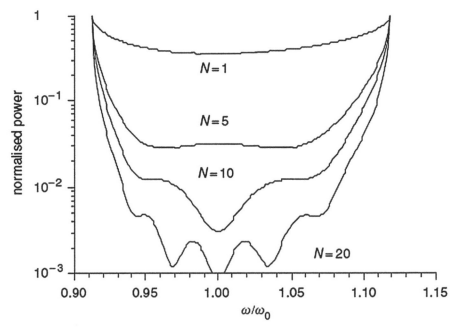

Figure 16.7 Spectrum of Reflected Power [1]

The amplitude of the reflected wave in the taper is given by the expression

$$R_T = -\frac{(i_{N+2} - i_{N+1}\exp(-jk_1a))}{(j_{N+2} - j_{N+1}\exp(-jk_1a))} \qquad (16.33)$$

At loop $N + 1$, the transmission coefficient is

$$T_T = i_{N+1} + R_T\, j_{N+1} \qquad (16.34)$$

Although the power reflection coefficient is the same as ($|\Re|^2$) but the power transmission coefficient becomes

$$|\Im|^2 = \mu_1 |T_T|^2 \sin(k_1a)/\sin(ka) \qquad (16.35)$$

The spectrum of the reflected power is shown in Figure 16.7.

This figure is for $\kappa = 0.2$ and $\mu_1 = 0.4$, for 1, 5, 10 and 20 elements.

16.1.5 Three-port devices

MI can be used also to create three-port devices and the theory presented in [1] relates to the theory of similar optical devices. The three-port devices discussed in this section include flux couplers and splitters.

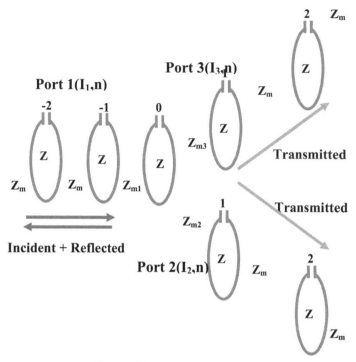

Figure 16.8 Three-Port Splitters [1]

16.1.5.1 General three-port splitters

Of the several configurations possible, this section discusses a one input and two output 3-port devices as shown in Figure 16.8.

An MI splitter is used to split an input power and transmit it through multiple channels towards different receivers. In the example given here the three ports are oriented at 120^0 apart. In this figure, four equations need to be identified for the three waveguides with indexes $i = 1, 2$ and 3 and one extra equation for the voltages in the loop identified with the index 0. The mutual inductances and currents in the three waveguides are given by the relations $Z_{mi} = j\omega M_i$; $I_{i,n}$; $i = 1, 2, 3$. The Kirchoff's voltage equations for the 3-port splitter are:

$$Z_{m1}I_0 + ZI_{1,-1} + Z_m I_{1,-2} = 0 \tag{16.36}$$

$$ZI_0 + Z_{m1}I_{1-1} + Z_{m2}I_{2,1} + Z_{m2}I_{3,1} = 0 \tag{16.37}$$

$$Z_{m2}I_0 + ZI_{2,1} + Z_m I_{2,2} = 0 \tag{16.38}$$

$$Z_{m3}I_0 + ZI_{3,1} + Z_m I_{3,2} = 0 \tag{16.39}$$

The current I_0 is unknown. The current in the input waveguide is assumed to consist of both an incident and reflected waves and the currents in the output waveguide are transmitted currents. Therefore

$$I_{1,n} = I \exp(-jkna) + R \exp(+jkna); \quad n \le -1 \tag{16.40}$$

$$I_{2,n} = T_2 \exp(-jkna) \quad n \ge 1 \tag{16.41}$$

$$I_{3,n} = T_3 \exp(-jkna) \quad n \ge 1 \tag{16.42}$$

To derive the expressions for the transmission and reflection coefficients and also the reflected power and transmitted power coefficients, the methods used in the previous sections are adopted with the definition that $\mu_i = Z_{mi}/Z_m$; $i = 1, 2, 3$. This leads to the relationships

$$\Re = \frac{R_{FP}}{I} = \frac{\left(1 - \mu_1^2\right)\exp(jka) + \left(1 - \mu_2^2 - \mu_3^2\right)\exp(-jka)}{\left(\mu_1^2 + \mu_2^2 + \mu_3^2 - 1\right)\exp(-jka) - \exp(jka)} \tag{16.43}$$

$$I_0 = \frac{I_0}{I} = \frac{\mu_1(\exp(-jka) - \exp(jka))}{\left(\mu_1^2 + \mu_2^2 + \mu_3^2 - 1\right)\exp(-jka) - \exp(jka)} \tag{16.44}$$

$$\Im_2 = \frac{T_2}{I} = \frac{\mu_1\mu_2(\exp(-jka) - \exp(jka))}{\left(\mu_1^2 + \mu_2^2 + \mu_3^2 - 1\right)\exp(-jka) - \exp(jka)} \tag{16.45}$$

$$\Im_3 = \frac{T_3}{I} = \frac{\mu_1\mu_3(\exp(-jka) - \exp(jka))}{\left(\mu_1^2 + \mu_2^2 + \mu_3^2 - 1\right)\exp(-jka) - \exp(jka)} \tag{16.46}$$

The power reflection and transmission coefficients for the splitter are

$$|\Re|^2 = \frac{[(\mu_1^2 - 1)^2 + (\mu_2^2 + \mu_3^2 - 1)^2] - 2(1 - \mu_1^2)(\mu_2^2 + \mu_3^2 - 1)\cos(2ka)}{[1 + (\mu_1^2 + \mu_2^2 + \mu_3^2 - 1)^2] - 2(\mu_1^2 + \mu_2^2 + \mu_3^2 - 1)\cos(2ka)} \tag{16.47}$$

$$|\Im_2|^2 = \frac{2\mu_1^2\mu_2^2\{1 - \cos(2ka)\}}{[1 + (\mu_1^2 + \mu_2^2 + \mu_3^2 - 1)^2] - 2(\mu_1^2 + \mu_2^2 + \mu_3^2 - 1)\cos(2ka)} \tag{16.48}$$

$$|\Im_3|^2 = \frac{2\mu_1^2\mu_3^2\{1 - \cos(2ka)\}}{[1 + (\mu_1^2 + \mu_2^2 + \mu_3^2 - 1)^2] - 2(\mu_1^2 + \mu_2^2 + \mu_3^2 - 1)\cos(2ka)} \tag{16.49}$$

It can be shown by adding these equations together that

$$|\Re|^2 + |\Im_2|^2 + |\Im_3|^2 = 1 \tag{16.50}$$

In other words, the input power is distributed into the three components consisting of a reflected power and two transmitted power components. The next section describes two specific cases of splitters, the trebly symmetric splitter and matched doubly symmetric splitter.

16.1.5.2 Trebly symmetric splitters

Consider the case when the mutual inductances between the sections in the three waveguides are all equal ($\mu_1 = \mu_2 = \mu_3 = 1$) and

$$\Re = \frac{-\exp(-jka)}{2\exp(-jka) - \exp(+jka)} \tag{16.51}$$

$$\Im_2 = \Im_3 = T = \frac{(\exp(-jka) - \exp(jka))}{2\exp(-jka) - \exp(+jka)} \tag{16.52}$$

The power reflection coefficients also reduce to

$$|\Re|^2 = \frac{1}{5 - 4\cos(2ka)} \tag{16.53}$$

$$|\Im_2|^2 = |\Im_3|^2 = \frac{2\{1 - \cos(2ka)\}}{5 - 4\cos(2ka)} \tag{16.54}$$

Since the device is symmetric, the scattering matrix for the device is

$$S = \begin{bmatrix} R & T & T \\ T & R & T \\ T & T & R \end{bmatrix}$$

If the device is lossless and reciprocal, the following relations hold

$$S \cdot S^{*T} = I \tag{16.55}$$

Where I is the identity matrix S^* and S^T are the complex conjugate and transpose of the scattering matrix.

The power conservation equation also holds:

$$R \cdot R^* + 2T \cdot T^* = 1 \tag{16.56}$$

The matrix coefficients also satisfy the equation

$$R \cdot T^* + T \cdot T^* = 0 \tag{16.57}$$

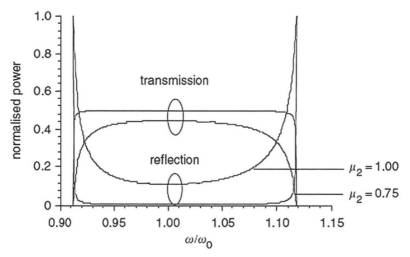

Figure 16.9 Power Spectrum of Three Port Splitters [1]

Figure 16.9 shows the reflected and transmitter power spectrum for the symmetric splitter. It shows the variation of $|\Re|^2$ and $|\Im|^2$ against ω/ω_0 for $\kappa = 0.2$ and $\mu_1 = 1$ and different values of $\mu_2(= \mu_3)$. The junction reflects everything at the band edge and at mid band when $ka \approx \pi/2$, the power reflection coefficient tends to 1/9. This is similar to the microwave symmetric splitter case when each line has impedance Z_0.

16.1.5.3 Matched doubly symmetric splitters

Consider the condition in the splitter when $\mu_1 = 1$ and $\mu_2 = \mu_3 \neq 1$, the reflection and transmission coefficients become:

$$\Re = \frac{-\left(2\mu_2^2 - 1\right)\exp(-jka)}{2\mu_2^2 \exp(-jka) - \exp(+jka)} \tag{16.58}$$

$$\Im_2 = \Im_3 = T = \frac{\mu_2(\exp(-jka) - \exp(jka))}{2\mu_2^2 \exp(-jka) - \exp(+jka)} \tag{16.59}$$

Let $2\mu_2^2 - 1 = 0$ which makes $\Re = 0$, or no reflections exist in the device and $\mu_2 = 1/\sqrt{2}$. This also implies from the second equation above that $\Im_2 = \Im_3 = T = 1/\sqrt{2}$. Because this result is independent of the term ka, the performance of the device is also independent of frequency and it performs as a matched filter. Figure 16.9 also shows the variation of $|\Re|^2$ and $|\Im|^2$ against ω/ω_0 for $\kappa = 0.2$ and $\mu_2 = 0.75$. When $\mu_2 \rightarrow 1/\sqrt{2}$, the power transmission approaches 0.5 in the MI band and the scattering matrix becomes

$$S = \begin{bmatrix} 0 & 1/\sqrt{2} & 1/\sqrt{2} \\ 1/\sqrt{2} & -1/\sqrt{2} & 1/2 \\ 1/\sqrt{2} & 1/2 & -1/\sqrt{2} \end{bmatrix}$$

The above equations give the first row in the scattering matrix. The remaining rows are obtained by permuting the ports and coupling terms. The matrix shows that any design which minimizes reflections in port 1 worsens the system when the inputs are applied to either port 2 or 3 because the relevant coefficients have become 1/2 rather than 1/3.

16.1.5.4 Matched asymmetric splitters

Let us generalize the discussions in the previous sections by assuming that the splitter is asymmetric and $\mu_1 = 1$, $\mu_2 \neq \mu_3$ and $\mu_2^2 + \mu_3^2 = 1$. This leads to the results: $R = 0$; $T_2 = \mu_2$; $T_3 = \mu_3$. The scattering matrix becomes

$$S = \begin{bmatrix} 0 & \mu_2 & \mu_3 \\ \mu_2 & -\mu_3^2 & \mu_3 \\ \mu_3 & \mu_2\mu_3 & -\mu_2^2 \end{bmatrix}$$

This leads to a reflectionless splitter with arbitrary power ratios in the waveguides. The relationship $S \cdot S^{*T} = I$ still holds in all the splitters covered in this section.

16.1.5.5 N-port splitters

The theory of 3-port splitters can be extended to N-ports. An N-port splitter has N mutual inductance ratios. The reflectionless power is divided into N ports in proportion to the square of their mutual inductance ratios μ_i^2.

$$\mu_2^2 : \quad \mu_3^2 : \quad \ldots : \quad \mu_N^2 \tag{16.60}$$

Assuming that the input power is into port 1 and $\mu_1 = 1$ then

$$\mu_2^2 + \mu_3^2 + \cdots + \mu_N^2 = 1 \tag{16.61}$$

In an N-port system, the angle between the ports is small. Hence the coupling between the branches increases causing interference which reduces the available power in the individual branches.

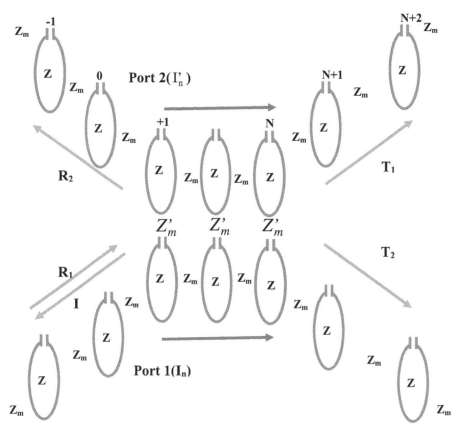

Figure 16.10 Directional Coupler

16.1.6 Directional couplers

The symmetric directional coupler is shown in Figure 16.10. It consists of four waveguides, two identical lines of infinite lengths that are coupled transvesely by nearest neighbour interactions, and two waveguides supporting forward transmissions only. Reflections occur in the infinite lines. The coupler region consists of N elements.

The impedance structure in the directional coupler consists of four impedances:

1. The loop impedance: $Z = j(\omega L + 1/\omega C)$
2. The coaxial impedance: $Z_m = j\omega M$
3. The transverse coupling impedance: $Z'_m = j\omega M'$
4. The currents in the lines are I_n and I'_n.

There are two sets of governing equations. The first set is in the input and output regions and in the second in the coupler region.

The governing equations in the input and output regions are

$$ZI_n + Z_m \{I_{n+1} + I_{n-1}\} = 0 \quad \text{when} \quad n < 1 \text{ and } n > N \qquad (16.62)$$

$$ZI'_n + Z_m \{I'_{n+1} + I'_{n-1}\} = 0 \quad \text{when} \quad n < 1 \text{ and } n > N \qquad (16.63)$$

These equations lead to the usual dispersion equation when we assume traveling MI currents I_n and I'_n.

The governing equations in the coupler region are

$$ZI_n + Z_m \{I_{n+1} + I_{n-1}\} + Z'_m I'_n = 0 \quad \text{when} \quad 1 \le n \le N \qquad (16.64)$$

$$ZI'_n + Z_m \{I'_{n+1} + I'_{n-1}\} + Z'_m I_n = 0 \quad \text{when} \quad 1 \le n \le N \qquad (16.65)$$

These equations have two modes of solutions called symmetric and anti-symmetric modes. The symmetric mode identical currents flow in the ports 1 and 2 ($I_n = I'_n$) and assuming $I_n = I_s \exp(-jk_s na)$ with dispersion equation solution

$$Z + 2Z_m \cos(k_s a) + Z'_m = 0 \qquad (16.66)$$

Or

$$\cos(k_s a) = -\frac{\{(\omega/\omega_0)^2 - 1\}}{\kappa} - \frac{\mu'}{2} \qquad (16.67)$$

Where $\mu' = Z'_m / Z_m$.

In the anti-symmetric mode the currents are at 180° flowing in the ports 1 and 2 $(I_n = -I'_n)$. The dispersion equation is

$$Z + 2Z_m \cos(k_a a) - Z'_m = 0 \qquad (16.68)$$

Or

$$\cos(k_a a) = -\frac{\{(\omega/\omega_0)^2 - 1\}}{\kappa} + \frac{\mu'}{2} \qquad (16.69)$$

For parts of the MI wave band the point at which the characteristic modes of the coupler are at cutoff depends on the sign of μ'.

How $k_a a - k_s a$ vary with ω/ω_0 is shown in Figure 8.1 for $\mu' = -0.2$ and $\kappa = 0.2$. The dispersion equations are also shown in Figure 16.12.

16.1.7 Mode excitation and propagation

Consider that the input to the coupler is an excitation current of one unit applied to port 1 and that none of the modes is at cutoff. Therefore two modes

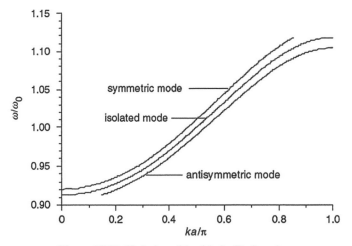

Figure 16.11 Variation of k with the Ratio ω/ω_0

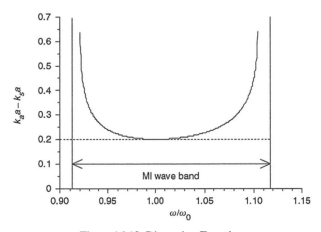

Figure 16.12 Dispersion Equation

will be created each of amplitude 0.5 and propagate in the coupler. These two modes will mix with each other as they propagate through the N sections. The total phase change after N sections is

$$\Phi = (k_s - k_a)Na \tag{16.70}$$

The two modes will be in anti-phase when $\Phi = \pi$ at which point the whole signal energy is transferred to port 2.

When $\omega = \omega_0$, the dispersion equations become:

$$\cos(k_s a) = -\frac{\mu'}{2} \quad \text{and} \quad \cos(k_a a) = +\frac{\mu'}{2} \tag{16.71}$$

$$k_s a = \frac{\pi}{2} + \frac{\mu'}{2} \quad \text{and} \quad k_s a = \frac{\pi}{2} - \frac{\mu'}{2} \tag{16.72}$$

Under this condition power transfer occurs when $(k_s - k_a) = \mu'$. This means that full power transfer first takes place when

$$\mu' N = \pm\pi \tag{16.73}$$

By making μ' small, reflections due to transitions between isolated and coupled guides can be minimized.

16.1.7.1 MI wave coupler

The general algorithms proposed earlier are used to create MI wave couplers. The essential currents are

$$I_{n+1} = -\left\{\frac{ZI_n + Z_m I_{n-1} + Z'_m I'_n}{Z_m}\right\} \tag{16.74}$$

$$I'_{n+1} = -\left\{\frac{ZI'_n + Z_m I'_{n-1} + Z'_m I_n}{Z_m}\right\} \tag{16.75}$$

These equations suffice with the requirement that $Z'_m = 0$ for $n < 1$ and $n > N$. For its operation an incident wave is launched in a port (eg. port 1) to obtain two sets of currents i_n and i'_n. To satisfy the boundary conditions the two transmitted outputs should be pure traveling waves. Also the boundary condition requires a unit amplitude backward travelling waves j_n and j'_n introduced into port 1. A similar wave should be inserted into port 2 to obtain two more final sets of currents k_n and k'_n.

The solutions are linear combinations of these currents as follows:

$$I_n = i_n + R_1 j_n + R_2 k_n \tag{16.76}$$
$$I'_n = i_n + R_1 j'_n + R_2 k'_n \tag{16.77}$$

In this expression R_1 and R_2 are the reflection coefficients at ports 1 and 2 respectively. Boundary conditions are satisfied when the following relations hold:

$$I_{N+1} = I_N \exp(-jka) \tag{16.78}$$
$$I'_{N+1} = I'_N \exp(-jka) \tag{16.79}$$

Therefore

$$R_1 \left[j_{n+1} - j_n \exp(-jka) \right] + R_2 \left[k_{n+1} - k_n \exp(-jka) \right]$$
$$= - \left[i_{n+1} - i_n \exp(-jka) \right] \tag{16.80}$$

and

$$R_1 \left[j'_{n+1} - j'_n \exp(-jka) \right] + R_2 \left[k'_{n+1} - k'_n \exp(-jka) \right]$$
$$= - \left[i'_{n+1} - i'_n \exp(-jka) \right] \tag{16.81}$$

By solving these equations, the transmission coefficients are:

$$T_1 = i_{N+1} + R_1 j_{N+1} + R_2 k_{N+1} \tag{16.82}$$

and

$$T_2 = i'_{N+1} + R_1 j'_{N+1} + R_2 k'_{N+1} \tag{16.83}$$

The reflection and transmission coefficients satisfy the relationship:

$$|R_1|^2 + |R_2|^2 + |T_1|^2 + |T_2|^2 = 1 \tag{16.84}$$

The variation of the coupling power when $\omega = \omega_0$ as a function of $-\mu' N / \pi$ is shown in Figure 16.13 for the values $\kappa = 0.2$ and $N = 50$. The figure shows a weak sinusoidal power exchange and maximum power is transferred when

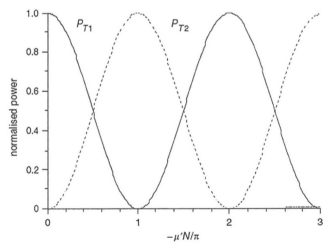

Figure 16.13 Normalised Output Power for MI Waveguide Couplers [1]

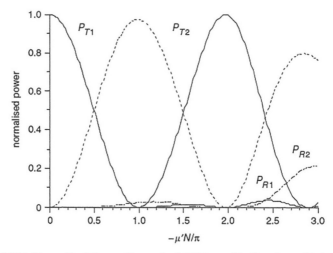

Figure 16.14 Normalised Output Power for MI Waveguide Couplers with Losses [1]

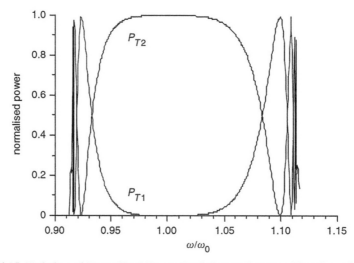

Figure 16.15 Variation of Normalised Transmitted Output Power as Function of ω/ω_0 [1] with $\kappa = 0.2$ ($N = 10$)

$-\mu'N/\pi = 1$ with little reflection. In Figure 16.14, a shorter waveguide with $N = 10$ shows power losses. In this case although coupling is stronger, reflections prevent maximum power transfer.

Figures 16.15 and 16.16 present these results in a different form as a function of normalized frequency for both when $N = 50$ (Figure 16.15) and $N = 10$ (Figure 16.16). Although Figure 16.16 represents weak coupling

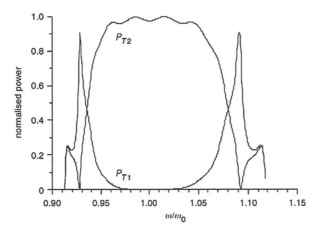

Figure 16.16 Variation of Normalised Transmitted Output Power as Function of ω/ω_0 [1] with $k = 0.2$

and ripples have appeared in the normalized coupled power, it is apparent that efficient power transfer has taken place for the values $\kappa = 0.2$ and when $-\mu'N/\pi = 1$.

All the previous analyses have not taken into account the effects of any resistive losses in the loops themselves. These effects are considered in the next section.

16.1.8 Losses in the MI loops

In general the magneto-inductive loops (wires) manifest resistive losses in their operations. These losses can be accounted for by including the resistance of the loop wire in the expression for the loop impedance and Q-value as follows:

$$Z = R_L + j\omega L - \frac{j}{\omega C} \quad \text{and} \quad Q = \frac{\omega L}{R_L} \tag{16.85}$$

R_L is the loop resistance. The loop resistance reduces the Q-value of the loop. The higher the loop resistance, the smaller the loop Q is. In a nutshell, these effects are reflected in the dispersion equation:

$$\left\{1 - \omega_0^2/\omega^2 - j/Q\right\} + \kappa \cos(ka) = 0 \tag{16.86}$$

To perceive this effect, k is assumed as a complex number is expressed clearly as $k = k' - jk''$ where k' is the propagation factor and k'' is the attenuation factor. The complex k value is substituted into the dispersion equation which can now be used to reveal the effects of the reduced loop Q as follows:

$$\left\{1 - \omega_0^2/\omega^2 - j/Q\right\} + \kappa \cos((k' - jk'')a) = 0 \tag{16.87}$$

$$\left\{1 - \omega_0^2/\omega^2 - j/Q\right\} + \kappa \cos(k'a)\cos(jk''a) + \kappa \sin(k'a)\sin(jk''a) = 0 \tag{16.88}$$

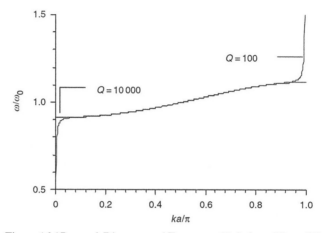

Figure 16.17 $\omega - k$ Diagram and Frequency Variation of Loss [1]

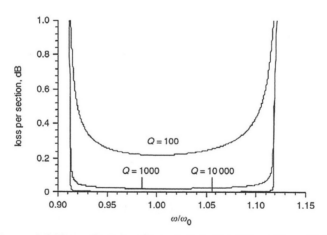

Figure 16.18 $\omega - k$ Diagram for Lossy Magneto-Inductive Waveguides with Different Q-Factors [1]

For this expression to be zero, both the real and imaginary terms are collected and set to zero. Or

$$l\left(1 - \omega_0^2/\omega^2\right) + \kappa \cos(k'a)\cosh(k''a) = 0 \qquad (16.89)$$
$$- 1/Q + \kappa \sin(k'a)\sinh(k''a) = 0 \qquad (16.90)$$

This section considers the approximate solution to this equation by assuming small loses in the loop, meaning that $k''a \ll 1$, so that $\cos(k'a) \approx 1$ and

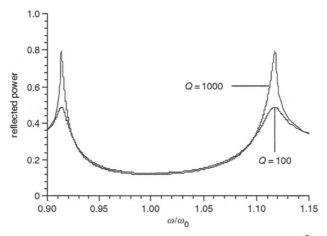

Figure 16.19 Lossy Magneto-Inductive Waveguide Mirror: Variation of $|\mathfrak{R}|^2$ With ω/ω_0 for $\kappa = 0.2$, $\mu = 0.7$ and Different Values of Q

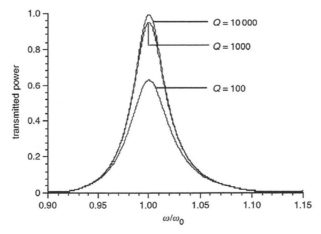

Figure 16.20 Lossy Magneto-Inductive Waveguide Fabry–Perot Interferometer: Variation of ω/ω_0 with $|\mathfrak{I}|^2$ with ω/ω_0 for $\kappa = 0.2$, $\mu_1 = \mu_2 = 0.4$ and Different Values of Q

$\sinh(k''a) \approx k''a$. The approximate solution is

$$\left(1 - \omega_0^2/\omega^2\right) + \kappa \cosh(k''a) = 0 \qquad (16.91)$$

$$k''a = 1/\kappa Q \sin(k'a) \qquad (16.92)$$

While the lower equation shows the effects of the losses as propagation losses, the upper equation is the dispersion equation for the lossless case. In the propagation loss equation, $k''a = 1/\kappa Q \sin(k'a)$, a low Q increases the propagation losses ($k''a$ is high) and high Q means low losses ($k''a$ is low). These losses are minimized at mid-band when $k'a \approx \pi/2$ and inversely proportional to the values of Q and κ. Therefore to obtain strongly coupled MI couplers, high Q materials or resonators should be used. Furthermore, Figures 16.17 and 16.18 demonstrate these conclusions over several Q regimes.

The loss equation permits the performances of various MI devices to be investigated for the MI mirror (Figure 16.19) power reflectance) and Fabry-Perot (Figure 16.20, transmission power coefficient) when $\kappa = 0.2$.

References

[1] R.R.A. Syms, E. Shamonina and L. Solymar, "Magneto Inductive Waveguide Devices", IEE Proc. – Microw. Antennas Propagat. Vol. 153, No. 2, Apr. 2006, pp. 111–121.

[2] O. Zhuromskyy, O. Sydoruk, E. Shamonina and L. Solymar; "Slow waves on magnetic metamaterials and on chains of plasmonic nanoparticles: Driven solutions in the presence of retardation", Journal of Applied Physics, Vol. 106, 104908, 2009.

[3] R.R.A. Syms, I.R. Young and L. Solymar, "Low-Loss Magneto-Inductive Waveguides", J. Phys D: Appl. Phys, Vol. 39, 2006, pp. 3945–395.

17

NFC Applications

In this chapter and in Chapter 19, various applications of Near Field Communication (NFC) have been studied. Our approach is to first provide an overview to Near Field Communication. After that, the capabilities of NFC are described.

17.1 Introduction

In the latter sections of this chaper, Near Field Communication applications are described. These applications are categorized into a few different groups which are as follows: *Biomedical Monitoring, Mobile Phones, Contact-less payment, Wireless Power Transfer, Substitute to UHF-RFID, MP3 Players, Battery charger, Military Devices.* The general description of some of these applications is provided in this chapter. Covers the detailed description of some other NFC applications.

An Overview to Near Field Communication (NFC)

NFC is a new interconnection technology launched by Sony, Nokia and Phillips who established the NFC Forum in March 2004. Which is a non-profit industry association with the goal of advancing the use of NFC short-range wireless interaction among mobile devices, consumer electronics and PCs [1]. It also aims to promote the standardization and implementation of the NFC technology to verify the interoperability between devices and services. NFC by itself, is a new technology which was evolved from a few RFID technologies (FeliCa/MIFARE/ISO14443A) providing wireless point to point interconnection in a very short range (max. 20 cm). Moreover, the ISO 14443RFID standard was extended by the use of NFC which provides inductively coupled communication and power delivery protocol [1]. The communication in this technology is based on inductive coupling using the

---

Co-authored by Samaneh Movassaghi & Agbinya JI, University of Technology, Sydney

J. Ihyeh Agbinya (PhD), Principles of Inductive Near Field Communications for Internet of Things, 251–279.

13.56 MHz carrier frequency providing a maximum data rate of 424 Kbps. This frequency happens to have the least interaction with animal and human tissues [2]. It has the capability of enabling users to access contents and services in touching smart objects by just holding them next to each other. It is also compatible with existing RFID applications in access control or public transport ticketing. However, it is more efficient than them in terms of security as it can provide a secure element for critical applications like payments. As an example, a credit card can be integrated in a mobile phone and used with NFC [3]. As it is cheaper and less complex with better usability and lower power consumption compared to alternative technologies, it can be a good solution for health monitoring devices. NFC also has the advantage of supporting the TouchMe interaction paradigm for mobile terminals. TouchMe is a user-friendly way of building connections and exchanging information between mobile terminals and other devices. TouchMe can transfer data among devices using NFC, however it is also compatible with other wireless technologies such as Bluetooth for creating a connection. It can also trigger predefined functions like on-off switching or labelling measurement [4].

In general, according to the facilitated measurements at real-life settings due to wireless communication, they are much more preferable than wired communications. However, their vulnerability to eavesdropping and high power consumption are drawbacks which require further improvement. Future communication devices should be simpler to use, cheaper and able to provide shorter operation time than nowadays devices. However, Bluetooth and IrDA are not as efficient compared to the benefits provided by NFC. Compared to these wireless technologies, NFC is more cost efficient, energy efficient and much more immune to eavesdropping. However, its short range communication limits it from real-time patient monitoring applications as the patients need continuous monitoring of their activities with a health monitoring device. Based on the high potential of this technology compared to other competing technologies, it can be used for short range connectivity between mobile terminals and health monitoring devices. Cost wise, Bluetooth costs several dollars per node which is very expensive for simple devices. Bluetooth has a power consumption of tens of miliwatts which is relatively too much for many simple devices. These problems can be solved if wired communication is used instead. However, they have their own limitations such as often missing when needed, not being feasible in everyday life, wearing out in use and being cumbersome in connection. Additionally, due to the resource issues in many countries, there has been an increase in the need of health care systems for self-care patients to manage their health on their own. Therefore a robust, easy to use, and maintenance-free technology is required to combine data from variant sources, provide accurate measurement devices, analyse data and output reports on mobile devices so that health care specialists can collect and analyse information efficiently [1].

NFC is suitable for relatively simple devices such as blood glucose monitors, blood pressure monitors, scales and heart rate monitors which are modern electrical monitoring devices that work with normal batteries, costing about tens to hundreds of dollars. Some are the stand-alone types with onboard display. The others are high-end devices with wireless or wired communication interfaces. As an example, the NFC device in a cell phone can control an insulin pump remotely, activate an implanted neural stimulation system, and relay collected measurements from a defibrillator-pacemaker to a monitoring physician. NFC can also be used in Digital Angel's implantable glucometer as a power and communication source. An NFC responder can also be embedded in the skull and communicate with probes within the brain which monitor brain functions. Characteristics such as low power consumption and rapid power decay, with distance make NFC a much more convenient technology for MP3 cord-off applications rather than other technologies such as Bluetooth [5].

NFC can also be convenient for providing keys for office and residence. This can be done with the configuration of a sensor in front of the door, capable of identifying who are permitted to enter the office. Additionally, the person who is given this permission can be listed via the sensor which leads to a safer and higher level of security [6]. It is also estimated that low cost NFC scanner subsystems will be widely used in commodity cellular phones. Due to its low cost, convenient power distribution and communication, it can be widely used in various medical devices [7].

Additionally, NFC can be applicable in the design of microchip antennas which is then used in cell phones. Via this design a person can simply identify the amount of money he has saved without the need for taking out the passing card and touching the sensor [6].

NFC can be also be used for sending SMS messages, calls and browsing by simply touching a RFID tag for storing a fixed data record. However, it is also used for connectivity between devices and objects, a variety of payment, and ticketing and access control applications. In Figure 17.1, some terminals that are NFC-enabled to transmit application measurements over the internet to a web-service have been shown. The measurement configurations and results are transferred via NFC enabled terminals and measurement devices.

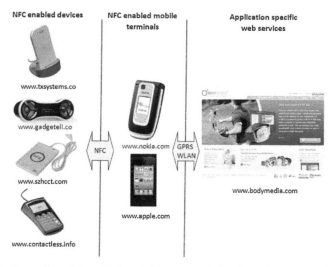

Figure 17.1 Examples of Terminal and Measurement Devices that can be NFC-Enabled (Adapted from [9])

Capabilities and Benefits of NFC

The NFC technology has many benefits which make it distinct from other wireless technologies and are given as follows [8]:

Inherent security: In the NFC design, their far-field power density is up to 50 dBm less than a typical RF device, intentionally designed for emission propagating, far-field, and electromagnetic waves. The power density in near-field communication attenuates 60 dB per decade or $1/(distance)^6$. Additionally, with the increase of distance from NFC systems, their emission level rapidly falls below ambient noise floor. Moreover, the risk of interference between the short-range communication system and the close by electronic devices is reduced.

Power consumption: The far-field RF systems are designed in a way that all their transmission energy leaves the antenna and radiates into free space. As a result, there is no power reuse. However, in NFC systems there is a resonating magnetic field around an antenna because of which the transmission energy is either around or inside the transmitter circuitry. Due to the transmission energy reuse in NFC, it has less power consumption compared to other RF communication systems.

Frequency contention and spectrum allocation: A majority of far-field RF systems are capable of sharing bandwidth via frequency allocation or time division due to the long range of RF signal propagation. Meanwhile, a large number of NFC systems can be co-located via the well defined communication bubble of magnetic energy.

Electromagnetic immunity: As NFC systems do not require propagated, far-field, electromagnetic waves for communication, they can be shielded from electromagnetic induction. This even improves their immunity to electromagnetic interference to ensure Low Probability of Interception (LPI) and Low Probability of Detection (LPD).

One of the most important capabilities of NFC is that it is useful for field-powered or implanted devices with larger power budgets by providing a data communication link to external computers. It also provides useful data links for the coordination of a range of cooperative systems (e.g. components measuring glucose and dispense insulin). It can be further used for the communication of self-powered devices like communication subsystems which are mainly field-powered hence not imposing additional drainage on battery limited resources which allows it to set operating parameters to ICDs (internal cardiac defibrillators) and collect historical data from them.

NFC Applications

The methodology for the application of Near Field Communication (NFC) is described in general in this chapter. The details of a selective range of applications which consists of mathematical formulas and algorithms used for their implementations have been provided in Chapter 19.

The application of the wireless NFC technology is rapidly growing and includes hands-free mobile phone earpieces, some hearing instruments and two-way radios. NFC is somehow similar to a telecoil as they both work with the principle of magnetic induction. However, with the use of NFC, the generated signal is encoded as well as allowing a two way communication. More specifically, both coils of the telecoil, which are further shown as we go through, create a magnetic field between which the communication takes place [9].

The NFC applications described in this chapter are divided into a few different categories, each of which consisting of a few sub-categories. General information is provided in the main categories which are as follows: *biomedical monitoring, mobile phones, wireless power transfer, being a substitute to UHF-RFID* and its use *in the military.* However, the detailed description of some of the sub-categories of these applications has been provided in Chapter 19.

17.1.1 Biomedical monitoring

Due to the resource issues in many countries, there is an increasing force for disease management in self care patients which provides them with the capability to manage their disease on their own. For such an aim, technology plays an important role via combining data from many sources, providing accurate measurement devices, analysing and providing reports on the data, and providing supervising healthcare specialists with collected and analysed information. However, this requires the technology to be easy to- use, robust and maintenance-free. Additionally, in order to have the capability of short and long term remote patient-monitoring, a real-time wireless communication system is to be designed which also applies a wireless protocol for this aim. This system monitors health data for pulse, blood pressure, cardiogram readings, body temperature and a number of other factors to keep track of changes in vital signs. The data is then wirelessly sent to a server which tracks patient information to check for any changes that represent an abnormal condition. In cases of dramatic change in the health status, the system will alert family, physicians and anyone else who is scheduled to be notified in an emergency. As part of its benefits, it can be worn outside a medical facility with the capability of automatically notifying the doctor about any abnormal symptoms through its wireless monitors. One other advantage is its capability for keeping track of the health of elderly people living in a faraway distance, out of reach of their relatives in case of assistance. Via the health monitoring system, their well being can be checked and a method of insurance is provided in case of a heart attack or other life threatening illness. Another interesting scenario is its capability of keeping track of vital signs and notifying the patient about the time for them to take a medication to prevent health complications.

However, a convenient, power efficient and fast enough technology is required to fulfil these aims.

The following features of NFC improve its usability for most health monitoring devices and therefore make it capable of being considered as a convenient technology in many applications:

1) It uses a low-power communication based on the TouchMe paradigm, and considerably short latency time compared to cables or a Bluetooth connection via a mobile terminal. As a result, these measurements can immediately be picked up and stored in the personal mobile terminal [10].
2) There is no need for finding a correct menu of configuration parameters and items, as the communication can easily be established once the two

devices are placed close to each other as in data-intensive applications and real-time monitoring.

3) The extremely low power consumption of the target device in NFCIP-1 mode enables it to be used for triggering and on-off switching by simply bringing a NFC-enabled device close to the health monitoring device which can also be done through clothing.

Based on the reasons mentioned above, NFC is widely used for biomedical monitoring. Its application in biomedical monitoring is sub-categorized into a few categories. The classification of medical devices is based on whether or not they are implanted within the body. These devices can then fit in one of the following categories: a) totally external, b) totally internal or c) combined external and internal. The following sub-sections are based on these configurations and provide information about implanted medical devices, health or disease management, real-time patient monitoring, offline health monitoring, heart rate monitoring, glucose, weight, blood pressure management [11].

Implanted medical devices

The implantable medical device which is either partially or totally inserted in a natural orifice or human body and is to stay there for 30 days or more. They are implanted via medical or surgical procedures and must be removed via the same way [12]. These devices use either electrical energy or some other type of energy. In case of electrical energy usage, the electromagnetic wave or magnetic coupling outside a human body communicates with the telemetry/programming apparatus inside the implantable medical devices. Due to the antenna size limitation, metallic container of titanium and shielding function of the tissue, there is a restriction in the use of the electromagnetic wave. It is important to note that via the use of mutual inductance coils working at low frequency the need for a large coil is cancelled out. In [13], this method has been widely studied and applied. The electromagnetic coupling coils being used in implantable medical devices provide the capability of transferring power and/or data to inner circuits. Additionally, due to the fact that NFC is not attenuated via tissue, they are suitable for the communication of implanted devices [14]. Coaxial coils are used for their overall power transfer efficiency to the load, output impedance of the receiver, voltage transfer ratio, displacement tolerance, size and bulk of the coils, phase angle between voltages across the coils, as well as the bandwidth [13].

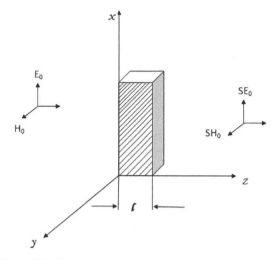

Figure 17.2 Electromagnetic Shielding (Adapted from [13])

Electromagnetic shielding: For devices to be implanted in the human body, a tissue and metallic container (made of titanium) exists between the coils of the transmitter and receiver. For a fine electric shielding, metal provides the container with fine conduction. As titanium is paramagnetic, its container has magnetic shielding. Moreover, in order to evaluate these shielding lamellas, a few parameters have been defined which are named after the *shielding efficiency* and *attenuation coefficient*. Lamellas are fine sheets of material with gill-shaped structure held adjacent to one another, with fluid in-between. The lamella shielding effect is shown in Fig. 17.2.

Parameters of Fig. 17.2 are defined as follows:

E_0: Electric field
H_0: Magnetic field
SE_0: Electric field inside the shield
SH_0: Magnetic field inside the shield

Equation (17.1), expresses the electromagnetic shielding efficiency defined with S.

$$S = \frac{4Z_d Z_M}{(Z_d + Z_M)^2 e^{\gamma t}\left[1 - \left(\frac{Z_d - Z_M}{Z_d + Z_M}\right)^2 \cdot e^{-2\gamma t}\right]} \tag{17.1}$$

The Attenuation coefficient is calculated via equation (17.2).

$$A = 20 \log \left| \frac{1}{S} \right| = 20 \log \left| \frac{(Z_d + Z_M)^2}{4 Z_d Z_M} \right| + 20 \log |e^{\gamma}|$$

$$+ 20 \log \left| 1 - \left(\frac{Z_d - Z_M}{Z_d + Z_M} \right)^2 \cdot e^{-2\gamma t} \right| \tag{17.2}$$

Parameters at equations (17.1) and (17.2) are defined as follows:

γ: propagation constant $\gamma = \sqrt{j\omega\mu(\sigma + j\omega\varepsilon)}$, which may be approximated for a conductor by neglecting the second term in the bracket.

Z_d: wave impedance of medium, $Z_d = \sqrt{\frac{\mu_d}{\varepsilon_d}}$

Z_M: wave impedance of shield, $Z_M = \sqrt{\frac{j\omega\mu}{\sigma}}$

μ_d: magnetic permeability of medium

ε_d: electric permittivity of medium

μ: magnetic permeability of shield

σ: electric conductance of shield

Subscript d refers to medium

Subscript m refers to shield

Storage of electronic health devices: Due to the increase in bandwidth in NFC, the privacy and usage of cryptographic protocols appropriate for medical systems are facilitated. Some standardization activities such as HL7 have enabled NFC-enabled RFID tags with the capability of storage for electronic health devices [14]. These stored records may include a wide range of data in summary or comprehensive form, including medication and allergies, demographics, medical history, billing info, radiology images and laboratory tests [15]. Via the usage of magnetic coupling in NFC technology a range of 20 cm is provided around them. It accelerates the selection of patients in hospital and other "first responder" environments where the identification of patients with radio-equipped medical devices is important [14].

Internal electrical stimulation: Another useful biomedical application which can be provided by NFC is internal electrical stimulation which is useful for a wide range of medical conditions. This is due to its capability of providing a small amount of energy for batteries recharged via NFC. Basically, electrical current is used for electrical stimulation to cause contraction for a single muscle or a group of muscles. These electrodes can be placed on various locations in the skin which then recruits the appropriate muscle fibres. Due to muscle contraction via electrical stimulation, the muscle is strengthened. The current setting of muscle contraction from forceful to gentle can be changed by the physical therapist. Via the muscle contraction, the blood supply to that area will be increased together with an increase in muscle strength [16]. Based on the previously mentioned features and advantages in regard of the NFC technology, it is considered to be a convenient technology for stimulation in the following potential applications:

- High frequency stimulation for mitigation of diabetic gastroparesis [17]
- Prevention of excessive eating by gastric stimulation creating a feeling of satisfaction for patients with morbid obesity.
- Deep brain stimulation for reducing Parkinson's disease symptoms [18]
- Spinal chord stimulation for direct mitigation of chronic pain [19, 20]

However, due to the matter that detection of implanted devices is evidence to compromised health, device corruption is harmful to the implantee's health. Therefore, it is essential that engineering practices and safeguards are exactly taken into account [14].

In [14], 13.56 MHz RFID tags have been implanted in a human cadaver. The results show the communication range of 4 cm to be sufficient for communicating with a scanner and a distance of 10 cm to be a good communication range for all locations in the body. In this design, a number of transponders in-lay within the PCB printed antennas which have been implanted in various locations in a preserved human cadaver of an elderly male. Also a NFC-enabled device is used in case of emergency in order to provide health history to first responders. It can also be held in escrow via an infrastructure service like the Health vault service which has recently been introduced by Microsoft. In these experiments 13.56 RFID tags have been implanted in a human cadaver by implementing a minimal community NFC transponder as a flexible PCB having a spiral antenna printed. Based on greater scanner and experimental output and its large size, a longer communication range is provided. However, transponder antennas are uninsulated and so not capable of communication when they are in direct contact with the tissue. So they were insulated by bags of 6 μm plastic film from the tissue.

An NFC enabled medical device could also be potentially used to provide health history to first responders in an emergency. This information could either be held within the device itself, or instead be held in escrow by an infrastructure service such as Microsoft's recently announced health vault service. In either case, it is important that this sensitive information can only be disclosed to appropriate personnel and even so, only in appropriate situations. As a whole, based on the experiments done so far, the commodity transponders and scanners working at 13.56 MHz RFID are suitable for the implanted devices in personal medical devices. However, corruption of system integrity and unintended disclosure of private information are required. While NFC is a descendent of a communication protocol designed with the aim to expose the presence of tags and transmit their own contents, it can also be extended to do neither of these tasks. Furthermore, due to its limited battery power reserves, NFC is secured against denial of service attacks [14].

Real-time patient monitoring

The monitoring capability of this application requires a continuous and uninterrupted connection between the health monitoring device and the patient's mobile terminal without patient activity. ECG tele-monitoring over the phone in case of heart disease symptoms is an example of this application [10]. In [21], an affordable healthcare system has been provided in order to prolong the elderly lives in their homes as well as supporting healthy lives. In Fig. 17.3 the health monitoring system is shown. As can be seen the designed monitoring health system proposed in [21] can be run in a PDA or a PC. They use 433 MHz wireless connection and have a receiver connected to the PC via USB by which they send the readings.

As for the way the system works, family members identify themselves via an NFC-enabled object (such as a bracelet, mobile, watch, card, etc). Therefore, health readings are transmitted wirelessly and stored in a backend system. These readings can be immediately shown to the user via. The user

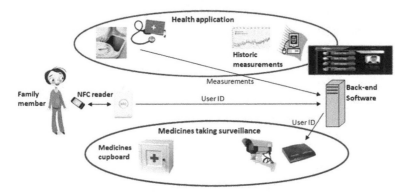

Figure 17.3 Health Monitoring Scenario (Adapted from [21])

can also access historical measurement via charts to simplify its understanding. However, based on the experimental graphs shown in [21], identification via a bracelet or card was considered much more convenient than a mobile phone. Once the user has identified himself to the system, his blood-pressure, weight and other health-related measurements are automatically associated with the identified user and stored in the back system. Historical measurements such as charts can also be provided in a easy way for the user to understand. Thereafter, this medical information can be checked by medical specialists and the user is also capable of sending them to healthcare professionals. As so the patient is guided to the best health status via the supervision of correct medication leading to an assurance of the patients' safety. This supervision is done by taking a video of the family member while they are taking their medicine. By the use of NFC technology, systems will become capable of saving information automatically in their tags without the need for it to be done manually by the patient as well as it being portable and so carried around easily.

Off-line health monitoring

Health is continuously monitored in a certain period of time in this application. After the monitoring period, the monitored data is stored in the monitoring device, from which it is sent to the mobile terminal or backend of the system [10].

Heart rate monitoring   In physical exercise and health care, heart rate needs to be continuously monitored in a long term which can be useful for diagnosis of stress and heart disease. These measurements can be done offline, after which the data is transferred to the terminal, visualized and analysed for the user's benefit. An ideal solution to this aim is the use of heart rate monitors with NFC communication. By simply touching the heart rate monitor, data could be transferred to a mobile terminal or PC [10].

In [22], the design of a small wearable Doppler radar unit for portable and mobile patient monitoring has been fully explained. The designed unit is embedded in a sort of clothing such as a garment or a patients' blanket which can also be seen in Fig. 17.4.

Due to the fact that it is directly placed behind or on the patient's torso extraneous, signals resulting from relative motion artifacts are greatly reduced. As it detects heart rate, based on Richard and Sarang's [22] assumptions they can be used for medical diagnostics and cardiology. The results

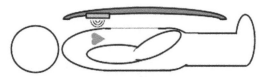

Figure 17.4 Illustration of a Doppler Radar Unit Embedded in Blanket. The Design Concept Indicates Relative Position of Doppler Sensor Module and Patients' Heart [22]

show the device to be a potential tool for cardiology and medical diagnostics specifically useful for cases where electrocardiogram ECG is too invasive, not practical and not available. Due to the short range operating system and quasi-near field nature of electro-magnetic communication, a reflected wave is produced at the air-body interface. Such a device is as small as 4 cm which can be embedded in clothing [22].

Health or disease management

The monitoring capability of this application is based on sparse measurements such as blood glucose, blood pressure or weight. However, there is a limitation on the amount of data and readings that must be immediately transferred to the mobile terminal or back-end. An example for this application is diabetes management [10].

17.1.2 Mobile phones

In [23], an online mobile application has been proposed for community members with the name Taggynet introducing new ways for exchanging information. It has chosen the NFC technology as the products with built-in NFC provide simple consumer interactions with one another, speed the connections, make secure and safe payments and receive and share secure payment. With the use of NFC, mobile devices can initialize transactions automatically by simply touching the NFC compliant transponder or NFC compliant device or a reader. Taggynet suggests a new way for bringing in NFC technology to provide a secure gateway with the capability of sharing and storing all kind of data to the world.

Fig. 17.5 represents Taggynet application, in which a mobile application using NFC technology retrieves commercial tag information when it gets close to a tag. Based on the application proposed in [23], a SCWS interface is built up to send an SMS consisting of specific data for expressing data's

Figure 17.5 Taggynet Application Presentation [23]

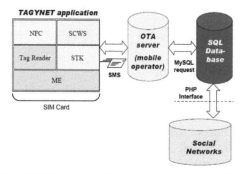

Figure 17.6 High Level Architecture of Taggynet Application [23]

feelings plus containing tag information. This data is then sent to an OTA (Over the Air) service which belongs to the service provider of the mobile phone. The high level architecture of Taggynet is shown in Fig. 17.6. As can be seen it is combined from four separate modules.

The main screen of the application of Taggynet is shown in Fig. 17.7. It has four usage scenarios which are as follows:

- A user wanting to share his opinion
- Retrieving community members experience
- Retrieve of location information
- Retrieve of customer feedbacks via Business parties

The use of NFC technology in Taggynet, guarantees reliability in exchange for information, because NFC uses the SIM card for communicating with tags. Therefore, a user wishing to share his feelings is always authenticated. The geographical and temporal information are also recorded each time the user tags an item.

Figure 17.7 Main Screen of Taggynet Application [23]

NFC in cellular handsets

In [10], the ABI research teams have done a forecast on NFC-enabled phone shipments for the next few years in regard of the deployment of NFC in cellular handsets. In 2005, market analysts forecasted the use of 500 million cellular handsets with the capability of NFC communication by 2011. This leads to an increase in the communication range and gives way to new services [21].

In Fig. 17.8 NFC related standards are shown which define the way of selecting the communication mode for the lower layer. The ECMA-340 standard has defined an NFC-specific communication mode for peer to peer data communication, namely NFCIP-1. It provides two modes for operation: passive mode and active mode. Both participants create a carrier for themselves in the active mode, during the data transmission process. However in the passive mode, the target device uses load communication for communicating with the initiator while the initiator only creates a carrier during communications. As so, it saves in power consumption which is a very good feature

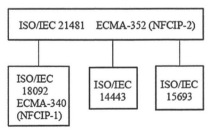

Figure 17.8 NFC Deployment in Cellular Handsets [10]

if the target device has a limited source of energy like a small battery. As so, a sensor which is readable over NFC can last for several years even by operating with a small lithium coin-cell battery.

One other standardization standard, namely ECMA-352 defines NFCIP-2 which specifies which operation mode to select automatically in the start of communications.

NFC for authentication of mobile e-identities

In [25], the MOBILSEC project is defined which uses NFC built in mobiles for authentication of mobile e-identities as they require strong authentication. However, it is important to note that the connection establishment should be as simple as possible for a wide range of devices as well as providing a secure connection in which the communicating parties are securely identified. Via the use of NFC technology, this establishment is made even simpler, causing a trend in the existing technologies. In essence, four usage scenarios of the NFC devices are provided in the following:

Password container:

The NFC technology could be used to provide a password container. Its goals and workflow are given as follows:

Goals: All passwords, PIN codes and paraphrases on your mobile can be secured via the password container. This function can provide other personal data like nickname, name, social-security number, ages, email, and date of birth on web forms. Using the NFC technology, your password and other personal data can be replaced. This stored data can be protected at different levels without requiring any protection, unique PIN, acknowledge container PIN.

Workflow:

1) At first the data input is logged in. Afterwhich the browser and Password container control application start by the user as well.
2) A URL is accessed by the user with a login form
3) The login form is detected by the PW Container Control application. It tries to obtain a number of appropriate passwords the phone the from the password store.
4) Then the phone receives an access PIN request from the user.
5) This password is then returned to the password target service and container control application to be accessed.

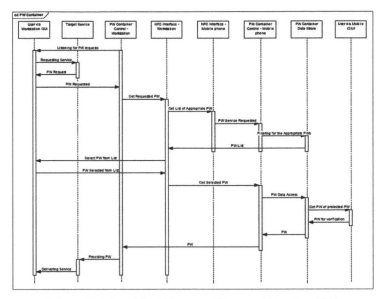

Figure 17.9 Data Flow Diagram of Password Container [25]

In Fig. 17.9 the data flow diagram of the password container is shown for the case of multiple accounts on a single URL. In the case where a wide range of appropriate passwords are available one of them is selected by the user.

Copy protection and secure licensing:
The NFC technology could be used in secure listening and copy protection. Its goals and workflow are given as follows:

Goals: Strong copy protection and secure licensing are one of the solutions which help multimedia piracy and addressing software. Based on their features, these applications can be globally accessed without the need for implementation of complex license schemes. Licenses are capable of being attached to natural persons. This solution need not support "wired" solution as it works as a hardware key. Via the use of NFC, end-user licenses are provided with personal protection and simple usage.

Workflow: It is noted that this usage scenario is somehow similar to the Password Container application; however the Licence Key Control application takes place of the PW container control. In the case of multimedia licensing, the browser and target service are merged to form a license software component. On all other cases the above data flow is applicable.

Virtual NFC tag embedded in to applications:

Goals: This solution allows the application and content providers to use virtual tag information in the service or application. Via the use of NFC, an intelligent component can be implemented which allows virtual NFC tags to be placed on the application GUI, web-pages and application data transfer via NFC exchange into mobile devices. This can then be transferred and used from the mobile device to another application [11-y].

Workflow:

1) The virtual tag comes up on the application GUI/html page.
2) This virtual tag is then selected via the user.
3) The application or service builds up this virtual tag.
4) The NFC interface of the workstation is then touched via the NFC enabled mobile devices.
5) The mobile application reads the information of the NFC tag and then processes it and stores it into its memory.
6) After that, the user touches the NFC interface of the virtual acceptor device
7) The acceptable virtual tag is then selected via the acceptor device
8) This selected virtual tag is then returned to the mobile device.

In Fig. 17.10 and Fig. 17.11 the data flow diagrams of basic workflows in case of PINs used for the protection of Virtual Tag operations in the mobile device is provided.

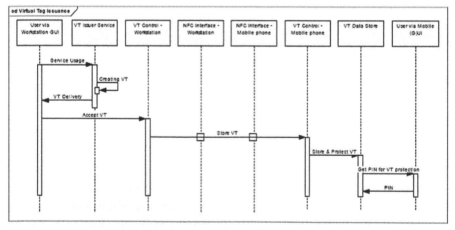

Figure 17.10 Data Flow Diagram of Virtual Tag Issuance [25]

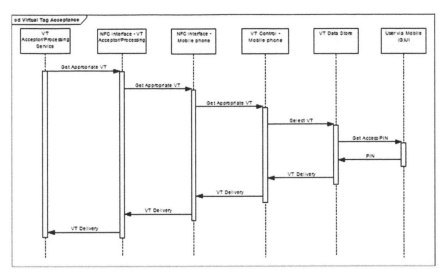

Figure 17.11 Data Flow Diagram: Virtual Tag Acceptance [25]

GPS coordinates in NFC tag:

Goals: Via this application, users can easily transfer and pickup route information as the service provides merchant geographical position or customer when the position is relevant. NFC makes the implementation of intelligent data exchange into the NFC enable mobile devices possible.

Workflow: This use case is similar to the virtual tag application in which the navigation data is then delivered to applications with the use of virtual tag data flows.

NFC gateway

In [26], a NFC gateway for network driven services has been proposed which allows network base service delivery and creation for operators and light-weight applications for end users.

The protocol is successfully implemented and evaluated in ENUM (Electronic Number) environment for conceptual proof. ENUM is a protocol that is the result of work of the Internet Engineering Task Force's (IETF's) Telephone Number Mapping Working Group. The charter of this working group was to define a Domain Name System (DNS)-based architecture and protocols for mapping a telephone number to a Uniform Resource Identifier (URI) which can be used to contact a resource associated with that number [27].

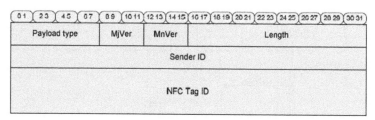

Figure 17.12 Request Protocol for NFC Gateway (Adapted from [26])

Figure 17.13 Response Protocol for NFC Gateway [26]

Via the usage of NFC gateway in this environment a flexible and convenient way of assigning services to certain customers and certain tags straight forwardly is provided. The NFC gateway protocol is an application level protocol with the aim of providing communication between NFC gateway and mobile phone. However, its major purpose is to enable network driven NFC services.

The format of the NFC gateway is shown in Fig. 17.12. As can be seen the header is of 32 bits length with 8 bits for payload type, 2 bits for major version, 2 bits for minor version and 16 bits for length. The format includes two ID fields for service selection: the sender ID and the identification field.

The NFC tag ID and sender ID build up the identification field. The NFC tag contains the canonical identification number (CID) of the read tag. However, the sender ID field identifies the sending party. Fig. 17.13 shows the format of the NFC gateway response protocol. It can be seen that it is only made up of a header field which includes payload types, major version, minor version, flags and length. However, the length is not currently used but will be used in the future. The payload, major version and minor version are very similar to the request protocol. The flags field consists of the results to the request status.

The key negotiation of the NFC gateway is shown in Fig. 17.14. As can be seen, TLS negotiation is used via the proposed design and Transport Layer Security (TLS) is only established in the first connection established to the gateway.

As the information being transferred between an NFC gateway and mobile application contain private information, there is a need for encryption. In the NFC gateway design proposed in [26], TLS is most probably used for

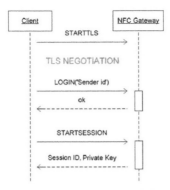

Figure 17.14 Key Negotiation of NFC Gateway [26]

providing communication privacy and endpoint authentication. Via this proposed design, TLS is only established in the first connection established to the gateway and establishes a connection whenever the gateway requires it. Once the TLS connection has established successfully, a new private key and session ID are provided via the user which are later used for user identification and message encryption with the AES symmetric encryption algorithm. Then the login function is called which uses the sender ID as a parameter. After the start session function is called, new values of Private key and session ID can then be used for a secure connection between the NFC gateway and the client. The TLS connection can then be disconnected after a new pair has been received. Additionally, a secured gateway request protocol has been proposed in [26] which is shown in Fig. 17.15. The AES is used for the encryption of the dashed areas. As can be seen the sender ID is removed from this protocol because of it redundant information.

0 1	2 3	4 5	6 7	8 9	10 11	12 13	14 15	16 17	18 19	20 21	22 23	24 25	26 27	28 29	30 31
Payload type				MjVer		MnVer		Length							
NFC Tag ID															
Sequence number															
Session Id															
Padding															
Session Id															

Figure 17.15 Secured NFC Request Protocol [26]

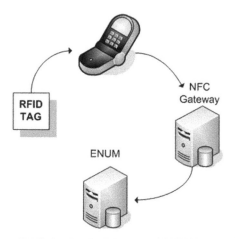

Figure 17.16 Service Architecture of NFC Gateway [26]

The architecture for the NFC gateway service is shown in Fig. 17.16. As can be seen it uses RFID tags for specifying the route of the incoming calls. These tags can then be attached to a car dashboard, home door or workplace desk. After that, the service request message is sent to the NFC gateway which is aware of the locations of all RFID tags. Once the RFID tag and sender ID are received by the NFC gateway, the route for incoming phone calls of that specific user will be specified. In the end the client reads the NFC tag and establishes a network connection to the ENUM to update the request of the NFC gateway .The sender identification and NFC tag identification are provided in the ENUM update request.

17.2 Wireless power transfer

In a system which is capable of wireless energy transfer, electrical energy is transmitted from a power source to an electrical load without interconnecting wires. The fact that this transmission link is wireless is useful in cases where instantaneous or continuous energy transfer is needed but interconnecting wires are inconvenient, hazardous, or impossible. However, it is less efficient than wired transfer through standardized cable conductors, a typical short range technology losing typically up to 60% of the input power over several meters [28]. This technology can be applicable in a battery charger. The details of its design in this specific application have been provided in the following section.

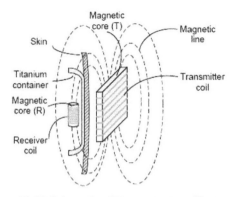

Figure 17.17 Schematic of Transcutaneous Charger [24]

Battery charger with NFC

NFC can also be useful for the communication of self powered devices. Some potential applications are the collection of historical data and arrangement of operating parameters for ICDs (internal cardiac defibrillators). In these sorts of configurations, NFC communication subsystems can be used which are capable of being field-powered and so not introducing additional drainage for limited battery resources [14].

One usage scenario of this type of usage is provided in [24] where two coupling coils are designed to charge the battery fit in the body. The schematic of it is shown in Fig. 17.17. The receiver coil is small and columned, encapsulated in a titanium container which is implanted under the skin. As can be seen in the design, the transmitter coil is flat and much bigger with the aim of stabilizing the recharge. It uses a charge current of 15 mA. The energy is transmitted from the outside into the coils. Due the current supply of the drive module, magnetic flux is generated and so induces a voltage on the receiver coil in the magnetic field which results in energy being transmitted into the body. The block diagram for this specific charger is shown in Fig. 17.18.

The power amplifier and frequency generator are the main parts of the drive module. Direct digital synthesis (DSS) is used to generate high frequency waveforms. A power MOSFET is used in the power amplifier. The transmission frequency is about tens of kHz which can also be changed by an AT89C51 MCU. A rectifier, filter circuit, lithium ion secondary battery and charging IC are used to build up the charge module.

Based on the model proposed in [24], the typical power requirement is said to be 100 μW with an average current of 30 μA. Based on the conducted

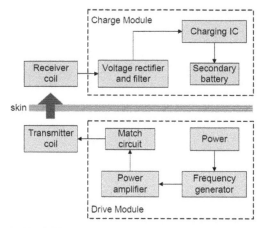

Figure 17.18 Block Diagram of the Designed Transcutaneous Charger [24]

calculations and experiments, it will take it 3 hours to be fully recharged and so the nerve simulation device can go on for more than one month.

The charging efficiency is calculated via equation (17.3):

$$\eta = \frac{V_{charge} \times I_{charge}}{W_{power} - W_{circuit}} \times 100\% \qquad (17.3)$$

V_{charge} = Voltage of the rechargeable battery (V)
I_{charge} = Current of the rechargeable battery (mA)
W_{power} = Power source output (W)
$W_{circuit}$ = Power consumption of circuit (W)

Substitute to UHF-RFID

RFID frequency ranges are divided into three separate bands which are as follows:

a) Low Frequency RFID ($\sim$134 kHz):
   Usage: implants identifying pets and livestock
b) MICS band (Medical Implant Communication Service) (402–405 MHz):
   Usage: Communication with medical devices and so field-powered communication with implanted devices.
c) UHF (900 MHz) (Ultra High Frequency) used in transponders

The UHF and MICS band offer higher data rates, however consume greater energy than LF RFID or 14 MHz which can eliminate the amount of safe power transfer through tissue due to thermal heating. In the case that short-range magnetically coupled UHF is used, highly efficient coupling through tissue is possible.

Issues in UHF:
UHF scanners and devices are difficult to be restricted in their range. The UHF RFID is prone to being reflected by small metallic objects causing the threat of eavesdropping and confusing emergency responder's efforts for identifying patients with radio-equipped medical devices. Moreover the inconsistent allocation of LF (Low Frequency) and UHF bands causes issues in internationalization. The NFC Forum has adopted protocols upon 13.56 MHz RFID.

Advantages:

- Provides bidirectional communication channel with higher data rates than UHF RFID (upto 400 Kbit/s from powered "scanner" devices and upto 100 kbit/s from field powered "passive" Devices.)
- Being internationally available
- The Short range HF scanners are not prone to UHF's distant phantom hot spot phenomena
- Provides increase in bandwidth
- Usable for a variety of data-intensive applications that facilitate the inclusion of privacy and integrity-preserving cryptographic protocols appropriate for medical systems.
- NFC-enabled RFID tags are also standardized (with standards such as HL7) enabling the storage of electronic health records in them.
- Uses magnetic coupling, which provides a short range of 20 cm which prevents eavesdropping.
- Reliable in patient selection identification in hospitals and other "first responder" environments with radio-equipped medical devices.
- Suitable for the communication of implanted medical devices as not being attenuated by tissue.
- Not requiring a battery or external electrical connection, making it compatible with other electronic devices.
- The NFC-enabled medical devices are capable of providing health history to first responders in an emergency.

Therefore based on the aforementioned issues and the advantages given above, the usage of NFC technology is much more efficient and does not get us involved with the mentioned issues.

17.3 NFC in the military

However RF communications are the best choice for long distances, the cordless communication of small devices over short range in military environment becomes an issue. In the case where one of the cords is disconnected, the entire system is inoperable. It is also important to note that features that make long range communications possible are disadvantageous to short range communication. Previously, RF-based communication technologies such as Bluetooth had been used for military use but due to issues such as security, spectrum allocation and reliability they can not be used nowadays. Currently, Freelinc has developed a secure, short-range wireless technology namely; Near Field Magnetic Induction (NFC) which is more than perfect for military use, as it's the only reliable, secure and sustainable non-RF wireless communication with no impact on other RF communications [8].

One of the major concerns in military communications is the control and assignment of RF frequency spectrum. The demand for available frequencies and clear channels is growing increasingly due to more vehicle mounted and infancy channels. Spectrum contention is more a problem in urban environments where soldiers and consumer devices must compete for frequency allocation. Also due to the matter that most available radio systems use the 2.4 GHz unlicensed frequency standard for Bluetooth, WiFi and other wireless communication systems, serious issues arise when a squad relies on its communication system in an urban environment where unexpected failure may occur due to over use of frequencies.

Additionally, the frequency spectrum should be kept clear for squad-to-command transmission which needs far-field communication for long range communications, not allowing short range squad level communication to utilize the frequency spectrum of these long range communications. There are also some security issues which require encryption for communication transmission. Therefore, in a body wireless network system where each soldier has their own equipment encryption keys, the loading process can be very complex and time consuming. However the Freelincs' NFC module is inherently secure and there will be no further requirement for encryption, allowing the short range communication to take place at squad level with a significant reduction in demodulation and interception threats.

The NFC technology is also applicable in military systems. Some examples of its applications in this regard are shown in Fig. 17.19.

NFC makes sense to service providers due to its rich media content driven by NFC-enabled device, eliminated cost for electronic issuance and the

Intercom system

Wireless Sensors and Signaling
Asset Health Monitoring
Fuselage/ Wing Integrity
Missile Health Status
Data Transmission
Thru Skin
Low Power

Local Intercom

Radio Interface

Robotic Controls
Denied Environment
Building Penetration

Dismounted perimeter communications

Enhanced mobility during
ingress and egress

Low probability of detection

Laptop Interface
Audio and Data Transfer
Programming Enclosed Devices
Low Spectrum Contention

Figure 17.19 NFC Applications in Wireless Vehicle Intercoms and UGS Control [8]

matter that NFC-enabled services are more preferable by consumers. However, the NFC technology is not a replacement for tactical radios, 802.xx or OTH satellite communication but a tool for removing the wires of short distance and military communication systems. Due to its inherent reliability and security, it has a wide usage and said to be "Bluetooth replacement". Other wireless communications can only reach the same level of benefit of NFC with numerous encryption methods, upgrades, add-ons and modifications. Freelincs' NFC is inexpensive, with high return-on-investment over time. In summary, NFC is estimated to have various market projections. Up to 700 million NFC- enabled mobile phones are to be sold by 2013 which means upto 25 percent of the market at that time. NFC mobile payments are said to exceed 30 billion US dollars by 2012.

References

[1] http://www.nfc-forum.org/home/

[2] Freudenthal, E., D. Herrera, et al. (2007). "Suitability of NFC for Medical Device Communication and Power Delivery". Engineering in Medicine and Biology Workshop, 2007 IEEE Dallas.

[3] Pui-Lam Siu; Chiu-Sing Choy; Chan, C.F.; Kong-Pan Pun; (2003). "A contactless smartcard designed with asynchronous circuit technique". Proceedings of the 29th European on Solid-State Circuits Conference, 2003. ESSCIRC '03.

[4] Pohjanheimo, L., Keränen, H., Ailisto, H. "Implementing TouchMe Paradigm with a Mobile Phone", Joint sOc-EUSAI conference, Grenoble, October 2005.

[5] http://www.3gtech.info/tag/nfc-vs-bluetooth

[6] Pasquet, M. Reynaud, J., Rosenberger, C. "Payment with mobile NFC phones: How to analyse the security problems", The International Symposium on Collaborative Technologies and Systems (CTS), 2008.

[7] Freudenthal, E., D. Herrera, et al. (2007). "Suitability of NFC for Medical Device Communication and Power Delivery". Engineering in Medicine and Biology Workshop, 2007 IEEE Dallas.

[8] "Use of near field magnetic induction in military communication environments", Technical Applications Group LLC, April 2010.

[9] G. Madlmayr, O. Dillinger, J. Langer, C. Schaffer, C. Kantner, J. Scharinger, "The benefit of using SIM application toolkit in the context of near field communication applications", International Conference on the Management of Mobile Business, 2007. ICMB 2007. Toronto, Ont. p. 5.

[10] Strommer, E., Kaartinen, J., Parkka, J., Ylisaukko-oja and Korhonen, I. (2006). "Application of Near Field Communication for Health Monitoring in Daily Life" 28th Annual International Conference of the IEEE Engineering in Medicine and Biology Society, 2006. EMBS '06.

[11] http://emedicine.medscape.com/article/1681045-overview

[12] Wu Ying; Yan Luguang; Xu Shangang;, "Modelling and performance analysis of the new contactless power supply system," *Proceedings of the Eighth International Conference on Electrical Machines and Systems*, 2005. ICEMS 2005, vol. 3, no., pp. 1983-7, vol. 3, 29-29, Sept. 2005, DOI: 10.1109/ICEMS.2005.202907.

[13] Xue, L., H. W. Hao, et al. (2005). "A new method of radio frequency links by coplanar coils for implantable medical devices". 27th Annual International Conference of the Engineering in Medicine and Biology Society, 2005. IEEE-EMBS 2005.

[14] Freudenthal, E., D. Herrera, et al. (2007). "Suitability of NFC for Medical Device Communication and Power Delivery". Engineering in Medicine and Biology Workshop, 2007 IEEE Dallas.

[15] http://en.wikipedia.org/wiki/Electronic_health_record
[16] http://www.determined2heal.org/exercise/functional-electrical-stimulation/
[17] D. Patterson, R. Thirlby, and M. Dobrio, "High-frequency gastric stimulation in a patient with diabetic gastroparesis," Diabetic Medicine, vol. 21, no. 2, p. 195, Feb 2004.
[18] M. Krausea, W. Fogela, A. Heckb, W. Hackea, M. Bonsantob, C. Trenkwalderc, and V. Tronnierb, "Deep brain stimulation for the treatment of parkinson's disease: subthalamic nucleus versus globus pallidus internu," J. Neurol Neurosurg, pp. 464–470, April 2001.
[19] http://en.wikipedia.org/wiki/electronic-health-record/
[20] K. Kumar, N.R, and G. Wyant., "Treatment of chronic pain by epidural crypto-graphic techniques in a manner that prevents such spinal chord stimulation; a 10-year experience," J. Neurosurg, vol. 75, no. 3, pp. 402–407, September 1991.
[21] Iglesias, R., Parra, J., Cruces, C., Segura, N.G. "Experiencing NFC-based touch for home healthcare". Petraacm (2009).
[22] Fletcher, R.R. and Kulkarni, S. "Wearable Doppler radar with integrated antenna for patient vital sign monitoring". Radio and Wireless Symposium (RWS), 2010 IEEE.
[23] Aziza, H. "NFC Technology in Mobile Phone Next-Generation Services". 2010 Second International Workshop on Near Field Communication (NFC).
[24] Chuansen, N., H. Hongwei, et al. (2006). "The Transcutaneous Charger for Implanted Nerve Stimulation Device". 28th Annual International Conference of the IEEE Engineering in Medicine and Biology Society, 2006. EMBS '06, pp. 4941–4944.
[25] M. Csapodi, A. Nagy, "New Applications for NFC Devices", Mobile and Wireless Communications Summit, 2007. 16th IST, Budapest.
[26] Ylinen, J., M. Koskela, Iso-Anttila, L., Loula, P. (2009). "Near Field Communication Network Services". Third International Conference on Digital Society, 2009. ICDS '09.
[27] http://computer.yourdictionary.com/enum
[28] "Wireless technologies are starting to power devices, 01.09.09, 06:25 PM EST". Forbes.com. http://www.forbes.com/2009/01/09/ces-wireless-power-tech-sciences-cx_tb_0109power.html. Retrieved 2009-06-04.

18

Wireless Power Transfer

Nowadays using electronic devices is an inevitable part of our life. They make life a lot easier for people. About 50 years ago, before the invention of mobile phones no one could imagine communicating with someone thousands of miles a way without the need for wires. Thanks to laptops and wireless modems, people can explore the web any time any where. These days, industrial robots are performing different tasks and reduce the required time and costs and the need for labour. They are even able to perform the tasks which might be harmful for humans such as tasks that military robots or space robots or robots for chemical analysis are performing. In a nutshell, all those electronic devices delivering a wide range of functions are here to improve our quality of life and to enhance our life experience. However they all have a common feature and that is they require a power source to fulfil a role.

18.1 Introduction

Those devices might be powered by wall mounted DC voltage source, which requires wires to connect the device to the power supply, or they may use rechargeable or disposable batteries. Although batteries facilitate mobility, they need to be recharged regularly using wire connections or to be exchanged with new battery set frequently. Using a cell phone or a laptop, users can easily communicate to any where in the world or to explore the web without the need for wires and regardless of where they are. This is possible as long as the battery in the device is alive. There has been a great interest to deliver wire-free power connection to electronic devices since WiFi was adopted by people worldwide for wireless communications [1]. Despite the interests of scientists in wireless power transmission (WPT), there has been little progress

Co-authored by Mehrnoush Masihpour & Agbinya JI, University of Technology, Sydney

J. Ihyeh Agbinya (PhD), Principles of Inductive Near Field Communications for Internet of Things, 281–300.

in this area and a convenient energy supply is a major challenge [2]. There-fore, delivering energy through a wireless system comparable to WiFi, to automatically charge electronic devices without the need for wires and in a large coverage range is still a dream of mankind. However, introducing novel wireless technologies in one hand and reduction of power consumption of electronics on the other hand allows this man's dream to be possible in some applications.

Research on WPT started since the early years of 20th century [3] and there have been different attempts to eliminate wires in power transmission in applications such as cell phones, laptops, robots, implantable devices and many more [3].

There are two main types of WPT [2]; the first is to use radio waves for WPT, which targets medium to long distance transmission. However this transmission type suffers from inefficient power delivery. The power trans-ferred is also quite small, usually from 1 to 100 mW [2]. WPT using micro-waves has been used in space solar powered satellites [4]. The main concept in such application is to transmit power from the sun to the earth through radio waves. Large satellites are built to generate the electricity from the sun energy and transfer it to the ground for instance to charge cell phones [4]. In the past decades many space solar power satellites have been designed and built in Japan [4]. However, this type of WPT is not the focus of this chapter.

Another type of WPT technique is achieved through one of the well known physical principles: magnetic field coupling [2]. This technique, which is the main focus of this chapter, can be used for short range WPT, but offers high efficiency up to 80% [2]. Non-contact magnetic coupling is capable of WPT range from several mW to several kW [1]. However it lacks enough analysis of efficiency and range enhancement, since the efficiency highly decreases as the distance increases. The basic concept of this technology is fairly simple. When two resonant circuits are located in close proximity, the current in the primary coil induces a current and voltage in the sec-ondary circuit; therefore energy can be transferred from one point to another without the need for wires. This principle has been used in different applica-tions to power up devices or to charge an internal battery such as to supply power for implantable devices or electric cars or different industrial applica-tions. However, very short coverage area is the main limitation of this tech-nique. Therefore it has not been used for applications where longer ranges are required for example to have the power coverage in an area as large as a room.

18.1.1 Applications of magnetically inductive WPT

In recent years electronic gadgets tend to be cheaper and smaller in size so they are widely adopted in people's everyday life. However, magnetically inductive (MI) WRT can be used to provide power for autonomous electronics such as sensors, robots, PDAs and many more [5]. One of the areas that MI-WRT can be beneficial is in different industrial applications, where a wide range of autonomous devices and machines are used and they can benefit from autonomous power supply [5]. According to [5], there are two main categories of energy supply that MI-WPT is required; firstly autonomous systems, where the key issue is to have a power supply at reasonable costs. Remote actuators and sensors in large chemical plants, industrial automation and control and temporary measurements (e.g oil exploration) are some examples which are classified in this category [5]. The second category is accessibility-limited systems, where the main issue is to transfer energy [5]. For instance systems with a very high voltage potential or with moving parts, aggressive atmospheric environment, robots, long term tests, mobile terminals, widely spread plants or fully automated production machines fall in this category [5]. Using WPT in such environments can be highly advantageous by being cost effective since engineering and wire cost are eliminated also by allowing easier extension and upgrade of the existing system [5].

To deploy WPT, requirements vary from application to application but it should address some common needs. For example WPT need to be usable for different positions and orientation of the target device also to deliver the power to the device while moving [5].

Electric cars are another field that MI-WPT can be used to recharge the battery automatically [6]. Although one can recharge the battery by plugging it to the wall mounted power supply, it is not necessary and it can be charged while it is parked on the street [6]. A simplified block diagram of such system is shown in Figure 18.1. As shown in Figure 18.1, a transmitting antenna dispenses the energy which is produced by a high frequency power source. Through magnetic resonance coupling, a receiving antenna receives the transmitted power and rectifies it to be able to charge the battery.

However, WPT for electric cars requires three key elements, which are large air gap (distance between the receiver and transmitter), high efficiency and large amount of power [6]. Authors in [6] claim that electromagnetic resonance coupling is the only technology that can cover the mentioned three requirements. Since the magnetic resonance coupling transfers the energy through near field coupling of two magnetic coils rather than radiating the

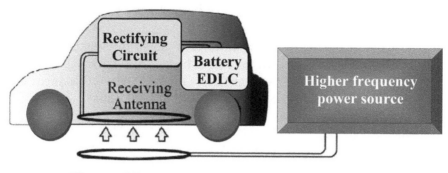

Transmitting antenna

Figure 18.1 Concept of Contact-Less Power Transfer System for Electric Car (Adopted from [6])

power (like microwaves), therefore, it conserves the energy meaning it is a lot more power efficient than microwave power transmission [6]. An experiment conducted in the University of Tokyo [6] shows that by using the same antenna coils at the receiving and transmitting side with radius 150mm, the efficiency of the WPT is approximately 95–97% for different distance separation 10, 15, 20 and 25 cm. However it is suggested in [6] that by increasing the air gap, the received power decreases therefore the efficiency drops.

Another important application of WPT is in implantable devices. Implants such as endoscope capsules, pace makers, blood pressure sensors or body temperature sensors all require a source of energy to be able to perform. Those devices might need a wire connection to the power source which results in problems such as lack of comfort and mobility for the patient [7]. There is a high risk of infection due to the wire crossing the skin [7]. However to eliminate wire, these devices use rechargeable or disposable batteries which also results in some issues such as large size of the device, limited life time and charging concerns [7]. Therefore, WPT has been seen as a promising solution to the mentioned issues that can eliminate the need for wires or large batteries and bring the comfort to the patients' life. There have been a number of studies that provide an insight into the WPT for medical devices [7, 8, 9, 10]. WPT system for implantable devices is discussed in detail in the next section.

18.1.2 Generalized WPT system

Although WPT is different in terms of configuration, specifications and required elements for different application types, the basic principle is almost the same. A generalized and simplified block diagram of a WPT system is shown in Figure 18.2 [2]. A typical system consists of a power source, transmitter and receiver circuits, which are tuned resonate circuits. In this simplified model the load is an LED and the aim is to light up the LED using WPT system through magnetic coupling of the transmitter and receiver antenna coils. In such system, energy source at the transmitting circuit creates a current in the circuit which in turn induces a current and AC voltage in the receiving circuit by means of transmitting and receiving coils. However, to be able to light up the LED, a DC voltage is required, which can be achieved by using a rectifying diode at receiver circuit [2]. A diode is required to have a large reverse voltage and fast switching speed to increase the efficiency of the system [2]. During the WPT process, there might be some undesirable changes which can result in decreasing the efficiency of the power transmission to the load, meaning that power losses are too high. One of the important and common problems in such system is changing the resonant frequency due to different parameters such as existence of obstacles within the inductive link, parasitic parameters, temperature rising in circuits and many more [2]. As a result of frequency variation, the efficiency decreases dramatically and usually the fast tracking of the frequency is a major issue [2]. Fast tracking is often more critical for higher frequencies (more than 1 MHz) [2].

A few studies have been done to address this drawback. In this section a method for matching the frequency is described [2]. Authors in [2], claim that by adding a protection circuit to the transmitter side, the efficiency will be optimized. The operating frequency of this circuit is retained close to the resonant frequency also on the right hand side of it and according to [2] it operates in the ZVS (zero-voltage switching) condition which is favourable in modern power converters to minimize the switching loss. The protection circuit is in fact a half-bridge converter which inhibits the operating frequency

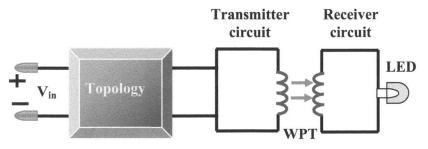

Figure 18.2 Generalized WPT System Model (Adopted from [2])

to transit to the left side of the resonant frequency from the right side in case of frequency variation [2]. Figure 18.3 shows a full diagram of a WPT system with addition of the protection circuit.

By varying the value of the load, which in this example is the LED, resonant frequency changes and makes the WPT system inefficient [2]. However by employing the protection circuit into the general design, 61% efficiency is achieved using a 16 cm × 18 cm spiral transmitting coil and 4 cm × 5 cm receiving coil at a frequency of 360 kHz [2]. This specific design is capable of providing 3.4 W at the load where the input voltage is 12 V [2].

18.1.3 WPT for medical applications

As mentioned earlier, MI-WPT for embedded medical devices is beneficial since it mitigates the need for wires, leaves the patient with more comfort

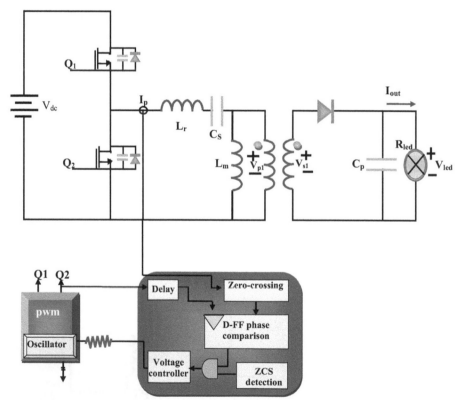

Figure 18.3 Full Diagram of the WPT with Protection Circuit (Adopted from [2])

and also decreases the device size by eliminating the battery or making them smaller. MI-WPT gained the interest of scholars in the field to provide an efficient and flexible power transmission system at lower costs. MI-WPT implementation is quite challenging mainly due to rapid efficiency drop with increasing the separation distance. However, MI-WPT has been considered to be a favourable power supply system for implantable devices such as retinal implant [8], endoscope robots [9] and many more. In this regard, the next section will discuss the concept of MI-WPT for a micro-robot for intestinal inspection [9], which is relevant to MI-WPT for different medical embedded devices and the same concept can be applied to different medical applications.

18.1.3.1 WPT for smart endoscopy devices

To diagnose and cure gastrointestinal diseases, it is required to insert endoscopy devices inside patient's body to record information through video or sampling tools. Also by using operational tools they can contribute to the treatment of patients [9]. However, such a process is painful, upsetting for the patient and also difficult for the doctors using conventional endoscopy devices. In this context it has been about two decades that researchers have been studying the design and implementation of micro-robots for this medical purpose to make the process of endoscopy easier and more reliable. However, those micro-robots are mainly based on trailing cables, which can result in resistance to the motion of robot inside the intestinal canal [9]. Trailing cables might also lead to intestinal tissue injury since they are quite rigid [9]. Recently a new endoscopy micro-robot prototype has been proposed by authors in [9], which is not trailing-cable-based so is very flexible. It uses MI-WPT to power it up. This micro-robot consists of six individual modules working consistently together [9]:

- wireless communication module
- personal computer (PC)
- micro-robot
- wireless power transfer control unit
- wireless power transfer coil

In fact the endoscopy micro-robot has a head cabin and three drivers or actuators, as shown in Figure 18.4. The head cabin includes the communication control unit as well as wireless power receiving module inside it [9]. The actuators are activated using a miniature DC motor with driving voltage of 3.2 V to provide a rectilinear earthworm like motion. However the output of the DC motor has consecutive rotary motion and it needs to be converted to

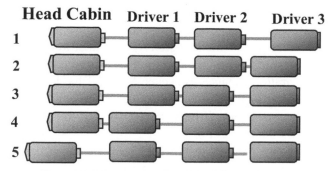

Figure 18.4 Locomotion Principle (Adopted from [9])

a rectilinear movement [9]. In the process of converting the rotary motion to rectilinear a lot of power will be lost therefore efficiency drops [9]. Hence high power transfer efficiency during this process is an important design parameter. However three drivers and head cabin are connected together via three gimbal joints. The reason for using the gimbals is that the robot can move even in a complex environment at the intestinal canal [9]. To prevent water penetration into different parts of the robot, tectorial membrane is used [9].

What makes this micro-robot very flexible is how it moves inside intestinal canal. The movement is based on a stretching and contracting action (Figure 18.4). To move forward, first the micro-robot requires a control signal from outside the body, then the last driver (driver 3) will be loaded by a positive voltage through WPT system and it moves toward driver 2 (left hand side) then it will stop since it is loaded with a negative voltage. After that driver 2 starts the same action as driver 3 after it is loaded by a positive voltage. The same process continues for driver 1 and finally head cabin which finally results is forward movement [9]. By repetition of this process the micro-robot can move as long as it is required. To perform a reverse movement (backward), the control signal should be inversed.

As mentioned earlier the movement needs to be controlled from outside the body. This can be done by a personal computer, which is managed by a person and input parameters can be entered. The input parameters need to reach the wireless communication module inside the head cabin. RS232-TTL is used to communicate the data form a PC to the wireless communication module [9]. After the input reached the wireless communication module, it sends the parameters to the communication control unit at the head cabin using radio frequency at 433 MHz and by using FSK modulation technique [9]. After that C8051F33 compiles the received parameters in order to create a specific waveform [9]. The generated signals then will be amplified and loaded on the drivers to control the direction, duration, speed and towing force of movement [9]. The process steps are shown in Figure 18.5.

The most important part of this design is that this micro-robot receives the power from outside using the wireless receiving module and the wireless power transfer controller and a transmitting antenna, which is located around the patient's trunk [9]. Transmitting and receiving antennas are two inductive solenoids, which are coupled through the magnetic field coupling concept to provide an inductive link between transmitter and receiver (inside the head cabin) [9].

It is well known that the coupling strength depends on the value of the coupling coefficient (k) value, which itself is a function of the mutual inductance of the two coils (M) as well as the self inductance of the transmitting (L_T) and receiving coils (L_R).

$$k = \frac{M}{\sqrt{L_T L_R}} \tag{18.1}$$

However the inductive link is almost the same for all applications and can be described as Figure 18.6. According to [9], transmitting coil can be excited by a voltage source and a switching circuit or a class E amplifier can be used to efficiently amplify the voltage. However a rectifier which has a half or full

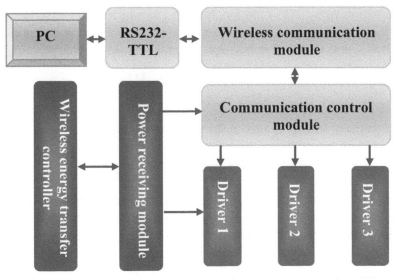

Figure 18.5 Functional Block Diagram of Micro-Robot (Adopted from [9])

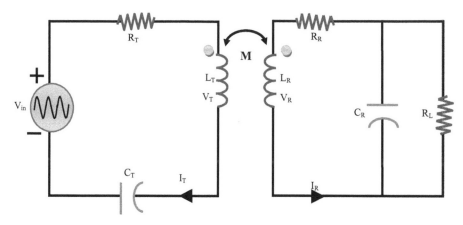

Figure 18.6 Equivalent Circuit of Wireless Power Transfer (Adopted from [9])

bridge configuration [9] is required to provide the DC voltage efficiently to optimize the losses due to conversion.

As it can be seen from Figure 18.6, the receiving coil is loaded with a parallel capacitor (C_R) to cancel the inductance at the receiving side [9]. However series-resistance of the transmitting and receiving coils are shown by R_T and R_R respectively. To achieve a sinusoidal high amplitude electric current I_T (maximum amplitude 2A in this case) at the transmitting circuit, transmitting coil is loaded with a series capacitor (C_T) [9]. In this specific application the excitation voltage (V_T) has amplitude 12 V, which is able to induce voltage V_R and current I_R at the receiving side via inductive link. According to Kirchhoff's law, relationship between system parameters holds as:

$$V_T = Z_T I_T - j\omega M I_R \tag{18.2}$$

$$j\omega M I_T = Z_R I_R \tag{18.3}$$

Where resonant frequency (ω), at the centre frequency f_0 is:

$$\omega = 2\pi f_0 = \frac{1}{\sqrt{L_T C_T}} = \frac{1}{\sqrt{L_R C_R}} \tag{18.4}$$

However Z_T and Z_R are the impedance of the transmitting and receiving circuit respectively. The voltage at the output which provides the power for the micro robot to operate is computed by multiplication of the excited current in the receiving circuit and the resistance of the load (R_L).

$$V_R = R_L I_R \tag{18.5}$$

By knowing that:

$$Z_T = R_T + j \left(\omega L_T - \frac{1}{\omega C_T} \right) \tag{18.6}$$

$$Z_R = R_L + R_R + j \left(\omega L_R - \frac{1}{\omega C_R} \right) \tag{18.7}$$

Also by substituting equation (18.2) and (18.3) in equation (18.5), the output voltage is calculated to be:

$$V_R = \frac{j\omega M R_L V_T}{R_T R_R + \omega^2 M^2 + j\omega(C_R R_R R_T R_L + L_R R_T + \omega^2 M^2 C_R R_L)} \tag{18.8}$$

Therefore the efficiency of the link can be expressed as:

$$\eta = \frac{V_R}{V_T} = \frac{\omega^2 M^2 R_L}{Z_R(Z_R Z_T + \omega^2 M^2)(1 + j\omega C_R R_L)^2} \tag{18.9}$$

As described in the earlier chapters (chapter 16), the quality factor of the transmitting and receiving resonant circuits are $Q_T = \omega L_T / R_T$ and $Q_R = \omega L_R / R_2$ respectively. Also by defining the ratio of the receiver load and transmitting coil resistance as $R_r = R_L / R_2$ the efficiency of the system is simplified to:

$$\eta = \frac{Q_R^2 R_r}{(Q_R^2 + R_r)^2} k^2 Q_R Q_T \tag{18.10}$$

From equation (18.10), it is derived that the maximum efficiency $\left[\frac{d\eta}{dR_r} = 0 \right]$ is achievable if $R_r = Q_R^2$. Therefore maximum efficiency is:

$$\eta_{\max} = \frac{k^2 Q_R Q_T}{4} \tag{18.11}$$

According to [9], the maximum efficiency is a function of the resonant frequency and by increasing the frequency, higher efficiency can be achieved. However, by increasing the resonant frequency the impedance of the two resonate circuits also increases [9]. According to this specific design of endoscope micro-robot, for the robot to work properly, coupling coefficient needs to be approximately 0.005 [9]. As mentioned in the earlier section by increasing the value of load resistance, efficiency decreases and there is an optimum value of load resistance. On the other hand, as micro-robot moves inside the intestinal channel, consumed power will change, therefore load resistance is designed to change in accordance to the power consumption during movement to maintain the efficiency of system [9].

Experiments show that the robot works properly as long as it is in power transmission limit [9]. However, voltage rapidly approaches zero when it is out of the limit. The transmission power by the transmitting coil is 25 W for this experiment, which 400 mW of that is dissipate by the locomotion unit [9].

18.1.4 Multi-voltage output MI-WPT system

With regards to providing an implantable device with MI-WPT system, it is often required to have different voltage levels [8]. High voltage is needed to provide the power for the stimulators and regular voltage to power up the other analog circuits and digital blocks [8]. There are two main types of multi-voltage systems; the first method is to drive a resonant circuit using a voltage oscillator and then rectifying and converting it to other required levels. However the second approach is to use both tapped and non-tapped coils, which is complex and requires larger size of coils since it increases total inductance [8]. The first is simpler to implement but it suffers from lower efficiency. To address this problem, a cascaded resonant tank circuit is used in [8], which improves the efficiency of the WPT (43% power efficiency) for multi-voltage WPT systems without the need for larger coils. The simplified circuit model is shown in Figure 18.7. The receiving circuit has two terminals T_1 and T_2 which provide the high voltage and regular voltage respectively [8].

There are two resonant frequencies (ω_1, ω_2) when the resonant tank is resonating [8]. To calculate the value of ω, R_a and R_b are replaced by testing voltage source V_a and V_b with corresponding current of i_{ta} and i_{tb} respectively [8]. The first resonant requency can be seen between the point b and the ground $(di_{tb}/d\omega = 0)$ and the second between points a and b $(di_{ta}/d\omega = 0$. By assigning the two resonant frequencies at the receiver to the resonant frequency of transmitter, two conditions can be referred to for computing the design components. The two resonant conditions are as follows [8]:

$$\frac{1}{\omega_1^2} \approx L\gamma^2(\frac{1}{\gamma} - 1)^2 C_3 \tag{18.12}$$

$$\frac{1}{\omega_2^2} \approx L(1 - \gamma)^2 C_3 + L\gamma^2 C_2 \tag{18.13}$$

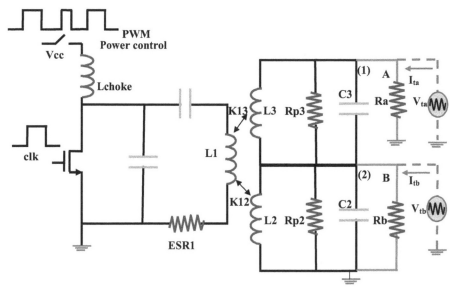

Figure 18.7 Simplified Model of Multiple Voltage Power Telemetry System (Adopted from [8])

Where L is the total inductance between the two terminals and $\gamma = V(\text{regular})/V(\text{high})$.

By using equation (18.12) and (18.13), system parameters can be computed.

However, the link efficiency is highly affected by the quality factor (Q) and coupling coefficient (k). Coupling coefficient has a value in the range 0.01 and 0.1 and is a function of the distance and the coil size [8]. For having an optimum design, the optimum Q needs to be determined. According to [8], a coil can be equivalent to a lumped capacitor in parallel with inductor at the frequency f and is derived using the following equation:

$$Q(f) \approx 2\pi f L \left(1 - \frac{f^2}{f_{self}^2}\right) \Big/ R_{DC} \left(1 + \frac{f^2}{f_h^2}\right) \qquad (18.14)$$

In this equation f_h represents the quantified impact of the proximity effect or so-called skin effect and is defined as [8]:

$$f_h = \frac{2\sqrt{2}}{\pi r_s^2 \mu_0 \sigma \sqrt{N_t N_s \eta A}} \qquad (18.15)$$

Parameters in equation (18.15) are defined as follows:

- r_s: wire radius used at the coils
- μ_0: magnetic permeability
- σ : magnetic conductivity
- N_t : number of turns
- N_s : number of strands per turn
- A : coil cross section area
- η : a value between 0.2 to 1 depending of the coil's geometry

f_{self} at equation (18.14) is the coil's self-resonant frequency and is computed using the following equation:

$$f_{self} = \frac{1}{2\pi\sqrt{LC_{self}}} \tag{18.16}$$

Where $C_{p,k}$ is the parasitic capacitance between turn p and k,

$$L = N_t^2 L_i \quad \text{and } C_{self} = \sum_{p<k} C_{p,k}(k-p)^2/N_t^2 \tag{18.17}$$

However, the frequency at which the coil has maximum $Q(f_{peak})$ can be derived by:

$$f_{peak}^2 \approx f_h^2 \left\| \frac{f_{peak}^2}{3} \right. \tag{18.18}$$

By using equations (18.12) to (18.18) not only the optimum Q is computed but also the maximum efficiency for multiple-voltage power telemetry system can be determined [8].

18.1.5 MI-WPT for embedded medical devices using spiral coils

One of the common types of inductive coils is the cylindrical configuration which provides good quality factor and simple modelling of the inductance and associated losses [7]. However, they are not optimum to be used for WPT of implantable devices since they can not be further optimized in terms of power efficiency [7]. There is another type of inductive coils known as spiral coils which can be optimized in terms of quality factor and coupling link strength and efficiency [7]. Spiral coils consist of a number of conductive rings and the strongest coupling can be achieved when they are coaxial and placed on the same plane [7]. The most important design parameters in spiral coils are the impact of internal and external radius, separation between the windings and the width of metal strips [7]. Spiral coils can be seen as a group of concentric rings. In this section a WPT model for implantable device using spiral micro-coils will be discussed.

18.1.5.1 Transmitter coil modelling

The model description in this section is based on a conventional model which has an inductance in series with the conductor resistance [7]. The self inductance of concentric rings, forming the spiral coil, is computed using equation (18.19). It is a function of average ring radius (b) and wire radius (R) and magnetic permeability of the material used to form the coil. However in free space, the wire radius is far smaller than the average ring radius [7].

$$L_P(b) = \mu_0 b \left(\ln \left(\frac{8b}{R} \right) - 2 \right)$$ (18.19)

However the mutual inductance between the windings of the coil (transmitting or receiving coil) is formulated as [7]:

$$Ma_1 a_2 = \pi \mu_0 \sqrt{a_1 a_2} \left(\frac{2}{\pi} \sqrt{\frac{a_1}{a_2}} \right) \cdot \left[K \left(\frac{a_2}{a_1} \right) - E \left(\frac{a_2}{a_1} \right) \right]$$ (18.20)

Where a_1 and a_2 are the mean radius of the two concentric rings, $K(x)$ and $E(x)$ are the complete elliptic integrals of the first and the second order [7]. Therefore the total inductance of coil is the result of summation of inductance of each ring and mutual inductance between different windings [7].

However, the resistance of a circular conductor is also formulated as [11]:

$$R_1(\omega) = \frac{\displaystyle\sum_{k=1}^{\infty} \frac{R_k}{R_k^2 + \omega^2 L^2}}{\left(\displaystyle\sum_{k=1}^{\infty} \frac{R_k}{R_k^2 + \omega^2 L^2} \right)^2 + \omega^2 \left(\displaystyle\sum_{k=1}^{\infty} \frac{L}{R_k^2 + \omega^2 L^2} \right)^2}$$ (18.21)

In this equation, $R_k = \xi_k^2 / 4\pi \sigma R^2$, and L is the self-inductance of the coil and is $L = \mu_0 \mu_r / 4\pi$. μ_r is the relative permeability and σ is the conductivity of the material used to form the coil windings. However, $\xi_k = (2k - 1)\pi / 2 + \pi / 4$ [7].

Mutual coupling between the transmitting and receiving coils depend on different parameters such as the size and shape of coils, side and angular alignment of them (i.e the orientation of coils). Inductive link is highly affected by the side and angular misalignment of the transmitting and receiving coils and such effect can decline the link strength significantly even by small misalignment. On the other hand in medical embedded devices, transceivers are not fixed and often they move and change their orientation therefore efficiency is highly influenced by motion or changing the location of the devices. The mutual inductance, taking into account all the mentioned functional parameters can be calculated by:

$$M(r_T, r_R, \Delta, d) = \pi \mu_0 \sqrt{r_T r_R} \int_0^\infty J_1\left(x\sqrt{\frac{r_T}{r_R}}\right) \cdot J_1\left(x\sqrt{\frac{r_R}{r_T}}\right)$$

$$\cdot J_0\left(x\frac{\Delta}{\sqrt{r_T r_R}}\right) \cdot e^{\left(-x\frac{d}{\sqrt{r_T r_R}}\right)} dx \qquad (18.22)$$

In this equation Δ is the side misalignment, r_T, r_R are the transmitting and receiving coils respectively, d is distance between the two coils and:

$$J(r) = \begin{cases} \dfrac{1}{\omega} f(r) & \text{when } \left(a - \dfrac{\omega}{2}\right) \leq r \leq \left(a + \dfrac{\omega}{2}\right) \\ 0 & \text{otherwise} \end{cases} \qquad (18.23)$$

where, $\int f(r) = \omega$ is the radial distribution [7].

However to improve the link quality and to decrease the sensitivity of link to misalignment, an approach has been proposed in [7]. In this method an array of spiral micro-coils are used at implanted device as the receiving antenna. However the transmitting unit only has one micro-coil to transmit the power to a multi-coil receiver. Figure 18.8 illustrates this configuration [7].

The following equation holds for the system shown in Figure (18.8) [7]:

$$V_1 = Z_1 I_1 + j\omega M I_2 + j\omega M I_3 + j\omega M I_4 + j\omega M I_5 \qquad (18.24)$$
$$0 = Z_i I_i + j\omega M I_i; \quad i = 1, 2, 3, 4, 5 \qquad (18.25)$$

Therefore the current in the transmitting circuit is:

$$|I_1| = \frac{|V_1|}{R_1 + \frac{(\omega M)^2}{Z_2} + \frac{(\omega M)^2}{Z_3} + \frac{(\omega M)^2}{Z_4} + \frac{(\omega M)^2}{Z_5}} \quad \text{where } (R_1 = Z_1)$$

$$(18.26)$$

However, when the two coils are coaxial they provide the maximum inductive coupling. Coupling coefficient and consequently the link quality decline since displacement happens. Therefore, as a result of misalignment, received

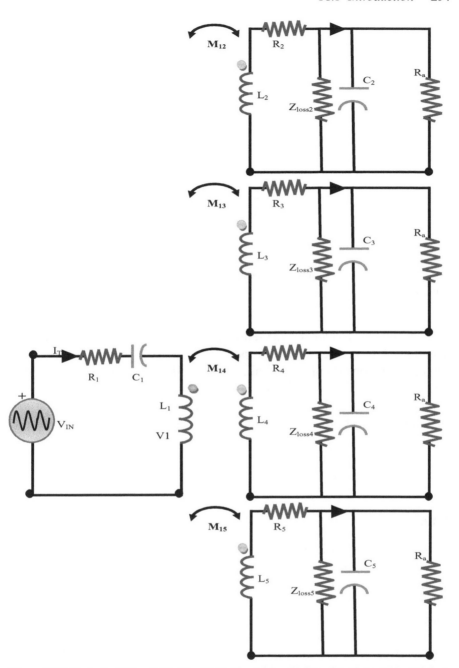

Figure 18.8 Equivalent Circuit Model of Inductive Link with Four Receivers (Adopted from [7])

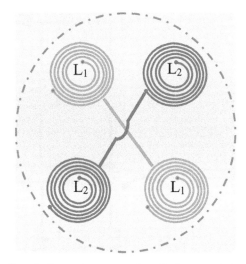

Figure 18.9 Schematic of diagonal connected coils in a multi-coil structure (adapted from [7]))

power decreases. To increase the sensitivity of the receiver, a favourable configuration of receiving coils is proposed in [7]. As shown in Figure 18.9, this topology consists of four similar spiral coils which are connected in a diagonal manner in series [7]. In case of displacement, while the received power increases at half of the receiving coil (L_1 or L_2), it decreases at the remaining half section [7]. Therefore the tolerance of inductive link between the external power transmission unit and the embedded receiving array of micro-coils to misalignment will be higher and less affected by both side and angular displacement [7].

18.1.6 Summary

In this chapter the basic principles of a wireless power transmission system is discussed. Different applications of WPT are introduced, such as charging the battery for electrical cars, WPT for industrial applications (robots, autonomous machines and etc) and WPT to power up implantable devices in medical applications. Then a simplified and generalized model for WPT is introduced. By discussing a wireless power transmission system for smart endoscopy devices, this concept has been studied in more depth. Different techniques have been discussed to improve the efficiency of wireless power transmission systems. Techniques such as multi-voltage output MI-WPT to provide different voltage levels (high and low voltage levels), for both stimulators and other analog circuits and digital blocks is introduced. It is also discussed that by using micro spiral coils and employing multiple coils at the receiver, efficiency will be enhanced and the receiver will be less sensitive to the side and angle misalignments.

References

[1] Zhen Ning Low; Chinga, R.A.; Tseng, R.; Jenshan Lin; "Design and Test of a High-Power High-Efficiency Loosely Coupled Planar Wireless Power Transfer System ", "Industrial Electronics" 2009 , issue 5, volume 56 pp. 1801–181.

[2] Jin-Ju Jang; Won-Yong Chae; Ho-Sung Kim; Dong-Gil Lee; Hee-Je Kim; "A Study on Optimization of the Wireless Power Transfer Using the HalfBridge Flyback Converter" , "Computer Research and Development", 2010, pp. 717–719.

[3] H. Mansor, M. A. A. Halim, M. Y. Mashor and M. A. Rahim; "Application on Wireless Power Transmission for Biomedical Implantable Organ", "IFMBE", 2008, pp. 40–43.

[4] Shinohara, N.; Kawasaki, S.; "Recent Wireless Power Transmission technologies in Japan for space solar power station/satellite", "Radio and Wireless Symposium", 2009, pp. 13–15.

[5] Scheible, G.; Schutz, J.; Apneseth, C.; "Novel wireless power supply system for wireless communication devices in industrial automation systems","IECON 02, Industrial Electronics Society", 2002, vol. 2, pp. 1358–1363.

[6] Imura, T.; Okabe, H.; Hori, Y.; "Basic experimental study on helical antennas of wireless power transfer for Electric Vehicles by using magnetic resonant couplings", "Vehicle Power and Propulsion Conference", 2009 , pp. 936–940.

[7] Mohamad Sawan, Saeid Hashemi, Mohamed Sehil, Falah Awwad and Mohamad Hajj-Hassan, et al; "Multicoils-based inductive links dedicated to power up implantable medical devices: modeling, design and experimental results", "Biomedical Microdevices", volume 11, no. 5, 2009, pp. 1059–1070.

[8] Lihsien Wu; Zhi Yang; Basham, E.; Wentai Liu; "An efficient wireless power link for high voltage retinal implant", "Biomedical Circuits and Systems Conference", 2008 , pp. 101–104.

[9] Guozheng Yan; Dongdong Ye; Peng Zan; Kundong Wang; Guanying Ma; "MicroRobot for Endoscope Based on Wireless Power Transfer", "Mechatronics and Automation", 2007 , pp. 3577–3581.

[10] Kiani, M.; Ghovanloo, M.; "A closed loop wireless power transmission system using a commercial RFID transceiver for biomedical applications", "Engineering in Medicine and Biology Society", 2009 , pp. 3841–3844.

[11] O.M.O. Gatous, J. Pissolato (2004). "Frequency-dependent skin-effect formulation for resistance and internal inductance of a solid cylindrical conductor", "IEEE Proc. Microwaves. Antennas", 2004.

19

NFC Applications (Part two)

This chapter looks into a detailed description of a few NFC application consisting of related formulas, block diagrams and circuit design. A general description in regard of these applications had been provided in Chapter 17 and is further extended in detail in this chapter.

19.1 Introduction

The chosen applications for discussion in the following secation include glucose, weight and blood pressure management, hearing aids, contactless payment and wireless power transfer.

19.1.1 Glucose, weight and blood pressure management

Nowadays, some blood pressure monitors and weights scales are capable of transferring data with other end devices. They have an internal memory in many of them which is used for the memory storage of these measurement devices. Due to the matter that these sort of devices are used by many persons in different families, sometimes it is even not possible to tell to whom these measurements belong as well as making them more difficult to use. Additionally, because of the measurement history of separate measurement devices their measurement history is spread over many places. Based on the aforementioned problems, the collection and analysis of data is somehow troublesome.

Via the use of NFC this problem can easily be solved by immediately transferring the measurement results to ones mobile terminal once the measurement has finished. Therefore, the health application can consist of the measurements of different sources which are then illustrated in the same user interface. The measured data can then be sent back to the information system in the back-end for medical use [1].

Co-authored by Samaneh Movassaghi & Agbinya JI, University of Technology, Sydney

J. Ihyeh Agbinya (PhD), Principles of Inductive Near Field Communications for Internet of Things, 301–320.
© 2011 *River Publishers. All rights reserved.*

The most typical sparse measurements for health management are glucose, weight and blood pressure. These results are normally documented manually for long-term monitoring. If the data management and electrical measurement visualizations are simple, patients are motivated to control and watch their situation by themselves which also enables a satisfactory cooperation between supervising medical specialists and patients.

In [2], the design of a device has been proposed for intraocular pressure monitoring. In this design, an external coil around an eyeglass produces a magnetic field which produces measurable changes of impedance in the external coil in response to variations on the pressure sensor. Fig. 19.1 shows the telemetry system by which pressure measurement is extracted in both types of sensors. As shown in the figure, the sensor is implanted between the orbital bones and the eyeball in a place where it can be fixed with easy surgical access. One external sensor is turned around an eyeglass and generates a magnetic field with lines which are perturbed by the sensor. As a consequence, the variations of pressure on the sensor cause measurable changes of impedance in the external coil [2].

In addition, a measurement subsystem is required to extract pressure data and store it in memory and a power subsystem is also required to complete the subsystem in terms of exciting the external coil in order to produce a magnetic field. However, the whole subsystem should be sufficiently small and provide portability with low power consumption.

The simplified diagram of this magnetic sensor is shown in Fig. 19.2 which consists of a using a sensing device capable of performing pressure, smart materials such as magnetostrictive materials and the variable which is to be directly measured, via a noticeable change in the average electromagnetive properties of the bulk sensor itself.

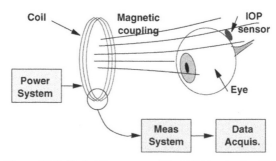

Figure 19.1 Basic Measurement for IOP Monitoring [2]

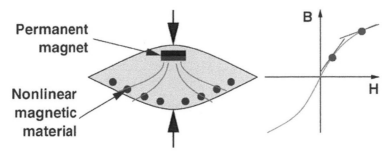

Figure 19.2 Simplified Diagram of a Magnetic Sensor [2]

These magnetic sensors can be illustrated via the circuit shown in Fig. 19.3. ΔL_1 expresses any increase in the magnetic permeability of the sensor. Additionally, the transformer circuit describes the eddy current effects. Due to simple calculations, the eddy effect is shown to be more dominant and the resistance variations show the loss to be dominant by the resistance component of the coil. The electric model for this magnetic sensor is shown in Fig. 19.3 which is equivalent to the circuit shown in Fig. 19.4.

The impedance of the primary coil is given in equation (19.1)

$$Z_{eq} = sL_1 \left(\frac{1 + sRC + s^2 L_2 C (1 - k^2)}{1 + sRC + s^2 L_2 C} \right) \tag{19.1}$$

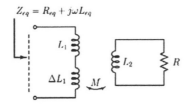

Figure 19.3 Electric Model for a Magnetic Sensor [2]

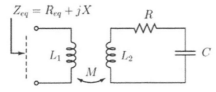

Figure 19.4 Equivalent Circuit for the Magnetic Sensor [2]

In case of small secondary losses: $R/L_2 \ll \omega_0$. Therefore, the input impedance will be as calculated in equation (19.2):

$$Z_{eq} = sL_1(1 - k^2)\frac{(s - z_1)(s - z_2)}{(s - p_1)(s - p_2)} \tag{19.2}$$

Which means, it has one zero at the origin and two conjugate complex zeros at:

$$z_{1,2} = -\frac{R}{L_2} \pm j(1 - k^2)\omega_0 \tag{19.3}$$

And a pair of conjugate complex poles at:

$$p_{1,2} = -\frac{R}{L_2} \pm j\omega_0 \tag{19.4}$$

Where $k = \frac{M^2}{L_1 L_2}$ and $\omega_0 = \frac{1}{\sqrt{L_2 C}}$

Based on these calculations, the input impedance has a peak at ω_0, independent to the magnetic coupling coefficient. In cases where the magnetic coupling coefficient becomes very small, imaginary zeros move towards the poles. In cases where the coupling coefficient is zero, there is a zero-pole cancellation, resulting on no loss in the circuit. A frequency peak will occur at a frequency dependent on the secondary losses and magnetic coupling coefficient.

The functional diagram proposed in [2] is made up of an analog front-end and a digital system. The analog part of the magnetic sensor prototype is shown in Fig. 19.5.

This circuit measures the real part of the impedence of the external coil and has been implemented via commercial integrated circuits. The major aim of this circuit is to measure the resonant frequency via isolating the real part

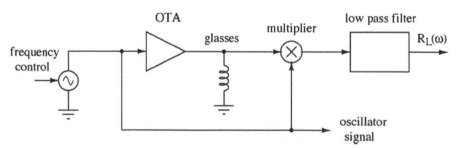

Figure 19.5 Function Diagram of Analog Selection of the Prototype [2]

of the external coil impedence. This is done via a voltage-controlled oscillator [(VCO), MAX038]. This voltage is then converted into a current via a transconductance amplifier in order to generate a voltage via the external coil [(OTA), OPA660].

Equation (19.5) provides the calculations for this voltage conversion.

$$V_L = A_0 R_L(\omega) \sin \omega_0 t + A_0 X_L(\omega) \cos \omega_0 t \tag{19.5}$$

The signal at the multiplier output (AD835) is then calculated via equation (19.6).

$$V_M = A_0^2 R_L(\omega) \frac{1 - \cos 2\omega_0 t}{2} + A_0^2 X_L(\omega) \frac{1 - \sin 2\omega_0 t}{2} \tag{19.6}$$

In the end, the average value of the multiplier output signal is extracted via the low pass filter which is given in equation (19.7).

$$V_{out} = \frac{A_0^2 R_L(\omega)}{2} \tag{19.7}$$

As can be seen, V_{out} is proportional to $R_L(\omega)$.

The block diagram for the digital part of the circuit is given is Fig. 19.6. Its main part is a microcontroller- the PIC16F676 and is in charge of doing various tasks such as controlling the frequency of the VCO, measuring the V_{out} of each frequency and finding its maximum and measuring the VCO frequency at the maximum found. The analog-digital (A/D) converter and the frequency meter are within the microcontroller and digital-analog converter (D/A) and the storage memory are external circuits.

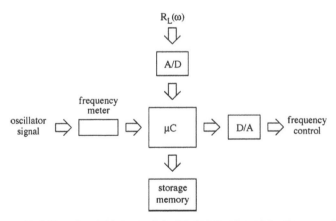

Figure 19.6 Functional Diagram of the Digital Section of the Prototype [2]

19.1.2 Hearing aids

A hearing aid is a device which increases the loudness of sounds in a user's ear. Its basic components are an earphone, a microphone and an amplifier. They are very much smaller, less conspicuous and fit within the ear canal or behind the earlobe. Some hearing aids have an automatic volume control which automatically varies the input amplification. A conventional acoustical hearing aid consists of two transducers. Each one, converting energy from one form to another. The flowchart in Fig. 19.7 describes the general steps by which most cochlear implants work. As can be seen, the input of the transducer is usually in the form of a diaphragm microphone which collects the sound waves on its diaphragm and then converts them to alternating current electrical signals. These are outputted from the input transducer and inputted to an electromagnetic coil setting up a magnetic field which changes in two directions. This magnetic field attracts and repels a permanent magnet attached to it alternately, and so repels the permanent magnet attached to which is also in attachment with the output diaphragm. This then results in the production of audible vibration via the output diaphragm that is a duplicate of the audible vibration impinging used in the input transducer at a much higher energy level which means it is done before the beginning the impinging over the behavior membrane amplification over the incoming sound waves. In a new invention for hearing aid devices, an electromagnetic transmitter which is operatively coupled to a remotely located receiver is surgically implanted on the ossicular chain without the need for interfering wires [3]. The disk preferably uses a magnetic material in the order of thickness of 10 microns with a positive magnetic coefficient of expansion which means it has positive

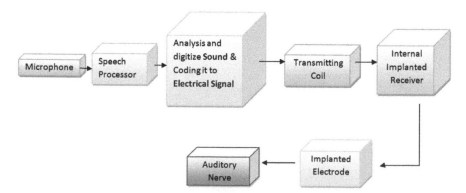

Figure 19.7 Flowchart for the Functionality of a Cochlear Implant (Adapted from [4])

magnetic features such as iron and in the case of a magnetic field it enlarges in direction. This is then attached to a second magnetic layer, with the same thickness of 10 microns however with negative magnet coefficient which shrinks when exposed to a magnetic field in planar direction.

The following components are used in most cochlear implants:

- **Microphone:**
  The microphone of a cochlear implant can be put up in a speech processor or a BTE unit. A directional microphone is capable to aid listening to speech in below adverse situations. This capability is important as it provides a more intelligible signal in noise to. Additionally, the placement and size of the microphone need to be effectively considered.
- **Speech processor:**
  The speech processor unit of a hearing aid needs to be durable, lightweight and as small as possible. However, in order to suit the needs of all cochlear implant wearers which also includes children; it should be simple to accept with large enough control.
- **Transmission Line:**
  There should be a great tolerance to movement in order to make the matter of coil alignment to be less critical. As so, the widest coaxial distance should easily be compatible with the most efficient transmission with the lowest power consumption.
- **Receiving Coil:**
  The number of independent transmission links, determines the number of coils for transcutaneous transmission. In cases were more than one electrode is required, each electrode must be provided with a separate transmission link.
- **Implanted electronics package:**
  For the electronics package to be placed in the mastoid bone of children and adults, it should be small in size and thin enough. Either Titanium or Ceramic can be used to increase implantable electronics.
- **Electrode array:**
  The electrode design needs to provide the following features: practical fabrication, traumatic insertion characteristic, biocompatibility and mechanical stability.
- **Batteries:**
  Cost, life and size are important design factors for a battery. As Ni-Cd batteries need to be totally discharged to achieve best recharging, the use of this type of battery has not been discouraging.

In [5], the design of implantable middle ear devices is fully discussed. This device uses transcutaneous magnetic induction in order to produce sound in a miniature speaker implanted in the mastoid region. A sound tube was then used to conduct the sound to the middle ear implant (MEI). Other contents use inductive coils or magnets in the middle ear to provide such an aim. Based on the experiments done in [6, 7], if direct energy coupling is provided the following advantages will be extended frequency response, improved transmission efficiency and elimination of airborne acoustic feedback. The basic components of the middle ear implant are as follows: amplifier, switch, battery, piezoelectric ceramic lead-titanate-zirconate vibrator and a microphone. In the case of the MEI, the ceramic vibrator is used in reverse to its previous design usage. By that we mean the excitation voltage to be converted to displacement. This design has been defined convenient for patients with bone conduction thresholds more than 20 dB hearing loss when there is a full implementation with an operational time of three years. However, for the case of partial implementations, candidates need to have bone conduction potentials of more than 30 dB. Fig 19.8 fully represents the place the where the MEI is implanted.

In Fig. 19.9 the output stage design of a hearing-aid is shown which is coupled to the receiver via an amplifier to achieve the largest output with a given "B" battery supply voltage.

In this circuit, the plate load of the output-tube V consists of a reciver of internal capacity C_R and a primary transformer T in parallel. The secondary winding of T is in parallel with a variable resistance of R_s.

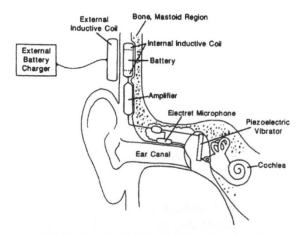

Figure 19.8 Implantable Middle Ear Piezoelectric Vibrator [5]

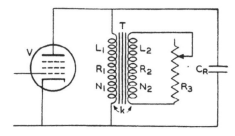

Figure 19.9 Circuit for Controlling Frequency Response [6]

If we consider the transformer by itself, the complex impedence looking on the primary side is given in equation (19.8).

$$Z_{12} = R_1 + \frac{(M\omega)^2 R_s}{R_s^2 + (L_s\omega)^2} + j\left[L_1\omega - \frac{(M\omega)^2 L_2\omega}{R_s^2 + (L_s\omega)^2}\right] = R_{12} + jL_2\omega$$

(19.8)

$R_s = R_2 + R_3$
R_{12}: *Apparent resistance of the primary*
L_{12}: *Apparent inductance of the primary*

$$k = \frac{M}{\sqrt{L_2 L_1}}$$

(19.9a)

$$N = \frac{N_1}{N_2} \approx \sqrt{L_1 L_2}$$

(19.9b)

Via the substitution of equations (19.9a) and (19.9b) in (19.8) equation (19.10) is obtained.

$$Z_{12} = R_1 + \frac{k^2 N^2 R_s}{1 + (R_s/L_2\omega)^2} + jL_1\omega\left[-\frac{k^2}{1 + (R_s/L_2\omega)^2}\right]$$

(19.10)

Via a turn ratio of N and a coupling coefficient of K, the apparent resistance looking in the primary is given in equation (19.11a).

$$R_{12} = R_1 + \frac{k^2 N^2 R_s}{1 + (R_s/L_2\omega)^2}$$

(19.11a)

Its apparent inductance is given in equation (19.11b). This equation is the basis of selective circuit discussion. In the case where the secondary winding T is short-circuited $R_S = R_2$. However, via a suitable transformer design,

$R_2/L_2\omega$ are smaller than unity.

$$L_{12} = L_1 \left[1 - \frac{k^2}{1 + (R_s/L_2\omega)^2} \right] \qquad (19.11b)$$

The highest resonating frequency of this circuit is calculated via equation (19.12).

$$f_h = \frac{1}{2\pi\sqrt{L_sC_R}} = \frac{1}{2\pi\sqrt{L_s(1-k^2)C_R}} \qquad (19.12)$$

The lowest resonating frequency is calculated via equation (19.13)

$$f_l = \frac{1}{2\pi\sqrt{L_1C_R}} = \frac{1}{2\pi\sqrt{L_1(1-k^2)C_R}} \qquad (19.13)$$

The ratio of these two frequencies is given in equation (19.14) which also provides a good coupling coeeficient. As a matter of which, the frequency response of the amplifier is capable of being controlled via R_S.

$$f_h/f_l = \frac{2\pi\sqrt{L_1C_R}}{2\pi\sqrt{L_1(1-k^2)C_R}} = \frac{1}{\sqrt{(1-k^2)}} \qquad (19.14)$$

Via this design, a particular way of selective amplification is provided for hearing aids which consists of the main part of hearing aids.

One other hearing aid device using magnetic induction uses electromagnetic transconductions of the sound at the oscillator chain. This device uses magnetic induction and permanently magnetic ossicular implants either in replacement to an ossicle or attachment to it. By direct coupling amplified to the middle ear, high fidelity in transconduction is achieved.

As can be seen in Fig. 19.10, a magnetic coil is externally coupled to a subcutaneous secondary coil. Additionally the magnetic ossicles are vibrated via a tertiary coil through which the current is created. There are magnets centrally placed in these coils are due to alignment and attachment purposes.

Based on the theoretical assumption in [8], no simple analytic approach can be done to estimate the force the electromagnetic transducer generates. However, considering the magnetic field on the axis of a coil with negligible wire diameter and vanishing length can provide some insight to the transducer. The magnetic field of a coil with negligible wire diameter and vanishing diameter gives some insight to the ehavior of the transducer. This

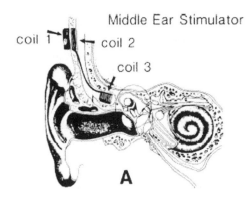

Figure 19.10 One Other Middle Ear Stimulator Design [5]

magnetic field is calculated with equation (19.15):

$$H = \frac{I \cdot N \cdot a^2}{8r^3}$$ (19.15)

H The magnetic field on the coil axis (A/m)
I current through the coil (A)
N number of turns in a coil
a diameter of the coil (in m)
d distance from the centre of the coil (m)
r distance to wire ($r^2 = d^2 + (a/2)^2$) (m)

In [8], it is noted that the coil should be as short as possible and its diameter should be optimized based on the intended gap between the permanent magnet and the coil. It is also important to note that the most important part of a hearing aid is its output transducer.

19.1.3 Contactless payment

The design of a contactless payment system consists of an embedded chip and antenna which provides consumers with the capability of waving their card over a reader at the point of sale. Some suppliers claim this type of transaction to be approximately twice as fast as a conventional debit card, cash, or credit purchase. It also does not require any signature for purchases of under US$25. However, Contactless payment has several limitations as they have no battery which makes them incapable of initializing the transaction and also having no keyboard or screen. In order to get over these limitations, a device with a screen, keyboard and batteries, such as the mobile phone provides a very interesting solution. For such a design, the NFC acronym for a mobile phone with contactless capability (ISO 18092 NFCIP-1 and ISO 21481 NFCIP-2 norms) and the RFID4 acronym which provides contactless cards technology (ISO 14 443 and ISO 7816 norms) are utilized [9].

NXP, a leading semiconductor company, sets four categories which can be compliant with NFC mobile phones. Its screen and keyboard allow the transaction confirmation to be done by accepting the transaction without any POS interaction or by simply entering a password. The contactless payment is made possible via the following two ways:

a) The first of which is a dual approach in which a NFC5 chip is used which includes a proper payment application and a separate chip which is the SIM is dedicated to the mobile usage (for sending and receiving SMS, MMS, call, . . .).

b) The second approach is a single chip approach with a SIM6 Card environment with payment application. In Fig. 19.11 such a configuration is shown. The NFC chip deals with RF exchange by the payment terminal. This configuration is made possible via merging a payment card and a SIM card in one. The end user with such an NFC mobile phone is capable of using his mobile phone for sending and receiving call, MMS and SMS and also launching contactless applications for payment similar to a means of payment such as a standard credit card.

Most NFC applications involve the usage of mobile phones. In these days the original communication aim of mobile phones have developed and evolved into portable multimedia systems. Multiple NFC applications such

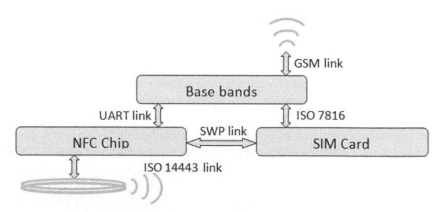

Figure 19.11 Mobile Phone with Single Chip (Adapted from [9])

as mobile payment, mobile ticketing, using electric keys, connecting to smart posters are done via mobile capabilities. In [10], an offline NFC based eVoucher payment system has been presented which uses the Nokia mobile phones, Nokia 6131 NFC and Nokia 6212 NFC. This highly secure system is capable of completion and performance of offline user-to-user transactions, meanwhile being able to consult their actual balance and other required task at anytime and anywhere free of cost as it does not require to be connected to an external server. However, a highly secure architecture has to be developed to achieve an offline payment system for mobile phones.

In addition, offline payments via NFC mobile phones are more likely to be based on phone rather than based on card. They also have no reliance or Mifare and provide the customers with the capability of offline transfer of value and provide offline payment support but no security determinations.

In Fig 19.12 the transaction steps between the actors are shown. The steps for such a transaction are explained as follows [11]:

1) First, eVouchers are received from the issuer. These eVouchers are SMS-based and provide an unconfirmed transaction between the user and the beneficiary.
2) In the next step, eVouchers are transferred from one device to another via sending one of the following:

   - A confirmation receipt via NFC.
   - An SMS from one phone to another.
   - An intermediate RFID tag.

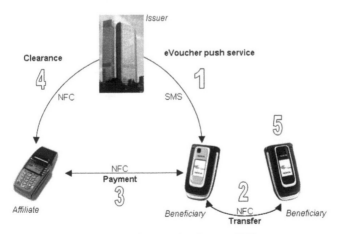

Figure 19.12 eVoucher System [11]

3) Payment: Next, the eVouchers are transferred from the beneficiary to the payment terminal (affiliate).
4) Clearance occurs between the issuer and the affiliate over a secured network.
5) eVouchers can be managed at all times for balance checking, retrieving the spreading history, expiration date checking of eVouchers.

As a contactless card has no any battery inside it, it utilizes electromagnetic induction for getting its power through when it is interacting with the reader. This received signal then goes through rectifying, filtering and regulation in order to supply the contactless card with a DC supply voltage. The communication interface between the contactless card and the reader is provided via an RF part which is shown in Fig 19.13.

The RF part is capable of demodulating the received signal and forwarding the data carried by the signal to the digital section later on. Via the same wireless coupling, the RF part is capable of modulating any outgoing signal which is then sent back to the reader. The RF interface circuit is shown in the Fig. 19.13. This circuit consists of the power on circuit (including filtering, rectifying and regulating circuits), reset circuit, modulation circuit and demodulation circuit.

The power on circuit of such a design is shown in Fig. 19.13. It is made up of the filtering, regulating and rectifying circuit. Signals get radiated via L_1 in the reader then after L_2 in the contactless card receives it. Via this circuit, the required DC voltage for all operations of the contactless card is derived. L_1 is the reader's antenna coil and L_2 is the contactless card's antenna coil.

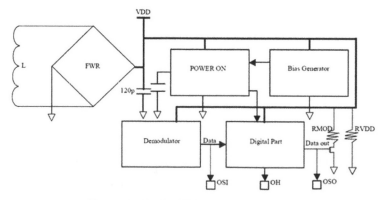

Figure 19.13 The RF Interface Circuit [12]

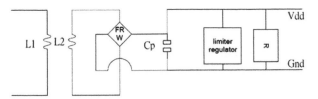

Figure 19.14 The Power on Circuit [12]

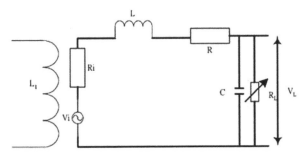

Figure 19.15 The Model of Antenna Circuit [12]

This RLC circuit shown in Fig. 19.15 is designed using the carrier frequency of 13.56 MHz and equations (19.16) to (19.19) with which the values of R, L and C are calculated.

$$\Omega = 2\pi f \qquad (19.16)$$

$$R = Z \cos \theta \qquad (19.17)$$

$$\omega L = Z \sin \theta \qquad (19.18)$$

$$\frac{1}{\omega C} = Z \sin \theta \qquad (19.19)$$

The voltage generated via this circuit is dependent on the distance between the card and the card reader which can be measured via a variable resistor across the capacitor also measuring the corresponding internal resistance.

Via this design, an asynchronous circuit technique has been used in the design of a contactless smartcard which has many advantages in terms of power consumption. This is due to the fact that contactless card remain inactive mostly until the data is to be processed which puts the asynchronous circuit into motion.

19.1.4 Wireless power transfer

Contactless Energy Transfer (CET) systems can be classified in two different categories which are **inductive coupling** and **magnetic coupling**. Inductive coupling allows power transfer from a few milliwatts to several kilowatts whereas capacitive coupling is utilized for low-power range. As an application toward NFC technology, magnetic coupling of CET systems are used.

In Fig. 19.16, the block diagram of the components in a wireless power transfer is given.

In [13] the design of a contactless energy transfer system using magnetic coupling has been proposed. The configuration of this method with Serial-to-Serial topology is considered. Fig.19.17 shows this configuration in more detail.

In the presented CET system, a rotatable transformer with adjustable air gap (up to 30 cm) is used. As can be seen, in the energy feeding input there are three phase diode rectifiers and a full IGBT converter. On the other side, a module R_0 with a bridge rectifier is presumed for the load. As in conventional applications, for the galvanic insulation between the source and the load, a transformer is used which provides a high magnetic coupling factor in between primary and secondary windings. Because of the air gap in between the two halve cores, the CET transformers operate under a much less magnetic coupling factor. Consequently, the main inductance is very small; however leakage inductances such as (L_{11}, L_{22}) are very large. Increase in the magnetizing current results in higher conduction losses due to large leakage inductance. One other problem related to the large gap is its strong radiation. Several power conversion topologies have been introduced to overcome these issues which are the resonant, quasi-resonant, flyback and self resonant system. The two methods proposed for inductance compensation are serial

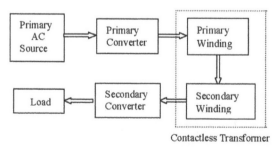

Figure 19.16 Block Diagram of Components in a Wireless Power System [12]

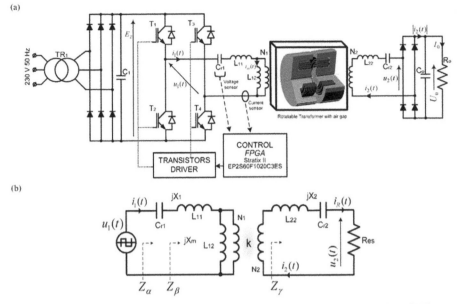

Figure 19.17 Configuration Design of a CET with SS Compensation Topology [13]

and parallel configurations giving: SP (series-parallel), SS (series-series), PS (parallel-series), PP (parallel-parallel).

If we consider the same number of windings for the primary and secondary ($N_1 = N_2$), we have the following equations:

$$L_1 = L_{11} + L_{22} \tag{19.20}$$

$$L_1 = L_{22} + n^2 L_{12} \tag{19.21}$$

$$M = n L_{12} \tag{19.22}$$

$$L_{11} = L_{22} = L - M = \frac{1 - k^2}{k} M, \quad k = M/L \tag{19.23}$$

The turns ratio is given by the expression

$$n = \frac{N_2}{N_1} \tag{19.24}$$

$$R_{es} = \frac{8}{\pi^2} R_0 = 0.8106 . R_0 \tag{19.25}$$

For SS compensation:

$$Z_\gamma = R_{es} + j X_2 \tag{19.26}$$

For SP compensation:

$$Z_\gamma = \left[j\omega L_{11} + \cfrac{1}{\cfrac{1}{R_{es}} + j\omega C_{r2}} \right] \tag{19.27}$$

$$Z_\beta = \frac{j X_m Z_\gamma}{j X_m + Z_\gamma} \tag{19.28}$$

$$Z_\alpha = j X_1 + Z_\beta \tag{19.29}$$

The impedance and reactance of the components shown in Fig. 19.32b are given in equations (19.25) to (19.31) where $\omega_s = 2\pi f_s$.

$$X_1 = \omega_s L_{11} - \frac{1}{\omega_s C_{r1}} \tag{19.30}$$

$$X_2 = \omega_s L_{22} - \frac{1}{\omega_s C_{r2}} \tag{19.31}$$

$$X_m = \omega_s M \tag{19.32}$$

The SS compensation topology of the CET system has a voltage transfer function is given in equation (19.35):

$$G_v = \frac{u_1}{u_2} = \left| \frac{Z_\beta R_{es}}{Z_\alpha Z_\gamma} \right| \tag{19.33}$$

The voltage transfer function is given as follows using equations (19.25) to (19.35).

$$G_v = \left[\left(1 + \frac{X_1}{X_m} \right)^2 + \left(\frac{X_1 + X_2 + \dfrac{X_1 X_2}{X_m}}{R_{es}} \right)^2 \right]^{-\frac{1}{2}} \tag{19.34}$$

Even with large value of the leakage inductances namely L_{11} and L_{22}, G_v is unity at the resonance (compensated) frequency which is $\omega_0 = 2\pi f_0$ for when $X_1 = X_2 = 0$.

$$\omega_0 = \frac{1}{\sqrt{L_r C_r}} = \frac{1}{\sqrt{L_{11} C_{r1}}} = \frac{1}{\sqrt{L_{22} C_{r2}}} \tag{19.35}$$

X_1 and X_2 can be rewritten as follows with the normalized frequency $\omega = \frac{\omega_s}{\omega_0}$:

$$X_1 = \omega_s L_{11} \left(1 - \frac{1}{\omega^2}\right) \tag{19.36}$$

$$X_2 = \omega_s L_{22} \left(1 - \frac{1}{\omega^2}\right) \tag{19.37}$$

In cases where the coupling factor varies in the operation, the circuit quality factor is given in equation (19.40) which is based on a voltage transfer given in (19. 39):

$$Q_{ac} = \frac{\omega \left(L_{11} + L_{22}\right)}{R_{es}} = \frac{\omega L_r}{R_{es}} \tag{19.38}$$

$$G_v = \left[\left(1 + \frac{1-k}{k}\left(1 - \frac{1}{\omega^2}\right)\right)^2 + \left(Q_{ac}\left(\omega - \frac{1}{\omega}\right)\right. \right.$$
$$\left. \left. \times \left(1 + \frac{1-k}{2k}\left(1 - \frac{1}{\omega^2}\right)\right)\right)^2 \right] \tag{19.39}$$

Via this design, a contactless power supply system is presented with an inductive CET as its core, operating at 60 kHz frequency. It is also important to note that this design is applicable to a wide class of contactless power supply with core or coreless transformers.

References

[1] Strommer, E., Kaartinen, J., Parkka,J., Ylisaukko-oja and Korhonen, I. (2006). "Application of Near Field Communication for Health Monitoring in Daily Life" 28th Annual International Conference of the IEEE Engineering in Medicine and Biology Society, 2006. EMBS '06.

[2] Lizon-Martinez, S., R. Giannetti, et al. (2005). "Design of a system for continuous intraocular pressure monitoring.", IEEE Transactions on Instrumentation and Measurement, **54**(4): 1534–1540.

[3] Miller, G.W., Janning, E.A., Janning, J.L. "Electromagnetic induction hearing aid device", United States Patent 5338287.

[4] Ylinen, J., M. Koskela, Iso-Anttila,L., Loula, P. (2009). "Near Field Communication Network Services". Third International Conference on Digital Society, 2009. ICDS '09.

[5] Dormer, K., Phillips, J. M.A. "Auditory Prostheses: Implantable and Vibrotactile Devices", Engineering in Medicine and Biology Magazine, IEEE, Vol. 6, Issue 2, pp. 36–41.

[6] Penn, W. D. "Fundmentals of Hearing-Aid Design", *Transactions journal*, Oct. 1944, vol. 63.

[7] Suzukia, J., Koderaa, K., Yanagiharab, N., "Middle Ear Implant for Humans", Acta Oto-Laryngologica, Volume 99, Issue 3 & 4 March 1985, pp. 313–7.

[8] Hough J, Vernon J, Dormer K, et al: "Experiences with implantable hearing devices and a presentation of a new device". Ann Otol Rhinol Laryngol 1986; 95: 60–5.

[9] Kompis, M., C. Kuhn, et al. (1998). "Design considerations for a contactless electromagnetic transducer for implantable hearing aids" Engineering in Medicine and Biology Society, 1998. Proceedings of the 20th Annual International Conference of the IEEE.

[10] Pasquet, M. Reynaud, J., Rosenberger, C. "Payment with mobile NFC phones: How to analyse the security problems", The International Symposium on Collaborative Technologies and Systems (CTS), 2008.

[11] Penn, W. D. (1944). "Fundamentals of Hearing-Aid Design." Transactions of the American Institute of Electrical Engineers, vol. 63, issue 10, pp. 744–49.

[12] G.V. Damme, K. Wouters, H. Karahan, B. Preneel, "Offline NFC Payments with Electronic Vouchers", Proceedings of the 1st ACM workshop on Networking, systems, and applications for mobile handhelds, Barcelona, Spain, 2009, pp. 25–30.

[13] Pui-Lam Siu; Chiu-Sing Choy; Chan, C.F.; Kong-Pan Pun; (2003). "A contactless smart-card designed with asynchronous circuit technique". Proceedings of the 29th European on Solid-State Circuits Conference, 2003. ESSCIRC '03.

[14] Moradewicz, A. J. and M. P. Kazmierkowski "Contactless Energy Transfer System With FPGA-Controlled Resonant Converter.", IEEE Transactions on Industrial Electronics **57**(9): 3181–3190.

20

NFMIC Simulator

Since simulating the wireless networks, using codes and so-called *computer language* is a challenging task and requires a lot of time and effort, a graphical user interface is developed by the authors, which enables the user to simulate different scenarios for an NFMIC system. This program evaluates NFMIC-link budget, based on *Agbinya-Masihpour* channel model, using MATLAB software.

20.1 Introduction

Network planning engineers and antenna designers can benefit from NFMIC-simulator by evaluating signal power at the receiver and communication range against different system parameters such as transmitter and receiver coil efficiency, quality factor, permeability, radius as well as the transmission power.

20.1.1 NFMIC-simulator

20.1.1.1 Home page
Firstly by running the program, the home page window as shown in Figure (20.1) appears. The home page is divided into two sections; the left column shows three push buttons, each operates a specific function while the right column, which can be seen in yellow, shows the brief explanation of the function of each push button. Therefore user can easily find the part of program that suits his/her intent. By clicking on either of push buttons, a new window will appear which their tasks are as follows:

20.1.1.2 First push button: "comparison between antennas using air-cored and ferrite-cored coils"
By clicking on the first push button, a window as shown in Figure (20.2) opens, which has two spots to show different axis for demonstrating the

Mehrnoush Masihpour (University of Technology, Sydney created the matlab codes) & co-authored the chapter with Agbinya JI

J. Ihyeh Agbinya (PhD), Principles of Inductive Near Field Communications for Internet of Things, 321–336.

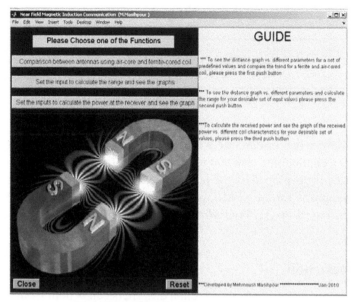

Figure 20.1 NFMIC-Simulator Home Page

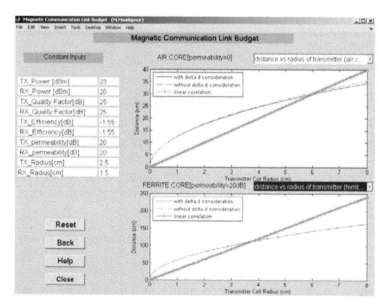

Figure 20.2 Air and Ferrite-Cored Coil Comparison

achievable distance against a range of antenna coil parameters that can impact the communication range, for two cases of using air and ferrite-cored antenna coils. Easily by choosing one of the desirable graphs from popup menu for each case, two graphs can be seen and compared for analysis. This part shows how the ferrite can improve communication range. Assumed values for different parameters are shown in the left column of window. In each graph, there are two curves, which show the communication range when considering Δd or neglecting it at the *Agbinya-Masihpour* link budget equation, therefore the difference between these two cases can be seen on the axis. There is a '*Reset*' button in this window, which can clear the axis.

20.1.1.3 Second push button: "set the values to calculate the range and see the graphs"

The second simulator, as shown in Figure (20.3) is even more dynamic and useful for network planning. Using this interface, each parameter's value is determined by the user at the edit boxes in the upper left column and simply by pushing one of the radio buttons the communication range against the desirable coil parameter on the axis for both ferrite-cored and air-cored antenna coil can be seen. To return the values to default and clear the axis,

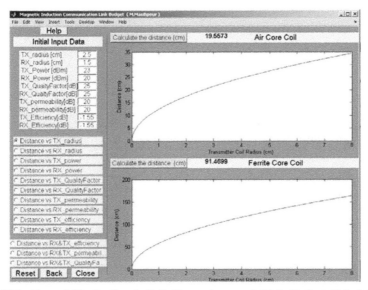

Figure 20.3 Air and Ferrite-Cored Coil Comparison and Link Budget Calculation for Desirable Input Values

'*Reset*' button needs to be pushed. Moreover, by clicking on '*Calculate the Distance (cm)*' button, the value of distance for entered values of each parameter will appear in front of the buttons for either case.

20.1.1.4 Third push button: "set the values to calculate the power at the receiver and see the graphs"

This part of the software is useful for estimation of received signal power for different scenarios. As it can be seen form Figure (20.4), there is a column at the upper left side of the window, indicating the input data in which default value can be changed to the desirable values as it is required for your network planning or antenna design. Then by choosing one of the radio button underneath, received power graph against the required parameters appears on the axis. Calculation of the power at receiver for your set of values also would be shown on the window by clicking on the push button below the axis. Calculated power can be seen in both units of dBm and mW. '*Reset*' button returns the default values back to the edit boxes, and program can be run for another set of values for system parameters.

However there is a '*Back*' button in each window, which returns user to the home page. '*Close*' button in each window will close NFMIC-simulator.

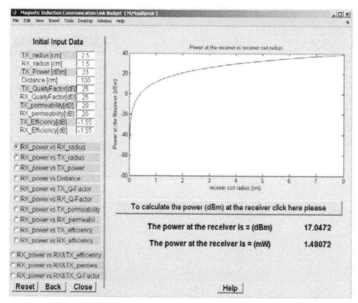

Figure 20.4 Received Power Calculation and Graphs vs. Antenna Coil Components for Inserted Input Data

To find out how each part of the program operates, simply click on the '*Help*' Button.

20.1.2 Deployment method

In this section, Matlab codes are provided to describe how NFMIC-Simulator is developed.

20.1.2.1 User interface

Each Matlab GUI has a set of codes that creates it. However, a GUI often shows a series of results such as graphs or numerical calculation of different functions. Therefore at least two set of codes are required to provide a GUI illustrating results. Each set of codes are required to be saved as an individual Matlab document ('.m' file). For instant a document is needed to create a GUI window and another document would contain the numerical calculations. Although the two documents can be combined as a single document, to prevent complexity and provide modularity it is preferred to create them in separated '.m' files. In fact by using different objects such as 'popup menu', 'push button' or 'radio button', the GUI refers to a different document to run a set of codes as required. In this process, selecting an option in a popup menu or clicking on a push button, a series of codes related to the GUI will run and calls a specific program to run its associated codes and show the result on GUI window. In this section, it is described how to create different objects within a GUI window and how to relate them to a desirable set of Matlab codes.

To develop a GUI window, the first step is to create a blank user interface using the following codes:

```
% Figure to create the user interface
figure1 = figure('Color',[0.2 0.2 0.4],'Name',...
'Near Field Magnetic Induction Communication_(C)Copyright Reserved
(University of Technology, Sydney, Australia)__Mehrnoush
Masihpour',...
'Position',[25 50 900 650]);
```

This set of codes creates a blank window in position [25 50 900 650]. The first number indicated the distance between the bottom of the computer screen and the GUI window. The second number is the distance difference between the left hand side of the screen and the GUI window. The third and the last number define the length and width of GUI window. To create a desirable colour for the GUI window, the numbers in front of the word 'color' need to be changed ('Color',[0.2 0.2 0.4],). The value needs to be between 0 and 1. However to name your software, you can type a desirable name for window in front of the word 'name' ('Name', 'Near Field Magnetic Induction Communication_(C)Copyright Reserved (University of Technology, Sydney, Australia)__Mehrnoush Masihpour'). This name is displayed on the top of the window.

Different object such as a text box, edit box and a lot more also can be defined in a window using a command called 'unicontrol'. Different characteristics of the object such as the object type, font size, font style, position, colour and etc are defined within the 'unicontrol' function, which are described in the next section for developing different objects.

20.1.2.2 Text box
'Unicontrol' function can be used to create a text box in GUI window using the following command:

```
% Unicontrol function to create a text box
Text_11=uicontrol(gcf,'Style','text','Position',[505 580 395 70],...
'String','GUIDE','FontSize',24,...
'HorizontalAlignment','center','BackgroundColor',[0.99 0.89 0.2]);
```

Firstly a name needs to be allocated to the specific function, which is 'Text-11' in the above example. To specify the style of your object the world 'text' needs to come after 'Style'. Also other styles can be defined by changing the word in front of the 'Style'. For instance to create a radio button the word 'radiobutton' should be replaced by the key word 'text'.

The followings are key words to define the style of your object:

- *'pushbutton': to create a push button*
- *'popupmenu': to create a popup menu*
- *'edit': to create edit box*

However the position in which the object is required to be displayed is defined by specifying four digits in brackets right after 'position'. As mentioned earlier the first and second digit shows the distance of object from length and width of GUI page respectively. The following two numbers are length and width of the object box itself.

To display a constant text on the GUI window, the text needs to be inside a pair of apostrophes after the keyword 'string' separating by a comma. In this example, the word 'GUIDE' will be displayed at location [505 580 395 70]. The font size also can be defined in front of the keyword 'FontSize'. However, the alignment of text also can be specified using keyword

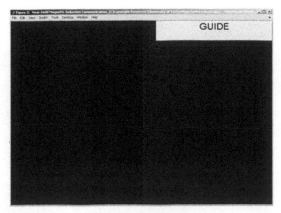

Figure 20.5 Text Box

'HorizontAlalignment'. It can be defined as 'centre', 'right' or 'left'. The same structure as mentioned in the previous section for defining GUI window colour can be applied to change the colour of text box background using keyword 'backgroundColor'. Therefore using the codes provided in this section and previous section creates a window as shown in Figure (20.5).

20.1.2.3 Image insertion

Images can be shown in a GUI window too. To insert an image into GUI window, 'subplot' function is used. However the function requires a name, which in this example is 'image1'. To show a picture in the window, the first step is to define a position for that image. Position is defined using four digits which are slightly different from the position definition in 'unicontrol' function. In this case, numbers should be a value between 0 and 1. This numbers are normalized values showing the distance of image from bottom and left hand side of GUI, length and width of the image respectively. For instance the value 0.56 in the third position means that the length of picture should be 56% the length of window. Similarly value 0.06 in the second position means the bottom of image should be located at a spot 6% of the width of GUI window, above the bottom of page. When position is defined, using another function: "image (imread('name of the image file', 'type of image file'))" the image can be shown in the window. This function loads a defined image into the GUI window. However command "axis off" is used to hide the default values on each axis. Therefore adding image1 function to the previous codes, a window as shown in Figure (20.6) will be deployed.

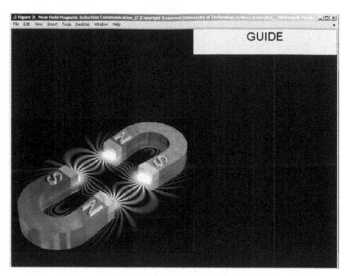

Figure 20.6 Image Insertion

```
% subplot function to insert an image into the GUI window
image1= subplot('Position',[0.0    0.06    0.56    0.56]);
image(imread('chain','jpg'));
axis off;
```

20.1.2.4 Push button

```
% Unicontrol function to create a push button)
p5 = uicontrol(gcf,'Style','pushbutton','Position',[260 290 200 25],...
    'String','Calculate the distance  (cm)','FontName','Arial',...
    'FontSize',12,'FontWeight','normal',...
    'HorizontalAlignment','center','CallBack','x=3; minoo');
```

'Unicontrol' function also can be used to create pushbuttons, simply by using the keyword 'pushbutton' after 'Style'. The following codes crate a push button at position [260 290 200 25], which includes the text 'calculate the distance (cm)' using font style/size Arial/12 with a normal weighting font at the centre of the push button. However, there is a new attributes added to the 'Unicintrol' function which is 'CallBack'. In fact what this command does is to refer to a Matlab document called "minoo.m" and it runs a part of the program where "$x = 3$". 'x' is a variable defined in an "if/else" function within a set of codes in minoo.m file, which are provided here:

```
% DISTANCE CALCULATION FOR FERRITE CORED COIL**************
elseif x==3
        d11=10.^((RXmu+TXmu+TXp+RXQ+TXQ+TXeff+RXeff+30.
            *log10(RXr.*TXr)-RXp)./60).*1.5;
        d22=10.^(-0.5.*log10(1+(TXr./d11).^2));
        distancef= d11.*d22;

        pText2=uicontrol(gcf,'Style','text','Position',
                [460 290 150 25],...
        'String',distancef,'FontSize',12,'FontWeight','bold',...
        'HorizontalAlignment','center','FontName','Arial',
        'BackgroundColor',[1 1 .5]);
end
```

The last part of above codes (pText2) will create a text box in position [460 290 150 25] but instead of a constant text, it returns the value of variable 'distancef' at the defined position. This value changes as the input values are modified. However all the inputs can be edited in the GUI window using 'edit' command, which is described in section (20.1.2.5).

However, more than one condition can be defined in apostrophes following the 'CallBack' command. The conditions need to be separated by a semicolon. For instance if it is required to run a part of codes (minoo.m) in which $x = 3$ and $y = 2$ the following codes can be used:

```
% Unicontrol function to create a push button)
p5 = uicontrol(gcf,'Style','pushbutton','Position',[260 290 200 25],...
        'String','Calculate the distance (cm)','FontName','Arial',...
        'FontSize',12,'FontWeight','normal',...
        'HorizontalAlignment','center','CallBack','x=3; y=2; minoo');
```

20.1.2.5 Edit box
To be able to define an input value of a variable on GUI window, 'Unicontrol' function is used. However, the style should be defined as edit and 'string' is not for defining texts any more but is a variable that can be changed by user. The set of codes in example below creates an edit box in position [175 570 50 20] which gets the inserted value and assign it to variable 'TXr' by using the function: 'Callback','TXr=str2num(get(Edit_1,''String''));minoo'. In fact it refers to the file "minoo.m" and assigns the defined value to the variable 'TXr' in that specific file. However, initial values have to be allocated to variable in the document consists of GUI codes, not the document that includes the mathematical calculation (minoo.m).

```
% Unicontrol function to create an edit box
Edit_1=uicontrol(gcf,'Style','edit','Position',[175 570 50 20],...
    'String',TXr,'FontSize',12,'BackgroundColor',[1 1 1],'Callback',...
    'TXr=str2num(get(Edit_1,''String''));minoo');
```

20.1.2.6 Pup up menu
```
% Unicontrol function to create a popup menu
sr=1;
pPopupmenuf=uicontrol(gcf,'Style','popupmenu','Position',
            [600 260 300 25],...
'String',{'distance vs radius of transmitter (ferrite core)',
         'distance vs radius of receiver (ferrite core)',...
    'distance vs transmission power (ferrite core)'
}......
,'value',sr,...
    'BackgroundColor',[1 1 1],'ForeGroundColor',[0 0 0],...
    'FontSize',12,'FontWeight','normal',...
    'HorizontalAlignment','center','FontName','Arial',...
    'CallBack','sr=get(pPopupmenuf,''value'');f=2; air');
```

The above set of codes shows how 'unicontrol' function can be set to create a popup menu. After 'style' and 'position' are defined, Matlab creates a popup menu at that specific spot on GUI window. When the arrow button of popup menu is clicked, it shows all the texts which are defined at 'string' command. Using a comma between the texts inside apostrophes, it can be defined that the first text is shown as the first option in menu and the following text is the second option and so on. However, to perform a specific function based on each option, 'callback' command is used: 'CallBack','sr=get(pPopupmenuf, ''value'');f=2; air');.

In this example, selecting either of available options in the menu, it sets the value of 'sr' first and refers to 'air.m' file where '$f = 2$':

```
%part of codes within the air.m file when f=2
elseif f==2
    subplot('Position',[0.38 0.06 0.6 0.33]); axis on;
    if sr==1
%Distance as a function of transmitter coil radius (ferrite core coil)
Ptx_dBm=23; % power at the transmitter {dBm}
Prx_dBm=20; % power at the receiver {dBm}
Qtx_dB=25;  % Quality factor of the transmitter {dB}
Qrx_dB=25; % Quality factor of the receiver {dB})
EFICIENCYtx_dB=-1.55;  %Efficiency of the transmitter {dB}
EFICIENCYrx_dB=-1.55;  %Efficiency of the receiver {dB}
MU_rx=20;  %permeability of receiver ferrite core coil in dB
```

```
MU_tx=20; %permeability of transmitter ferrite core coil in dB
radius_tx_cm=(0:0.005:8);        % radius of the transmitter coil {cm}
radius_rx_cm =1.5;% radius of the receiver coil {cm}
distance_prime=10.^((Ptx_dBm+Qtx_dB+Qrx_dB+EFICIENCYtx_dB
                +EFICIENCYrx_dB+30.*log10(radius_rx_cm.
                *radius_tx_cm)-Prx_dBm+MU_tx+MU_rx)./60).*1.5;
distance_delta=10.^(-0.5.*log10(1+(radius_tx_cm./distance_prime).^2));
distance_cm= distance_prime.*distance_delta;
line1=0:0.05:8;
line2=30.*line1;
plot(radius_tx_cm,distance_cm,'-',radius_tx_cm,distance_prime,
 '--',line1,line2,'-+'); %plot of distance as a function of transmitter
 ciol radius
xlabel('Transmitter Coil Radius (cm)')
ylabel('Distance (cm)')
legend('with delta d consideration','without delta d consideration',
 'linear correlation',2)

    end

if sr==2
    %Distance as a function of receiver coil radius(ferrite core coil)
Ptx_dBm=23; % power at the transmitter {dBm}
Prx_dBm=20; % power at the receiver {dBm}
Qtx_dB=25;  % Quality factor of the transmitter {dB}
Qrx_dB=25;  % Quality factor of the receiver {dB})
EFICIENCYtx_dB=-1.55;  %Efficiency of the transmitter {dB}
EFICIENCYrx_dB=-1.55;  %Efficiency of the receiver {dB}
MU_rx=20;  %permeability of reeciver ferrite core coil in dB
MU_tx=20; %permeability of transmitter ferrite core coil in dB
radius_tx_cm=2.5;        % radius of the transmitter coil {cm}
radius_rx_cm=(0:0.005:7);% radius of the receiver coil {cm}
distance_prime=10.^((Ptx_dBm+Qtx_dB+Qrx_dB+EFICIENCYtx_dB
                +EFICIENCYrx_dB+30.*log10(radius_rx_cm.*radius_tx_cm)
                -Prx_dBm+MU_tx+MU_rx)./60).*1.5;
distance_delta=10.^(-0.5.*log10(1+(radius_tx_cm./distance_prime).^2));
distance_cm= distance_prime.*distance_delta;
line1=0:0.05:7;
line2=30.*line1;
plot(radius_rx_cm,distance_cm,'-',radius_rx_cm,distance_prime,'--',
 line1,line2,'-+') %plot of distance as a function of receiver
  ciol radius
%text(4.5,124,'(X=6.2,Y=124),\rightarrow')
xlabel('Receiver Coil Radius (cm)')
ylabel('Distance (cm)')
```

```
legend('with delta d consideration','without delta d consideration',
 'linear correlation',2)

end
if sr==3

%Distance as a function of transmission power (ferrite core coil)
Prx_dBm=20; % power at the transmitter {dBm}
Ptx_dBm=(0.001:0.05:40); % power at the receiver {dBm}
Qrx_dB=25;  % Quality factor of the transmitter {dB}
Qtx_dB=25;  % Quality factor of the receiver {dB})
EFICIENCYrx_dB=-1.55;  %Efficiency of the transmitter {dB}
EFICIENCYtx_dB=-1.55;  %Efficiency of the receiver {dB}
MU_rx=20;  %permeability of receiver ferrite core coil in dB
MU_tx=20;  %permeability of transmitter ferrite core coil in dB
radius_tx_cm=2.5 ;   % radius of the transmitter coil {cm}
radius_rx_cm =1.5;% radius of the receiver coil {cm}
distance_prime=10.^((Ptx_dBm+Qtx_dB+Qrx_dB+EFICIENCYtx_dB
                +EFICIENCYrx_dB+30.*log10(radius_rx_cm.*radius_tx_cm)
                -Prx_dBm+MU_tx+MU_rx)./60).*1.5;
distance_delta=10.^(-0.5.*log10(1+(radius_tx_cm./distance_prime).^2));
distance_cm= distance_prime.*distance_delta;
plot(Ptx_dBm,distance_cm,'-',Ptx_dBm,distance_prime,'--')
 %plot of distance as a function of transmission power
xlabel('Transmission Power(dBm)')
ylabel('Distance (cm)')

legend('with delta d consideration','without delta d consideration',2)


end
end
```

'sr' is the variable that identifies which option (first, second, third and...) is selected from the menu. For instance when 'sr = 1' it implies that the first option is selected. Therefore the part of program in which '$f = 2$' and 'sr = 1' will be performed. In this example, when '$f = 2$' and '$sr = 1$', a graph will appear showing the communication distance versus transmitting coil radius at position [0.38 0.06 0.6 0.33] as defined in 'subplot' function.

20.1.2.7 Radio button
By defining the 'style' a 'radiobutton' in 'unicontrol' function, a radio button at defined position is created. Each radio button can have two values of one and zero. The value '1', indicates that the radio button is *on* whilst the value

'0' turns *off* the radio button. These values are defined using two keywords 'min' and 'max'. However, to perform a function by turning a radio button on, the relevant '.m' file has to be called through 'callback' command.

```
% Unicontrol function to create a radio button
option=1;

radiobutton1 = uicontrol(gcf,'Style','radiobutton','Value',0,'Position',
              [10 350 230 20] ,'Min',0,'Max',1,'BackgroundColor',
              [1 1 0.4],'String','Distance vs TX_radius','FontName',
              'Arial','FontSize',12,'FontWeight','normal',
              'HorizontalAlignment','left','CallBack',
              'x=1; Option=1; minoo');

radiobutton2 = uicontrol(gcf,'Style','radiobutton','Value',0,'Position',
              [10 325 230 20], 'Min',0,'Max',1,'BackgroundColor',
              [1 1 0.4],'String','Distance vs RX_radius','FontName',
              'Arial','FontSize',12,'FontWeight','normal',
              'HorizontalAlignment','left','CallBack',
              'x=1; Option=2; minoo');
```

In the above Matlab codes, two radio buttons are created. By clicking on the first radio button, it refers to minoo.m file where $x = 1$ and option $= 1$ (see below). When option is '1', meaning that the first radio button is on, all other radio buttons should be off using the commands: set (radiobutton1, 'Value', 1)and set (radiobutton2, 'Value', 0);

Similarly, when the second button is on the other one should be off using the same command as mentioned earlier but assigning different values to the radio buttons by their specific names. In this example when the first button is activated, Matlab plots the graph of distance versus transmitting coil radius. However, by clicking on the second pushbutton, a graph showing distance versus receiving coil radius will be plotted at the specified position.

```
% the part of codes within the minoo.m file when x=1 and option is 1
   and 2.
if x==1
    if Option==1
        set (radiobutton1, 'Value',1);
        set (radiobutton2, 'Value',0);

        r1=0.0001:0.005:8;        %range of transmitter coil radius

        d1=10.^((TXp+TXQ+RXQ+TXeff+RXeff+30.*log10(RXr.*r1)-RXp)./60).
           *1.5;
```

```
        d2=10.^(-0.5.*log10(1+(r1./d1).^2));
        da=d1.*d2;
        subplot('Position',[0.35 0.55 0.63 0.35]); axis on;
        plot(r1,da);
        xlabel('Transmitter Coil Radius (cm)')
        ylabel('Distance (cm)')
      d11=10.^((RXmu+TXmu+TXp+TXQ+RXQ+TXeff+RXeff+30.*log10(RXr.*r1)
            - RXp)./60).*1.5;
        d22=10.^(-0.5.*log10(1+(r1./d11).^2));
        df= d11.*d22;
        subplot('Position',[0.35 0.05 0.63 0.35]); axis on;
        plot(r1,df);
        xlabel('Transmitter Coil Radius (cm)')
        ylabel('Distance (cm)')


elseif Option==2
        set (radiobutton1, 'Value',0);
        set (radiobutton2, 'Value',1);
        r2=0.0001:0.005:7;      %range of receiver coil radius
        d1=10.^((TXp+TXQ+RXQ+TXeff+RXeff+30.*log10(r2.*TXr)-RXp)./60).
            *1.5;
        d2=10.^(-0.5.*log10(1+(TXr./d1).^2));
        da= d1.*d2;
        subplot('Position',[0.35 0.55 0.63 0.35]); axis on;
        plot(r2,da);
        xlabel('Receiver Coil Radius (cm)')
        ylabel('Distance (cm)')

        d11=10.^((RXmu+TXmu+TXp+TXQ+RXQ+TXeff+RXeff+30.*log10(r2.*TXr)
            -RXp)./60).*1.5;
        d22=10.^(-0.5.*log10(1+(TXr./d11).^2));
        df= d11.*d22;
        subplot('Position',[0.35 0.05 0.63 0.35]); axis on;
        plot(r2,df);
        xlabel('Receiver Coil Radius (cm)')
        ylabel('Distance (cm)')
```

20.1.2.8 Close button

The following set of codes can be used to create a close button. This is a simple pushbutton that refers to a Matlab built-in function called 'close'. By clicking on close button, GUI window will be closed.

```
%CLOSE BUTTON
p1 = uicontrol(gcf,'Style','pushbutton','Position',[10 10 70 25],...
```

```
'String','Close','BackgroundColor',[1 0.4 0.5],'FontName','Arial',...
'FontSize',14,'FontWeight','normal',...
'HorizontalAlignment','center','CallBack','close');
```

20.1.2.9 Reset button

Similar to the close button, 'Reset' button is also a simple pushbutton, which can be created using 'Unicontrol' function. However, to be able to reset a window to its default content, the program should refer to itself. It is achieved through using 'callback' command, in which it calls the same document (.m file) that includes the current set of codes.

```
%RESET BUTTON____this function is defined in chaingui.m
p2 = uicontrol('Style','pushbutton','position',[420 10 70 25],
    'FontName','Arial','FontSize',14,'FontWeight','normal',
    'BackgroundColor',[1 0.4 0.5],'HorizontalAlignment',
    'center','String','Reset');
set(p2,'Callback','chaingui')
```

20.1.2.10 Create axis in a specific position in the user interface window
```
%creating axis at a specific spot of the window permanently
plot1=subplot('Position',[0.35 0.05 0.63 0.35]); axis on;

plot2=subplot('Position',[0.35 0.55 0.63 0.35]); axis on;
```

The above code, create two blank axis in a GUI window at position [0.35 0.05 0.63 0.35] and [0.35 0.55 0.63 0.35]. However, they must be defined in the '.m' file which creates GUI window. Different graphs can be shown in these spots.

20.1.3 Acknowledgment

Authors would like to thank Dr. Behnam Fatahi for his help and guidance.

MATLAB Codes available for download:
http://riverpublishers.com/river_publisher/book_details.php?book_id=90

Reference

[1] Matlab® help document

Appendix I

```
%Accompanying Matlab Source Code
%In the book: Principles of Near Field Communications for Internet of Things
   (Agbinya, JI), River Publishers Denmark, 2011
%Developed by: Mehrnoush  Masihpour 30.12.2009
%Magnetic Communication Link Budget _ GUI(home page)
%file name: chainGUI.m
%sub-programs: air.m; twograph.m; minoo.m; minooGUI.m; NFMIC.m; NFMICGUI.m
clc
clear all
close all

% Figure to create the user interface
figure1 = figure('Color',[0.2 0.2 0.4],'Name',...
    'Near Field Magnetic Induction Communication_(C)Copyright Reserved
    (Univercity of Technology,Sydney,Australia)__Mehrnoush Masihpour',...
    'NumberTitle','off','Position',[25 50 900 650]);
% Image

image1= subplot('Position',[0.0 0.06 0.56 0.56]); image(imread('chain','jpg'));
%axis square;
axis off;

Text_11=uicontrol(gcf,'Style','text','Position',[505 580 395 70],...
    'String','GUIDE','FontSize',24,...
    'HorizontalAlignment','center','BackgroundColor',[0.99 0.89 0.2]);

Text_22=uicontrol(gcf,'Style','text','Position',[505 490 395 90],...
    'String','*** To see the distance graph vs. different parameters for a set
       of predefined values and compare the trend for a ferrite and air-cored
       coil, please press the first push button ','FontSize',10,...
    'HorizontalAlignment','left','BackgroundColor',[0.99 0.89 0.2]);

Text_33=uicontrol(gcf,'Style','text','Position',[505 410 395 90],...
    'String','*** To see the distance graph vs. different parameters and
       calculate the range for your desirable set of input values please press
       the second push button','FontSize',10,...
    'HorizontalAlignment','left','BackgroundColor',[0.99 0.89 0.2]);

Text_44=uicontrol(gcf,'Style','text','Position',[505 330 395 90],...
    'String','***To calculate the received power and see the graph of the
       received power vs. different coil characteristics for your desirable
       set of values, please press the third push button '...
    ,'FontSize',10,...
```

```
                         'HorizontalAlignment','left','BackgroundColor',[0.99 0.89 0.2]);

Text_55=uicontrol(gcf,'Style','text','Position',[505 260 395 90],...
    'String','','FontSize',10,...
    'HorizontalAlignment','left','BackgroundColor',[0.99 0.89 0.2]);

Text_66=uicontrol(gcf,'Style','text','Position',[505 190 395 90],...
    'String','','FontSize',10,...
    'HorizontalAlignment','left','BackgroundColor',[0.99 0.89 0.2]);

Text_77=uicontrol(gcf,'Style','text','Position',[505 120 395 90],...
    'String','','FontSize',10,...
    'HorizontalAlignment','left','BackgroundColor',[0.99 0.89 0.2]);

Text_88=uicontrol(gcf,'Style','text','Position',[505 30 395 90],...
    'String','','FontSize',10,...
    'HorizontalAlignment','left','BackgroundColor',[0.99 0.89 0.2]);

Text_99=uicontrol(gcf,'Style','text','Position',[505 0 395 30],...
    'String','***Developed by:Mehrnoush Masihpour ********************
        Jan-2010','FontSize',9,...
    'HorizontalAlignment','left','BackgroundColor',[0.99 0.89 0.2]);
%***************************************************************

hb1 = uicontrol('Style','pushbutton','position',[5 550 500 30],
        'FontName', 'Arial',...
        'FontSize',12,'FontWeight','normal','HorizontalAlignment','left',
        'BackgroundColor',[1 0.7 0.7],...
        'String','Comparison between antennas using air-core and
          ferrite-cored coil');
        set(hb1,'Callback','twogragh')

hb2 = uicontrol('Style','pushbutton','position',[5 500 500 30],
        'FontName','Arial',...
        'FontSize',12,'FontWeight','normal','BackgroundColor',[1 0.7 0.7],
        'HorizontalAlignment','left',...
        'String','Set the input to calculate the range and
        see the graphs');
        set(hb2,'Callback','minoogui')

hb3 = uicontrol('Style','pushbutton','position',[5 450 500 30],
        'FontName','Arial',...
        'FontSize',12,'FontWeight','normal','BackgroundColor',[1 0.7 0.7],
        'HorizontalAlignment','left',...
        'String','Set the inputs to calculate the power at the receiver and
        see the graph');
        set(hb3,'Callback','NFMICgui')

Text_0=uicontrol(gcf,'Style','text','Position',[60 600 380 30],...
    'String','Please Choose one of the Functions','fontSize',14,...
    'HorizontalAlignment','center','BackgroundColor',[0.8 0.8 1]);

%CLOSE BUTTON
```

```
p1 = uicontrol(gcf,'Style','pushbutton','Position',[10 10 70 25],...
    'String','Close','BackgroundColor',[1 0.4 0.5],'FontName','Arial',...
    'FontSize',14,'FontWeight','normal',...
    'HorizontalAlignment','center','CallBack','close');


p2 = uicontrol('Style','pushbutton','position',[420 10 70 25],
    'FontName','Arial',...
    'FontSize',14,'FontWeight','normal','BackgroundColor',[1 0.4 0.5],...
    'HorizontalAlignment','center',...
    'String','Reset');
     set(p2,'Callback','chaingui')


%***************************************************************
%***************************************************************
%                       THE END
%***************************************************************
%***************************************************************

%developed by: Mehrnoush  Masihpour 14.12.2009
%Magnetic Communication Link Budget
%file name: Twograph.m
%sub-programs: air.m

clear all
clc
close all

%creating the main window
%***************************************
f1 = figure('Color',[0.6 0.8 0.9],'Name',...
    'Magnetic Communication Link Budget_(C)Copyrigh Reserved(University of
      Technology, Sydney, Australia)__Mehrnoush Masihpour ',...
    'NumberTitle','off','Position',[30 40 900 650]);
%********************************************

Text_1=uicontrol(gcf,'Style','text','Position',[200 625 500 25],...
    'String','Magnetic Communication Link Budget',...
    'FontSize',14,'HorizontalAlignment',...
    'center','BackgroundColor',[.9 .5 .8]) ;

Text_2=uicontrol(gcf,'Style','text','Position',[300 575 300 25],...
    'String','AIR CORE[relative permeability=1]',...
    'FontSize',12,'HorizontalAlignment',...
    'center','BackgroundColor',[.6 .8 .95]) ;


Text_3=uicontrol(gcf,'Style','text','Position',[300 260 300 25],...
    'String','FERRITE CORE[relative perm.=10000]',...
    'FontSize',12,'HorizontalAlignment',...
    'center','BackgroundColor',[.6 .8 .95]) ;
```

```
Text_4=uicontrol(gcf,'Style','text','Position',[25 575 200 25],...
    'String','Constant Inputs',...
    'FontSize',12,'HorizontalAlignment',...
    'center','BackgroundColor',[.9 .5 .8]) ;
% line********************************************************
line1=uicontrol(gcf,'Style','text','Position',[270 25 3 540],...
    'String','','BackgroundColor',[0.2 1 1]);

%****************************************************************

Text_TxP=uicontrol(gcf,'Style','text','Position',[10 520 200 20],...
    'String','TX_Power [dBm] ','FontSize',13,...
    'HorizontalAlignment','left','BackgroundColor',[1 1 1]);

Text_Txpv=uicontrol(gcf,'Style','text','Position',[215 520 50 20],...
    'String','23','FontSize',13,...
    'HorizontalAlignment','left','BackgroundColor',[1 1 1]);

Text_RxP=uicontrol(gcf,'Style','text','Position',[10 495 200 20],...
    'String','RX_Power [dBm] ','FontSize',13,...
    'HorizontalAlignment','left','BackgroundColor',[1 1 1]);

Text_RxPv=uicontrol(gcf,'Style','text','Position',[215 495 50 20],...
    'String','0','FontSize',13,...
    'HorizontalAlignment','left','BackgroundColor',[1 1 1]);

Text_TxQ=uicontrol(gcf,'Style','text','Position',[10 470 200 20],...
    'String','TX_Quality Factor ','FontSize',13,...
    'HorizontalAlignment','left','BackgroundColor',[1 1 1]);

Text_TxQv=uicontrol(gcf,'Style','text','Position',[215 470 50 20],...
    'String','1000','FontSize',13,...
    'HorizontalAlignment','left','BackgroundColor',[1 1 1]);


Text_RxQ=uicontrol(gcf,'Style','text','Position',[10 445 200 20],...
    'String','RX_Quality Factor ','FontSize',13,...
    'HorizontalAlignment','left','BackgroundColor',[1 1 1]);
Text_RxQv=uicontrol(gcf,'Style','text','Position',[215 445 50 20],...
    'String','1000','FontSize',13,...
    'HorizontalAlignment','left','BackgroundColor',[1 1 1]);




Text_TxE=uicontrol(gcf,'Style','text','Position',[10 420 200 20],...
    'String','TX_Efficiency','FontSize',13,...
    'HorizontalAlignment','left','BackgroundColor',[1 1 1]);
Text_TxEv=uicontrol(gcf,'Style','text','Position',[215 420 50 20],...
    'String','0.98','FontSize',13,...
    'HorizontalAlignment','left','BackgroundColor',[1 1 1]);
```

```
Text_RxE=uicontrol(gcf,'Style','text','Position',[10 395 200 20],...
    'String','RX_Efficiency','FontSize',13,...
    'HorizontalAlignment','left','BackgroundColor',[1 1 1]);
Text_RxEv=uicontrol(gcf,'Style','text','Position',[215 395 50 20],...
    'String','0.98','FontSize',13,...
    'HorizontalAlignment','left','BackgroundColor',[1 1 1]);


Text_Txmu=uicontrol(gcf,'Style','text','Position',[10 370 200 20],...
    'String','TX_permeability','FontSize',13,...
    'HorizontalAlignment','left','BackgroundColor',[1 1 1]);
Text_Txmuv=uicontrol(gcf,'Style','text','Position',[215 370 50 20],...
    'String','10000','FontSize',13,...
    'HorizontalAlignment','left','BackgroundColor',[1 1 1]);


Text_Rxmu=uicontrol(gcf,'Style','text','Position',[10 345 200 20],...
    'String','RX_permeability','FontSize',13,...
    'HorizontalAlignment','left','BackgroundColor',[1 1 1]);
Text_Rxmuv=uicontrol(gcf,'Style','text','Position',[215 345 50 20],...
    'String','10000','FontSize',13,...
    'HorizontalAlignment','left','BackgroundColor',[1 1 1]);


Text_TXr=uicontrol(gcf,'Style','text','Position',[10 320 200 20],...
    'String','TX_Radius[cm]   ','FontSize',13,...
    'HorizontalAlignment','left','BackgroundColor',[1 1 1]);
Text_TXrv=uicontrol(gcf,'Style','text','Position',[215 320 50 20],...
    'String','2.5','FontSize',13,...
    'HorizontalAlignment','left','BackgroundColor',[1 1 1]);

Text_RXr=uicontrol(gcf,'Style','text','Position',[10 295 200 20],...
    'String','RX_Radius[cm]   ','FontSize',13,...
    'HorizontalAlignment','left','BackgroundColor',[1 1 1]);

Text_RXrv=uicontrol(gcf,'Style','text','Position',[215 295 50 20],...
    'String','1.5','FontSize',13,...
    'HorizontalAlignment','left','BackgroundColor',[1 1 1]);


%********************ferrite
subplot('Position',[0.38 0.06 0.6 0.33]); axis on;
sr=1; % (Fig. type)
pPopupmenuf=uicontrol(gcf,'Style','popupmenu','Position',[600 260 300 25],...
    'String',{'distance vs radius of transmitter (ferrite core)',
    'distance vs radius of receiver (ferrite core)',...
    'distance vs transmission power (ferrite core)',
    'distance vs receiver power(ferrite core)',...
'distance vs efficiency of transmitter and receiver (ferrite core)',...
'distance vs efficiency of receiver (ferrite core)',
'distance vs efficiency of transmitter (ferrite core)',...
'distance vs quality factor of transmitter and receiver (ferrite core)',...
'distance vs quality factor of transmitter (ferrite core)',...
'distance vs quality factor of receiver (ferrite core)',...
```

```
'distance vs permeability of transmitter and receiver (ferrite core)',...
'distance vs permeability of transmitter (ferrite core)',...
'distance vs permeability of receiver (ferrite core)'}......
,'value',sr,...
    'BackgroundColor',[1 1 1],'ForeGroundColor',[0 0 0],...
    'FontSize',12,'FontWeight','normal',...
    'HorizontalAlignment','center','FontName','Arial',...
    'CallBack','sr=get(pPopupmenuf,''value'');f=2; air');

%*********air
subplot('Position',[0.38 0.49 0.6 0.37]); axis on;
st=1; % (Fig. type)
pPopupmenu=uicontrol(gcf,'Style','popupmenu','Position',[600 575 300 25],...
    'String',{'distance vs radius of transmitter (air core)',
    'distance vs radius of receiver (air core)',...
    'distance vs transmission power (air core)','distance vs receiver
      power(air core)',...
'distance vs efficiency of transmitter and receiver (air core)',...
'distance vs efficiency of receiver (air core)',
'distance vs efficiency of transmitter (air core)',...
'distance vs quality factor of transmitter and receiver (air core)',...
'distance vs quality factor of transmitter (air core)',...
'distance vs quality factor of receiver (air core)'}...
,'value',st,...
    'BackgroundColor',[1 1 1],'ForeGroundColor',[0 0 0],...
    'FontSize',12,'FontWeight','normal',...
    'HorizontalAlignment','center','FontName','Arial',...
    'CallBack','st=get(pPopupmenu,''value''); f=1; air');

%close***********************
pushbutton_close= uicontrol(gcf,'Style','pushbutton','Position',...
    [80 25 100 30],'FontSize',12,'FontWeight','Bold','String',...
    'Close','CallBack','close');

hb000 = uicontrol('Style','pushbutton','position',[80 115 100 30],
        'FontName','Arial',...
        'FontSize',14,'FontWeight','normal','HorizontalAlignment','left',...
        'String','Back');
        set(hb000,'Callback','chaingui')

%***************************************************************
%***************************************************************
%                      THE END
%***************************************************************
%***************************************************************

%developed by: Mehrnoush  Masihpour 27.12.2009
%Magnetic Communication Link Budget _ GUI_ gets the value of each parameter
%and return the value of distance also the graphs of distance vs. different
%variable
%file name: minooGUI.m
% sub_program: minoo.m
```

```
clc
clear all
close all

% ***********************************************************************
% Fiure for GUI
figure1 = figure('Color',[0.6 0.5 0.9],'Name',...
'   Magnetic Induction Communication Link Budget_(C)Copyright Reserved
    (University of Technology,Sydney, Australia)__Mehrnoush Masihpour  ',...
    'NumberTitle','off','Position',[25 50 900 650]);



% input initial data***************************************************


RXQ=1000
TXQ=1000
RXmu=0.012
TXmu=0.012
RXp=0
TXp=23
RXr=1.5
TXr=2.5
TXeff=0.9
RXeff=0.9
mu0=log10(4.*pi.* 10.^-7);
TXmu1=log10(TXmu)
RXmu1=log10(RXmu)

% text box and variable box for each parameter ****************************
Text_1=uicontrol(gcf,'Style','text','Position',[25 570 150 20],...
    'String','TX_radius [cm] ','FontSize',12,...
    'HorizontalAlignment','left','BackgroundColor',[.9 0.9 1]);
Edit_1=uicontrol(gcf,'Style','edit','Position',[175 570 60 20],...
    'String',TXr,'FontSize',12,'BackgroundColor',[1 1 1],'Callback',...
    'TXr=str2num(get(Edit_1,''String''));');


Text_2=uicontrol(gcf,'Style','text','Position',[25 550 150 20],...
    'String','RX_radius [cm] ','FontSize',12,...
    'HorizontalAlignment','left','BackgroundColor',[.9 0.9 1]);
Edit_2=uicontrol(gcf,'Style','edit','Position',[175 550 60 20],...
    'String',RXr,'FontSize',12,'BackgroundColor',[1 1 1],'Callback',...
    'RXr=str2num(get(Edit_2,''String''));');


Text_3=uicontrol(gcf,'Style','text','Position',[25 530 150 20],...
    'String','TX_Power [dBm] ','FontSize',12,...
    'HorizontalAlignment','left','BackgroundColor',[.9 0.9 1]);
Edit_3=uicontrol(gcf,'Style','edit','Position',[175 530 60 20],...
    'String',TXp,'FontSize',12,'BackgroundColor',[1 1 1],'Callback',...
    'TXp=str2num(get(Edit_3,''String''));');
```

```
Text_4=uicontrol(gcf,'Style','text','Position',[25 510 150 20],...
    'String','RX_Sensitivity [dBm] ','FontSize',12,...
    'HorizontalAlignment','left','BackgroundColor',[.9 0.9 1]);
Edit_4=uicontrol(gcf,'Style','edit','Position',[175 510 60 20],...
    'String',RXp,'FontSize',12,'BackgroundColor',[1 1 1],'Callback',...
    'RXp=str2num(get(Edit_4,''String''));');


Text_5=uicontrol(gcf,'Style','text','Position',[25 490 150 20],...
    'String','TX_QuaityFactor','FontSize',12,...
    'HorizontalAlignment','left','BackgroundColor',[.9 0.9 1]);
Edit_5=uicontrol(gcf,'Style','edit','Position',[175 490 60 20],...
    'String',TXQ,'FontSize',12,'BackgroundColor',[1 1 1],'Callback',...
    'TXQ=str2num(get(Edit_5,''String''));');


Text_6=uicontrol(gcf,'Style','text','Position',[25 470 150 20],...
    'String','RX_QuaityFactor','FontSize',12,...
    'HorizontalAlignment','left','BackgroundColor',[.9 0.9 1]);
Edit_6=uicontrol(gcf,'Style','edit','Position',[175 470 60 20],...
    'String',RXQ,'FontSize',12,'BackgroundColor',[1 1 1],'Callback',...
    'RXQ=str2num(get(Edit_6,''String''));');


Text_7=uicontrol(gcf,'Style','text','Position',[25 450 150 20],...
    'String','TX_permeability','FontSize',12,...
    'HorizontalAlignment','left','BackgroundColor',[.9 0.9 1]);
Edit_7=uicontrol(gcf,'Style','edit','Position',[175 450 60 20],...
    'String',TXmu,'FontSize',12,'BackgroundColor',[1 1 1],'Callback',...
    'TXmu=str2num(get(Edit_7,''String''));');


Text_8=uicontrol(gcf,'Style','text','Position',[25 430 150 20],...
    'String','RX_permeability ','FontSize',12,...
    'HorizontalAlignment','left','BackgroundColor',[.9 0.9 1]);
Edit_8=uicontrol(gcf,'Style','edit','Position',[175 430 60 20],...
    'String',RXmu,'FontSize',12,'BackgroundColor',[1 1 1],'Callback',...
    'RXmu=str2num(get(Edit_8,''String''));');


Text_9=uicontrol(gcf,'Style','text','Position',[25 410 150 20],...
    'String','TX_Efficiency','FontSize',12,...
    'HorizontalAlignment','left','BackgroundColor',[.9 0.9 1]);
Edit_9=uicontrol(gcf,'Style','edit','Position',[175 410 60 20],...
    'String',TXeff,'FontSize',12,'BackgroundColor',[1 1 1],'Callback',...
    'TXeff=str2num(get(Edit_9,''String''));');


Text_10=uicontrol(gcf,'Style','text','Position',[25 390 150 20],...
    'String','RX_Efficiency','FontSize',12,...
    'HorizontalAlignment','left','BackgroundColor',[.9 0.9 1]);
Edit_10=uicontrol(gcf,'Style','edit','Position',[175 390 60 20],...
    'String',RXeff,'FontSize',12,'BackgroundColor',[1 1 1],'Callback',...
```

```
'RXeff=str2num(get(Edit_10,''String''));minoo');

%RADIO BUTTON_showing the graphs for air and ferrite core coils  *********
option=1;


radiobutton1 = uicontrol(gcf,'Style','radiobutton','Value',0,'Position',
              [10 350 230 20],...
    'Min',0,'Max',1,'BackgroundColor',[1 1 0.4],...
    'String','Distance vs TX_radius','FontName','Arial',...
    'FontSize',12,'FontWeight','normal','HorizontalAlignment','left',
    'CallBack','x=1; Option=1; minoo');

radiobutton2 = uicontrol(gcf,'Style','radiobutton','Value',0,'Position',
              [10 325 230 20],...
    'Min',0,'Max',1,'BackgroundColor',[1 1 0.4],...
    'String','Distance vs RX_radius','FontName','Arial',...
    'FontSize',12,'FontWeight','normal','HorizontalAlignment','left'
    'CallBack','x=1; Option=2; minoo');

radiobutton3 = uicontrol(gcf,'Style','radiobutton','Value',0,'Position',
              [10 300 230 20],...
    'Min',0,'Max',1,'BackgroundColor',[1 1 0.4],...
    'String','Distance vs TX_power','FontName','Arial',...
    'FontSize',12,'FontWeight','normal','HorizontalAlignment','left',
    'CallBack','x=1; Option=3; minoo');

radiobutton4 = uicontrol(gcf,'Style','radiobutton','Value',0,'Position',
              [10 275 230 20],...
    'Min',0,'Max',1,'BackgroundColor',[1 1 0.4],...
    'String','Distance vs RX_power','FontName','Arial',...
    'FontSize',12,'FontWeight','normal','HorizontalAlignment','left',
    'CallBack','x=1; Option=4; minoo');

radiobutton5 = uicontrol(gcf,'Style','radiobutton','Value',0,'Position',
              [10 250 230 20],...
    'Min',0,'Max',1,'BackgroundColor',[1 1 0.4],...
    'String','Distance vs TX_QualityFactor','FontName','Arial',...
    'FontSize',12,'FontWeight','normal','HorizontalAlignment','left',
    'CallBack','x=1; Option=5; minoo');

radiobutton6 = uicontrol(gcf,'Style','radiobutton','Value',0,'Position',
              [10 225 230 20],...
    'Min',0,'Max',1,'BackgroundColor',[1 1 0.4],...
    'String','Distance vs RX_QualityFactor','FontName','Arial',...
    'FontSize',12,'FontWeight','normal','HorizontalAlignment','left',
     'CallBack','x=1; Option=6; minoo');

radiobutton7 = uicontrol(gcf,'Style','radiobutton','Value',0,'Position',
              [10 200 230 20],...
    'Min',0,'Max',1,'BackgroundColor',[1 1 0.4],...
    'String','Distance vs TX_permeability','FontName','Arial',...
    'FontSize',12,'FontWeight','normal','HorizontalAlignment','left',
```

```
          'CallBack','x=1; Option=7; minoo');

radiobutton8 = uicontrol(gcf,'Style','radiobutton','Value',0,'Position',
              [10 175 230 20],...
    'Min',0,'Max',1,'BackgroundColor',[1 1 0.4],...
    'String','Distance vs RX_permeability','FontName','Arial',...
    'FontSize',12,'FontWeight','normal','HorizontalAlignment','left',
    'CallBack','x=1; Option=8; minoo');

radiobutton9 = uicontrol(gcf,'Style','radiobutton','Value',0,'Position',
              [10 150 230 20],...
    'Min',0,'Max',1,'BackgroundColor',[1 1 0.4],...
    'String','Distance vs TX_efficiency','FontName','Arial',...
    'FontSize',12,'FontWeight','normal','HorizontalAlignment','left',
    'CallBack','x=1; Option=9; minoo');

radiobutton10 = uicontrol(gcf,'Style','radiobutton','Value',0,'Position',
              [10 125 230 20],...
    'Min',0,'Max',1,'BackgroundColor',[1 1 0.4],...
    'String','Distance vs RX_efficiency','FontName','Arial',...
    'FontSize',12,'FontWeight','normal','HorizontalAlignment','left',
    'CallBack','x=1; Option=10; minoo');

radiobutton11 = uicontrol(gcf,'Style','radiobutton','Value',0,'Position',
              [1 90 250 20],...
    'Min',0,'Max',1,'BackgroundColor',[1 1 0.4],...
    'String','Distance vs RX&TX_efficiency','FontName','Arial',...
    'FontSize',12,'FontWeight','normal','HorizontalAlignment','left',
    'CallBack','x=1; Option=11; minoo');

radiobutton12 = uicontrol(gcf,'Style','radiobutton','Value',0,'Position',
              [1 65 250 20],...
    'Min',0,'Max',1,'BackgroundColor',[1 1 0.4],...
    'String','Distance vs RX&TX_permeability','FontName','Arial',...
    'FontSize',12,'FontWeight','normal','HorizontalAlignment','left',
    'CallBack','x=1; Option=12; minoo');

radiobutton13 = uicontrol(gcf,'Style','radiobutton','Value',0,'Position',
              [1 40 250 20],...
    'Min',0,'Max',1,'BackgroundColor',[1 1 0.4],...
    'String','Distance vs RX&TX_QualityFactor','FontName','Arial',...
    'FontSize',12,'FontWeight','normal','HorizontalAlignment','left',
    'CallBack','x=1; Option=13; minoo');

%CREATING THE LINE    ****************************************************
line1=uicontrol(gcf,'Style','text','Position',[250 20 3 606],...
    'String',' ','BackgroundColor',[1 0 0]);

%TEXT BOXES    *********************************************************
Text_11=uicontrol(gcf,'Style','text','Position',[10 600 230 25],...
    'String','Initial Input Data',...
    'FontSize',14,'HorizontalAlignment',...
    'center','BackgroundColor',[.9 0.9 1]);
```

```
Text_22=uicontrol(gcf,'Style','text','Position',[460 610 450 25],...
    'String','Air Core Coil',...
    'FontSize',14,'HorizontalAlignment',...
    'center','BackgroundColor',[1 1 .5]);

Text_33=uicontrol(gcf,'Style','text','Position',[460 290 450 25],...
    'String','Ferrite Core Coil',...
    'FontSize',14,'HorizontalAlignment',...
    'center','BackgroundColor',[1 1 .5]);

%Axis     **************************************************************
plot1=subplot('Position',[0.35 0.05 0.63 0.35]); axis on;

plot2=subplot('Position',[0.35 0.55 0.63 0.35]); axis on;

% Push buttons       **************************************************
%CLOSE BUTTON
p3 = uicontrol(gcf,'Style','pushbutton','Position',[160 10 70 25],...
    'String','Close','BackgroundColor',[.8 1 1],'FontName','Arial',...
    'FontSize',14,'FontWeight','normal',...
    'HorizontalAlignment','center','CallBack','close');

%CALCULATION BUTTON (AIR)
p4 = uicontrol(gcf,'Style','pushbutton','Position',[260 610 200 25],...
    'String','Calculate the distance (cm)','FontName','Arial',...
    'FontSize',12,'FontWeight','normal',...
    'HorizontalAlignment','center','CallBack','x=2; minoo');

%CALCULATION BUTTON (FERRITE)
p5 = uicontrol(gcf,'Style','pushbutton','Position',[260 290 200 25],...
    'String','Calculate the distance  (cm)','FontName','Arial',...
    'FontSize',12,'FontWeight','normal',...
    'HorizontalAlignment','center','CallBack','x=3; minoo');

hb11 = uicontrol('Style','pushbutton','position',[80 10 70 25],
       'FontName','Arial',...
        'FontSize',14,'FontWeight','normal','BackgroundColor',[.8 1 1],
        'HorizontalAlignment','left',...
        'String','Back');
        set(hb11,'Callback','chaingui')
%*********************************************************************
%*********************************************************************
%                       THE END
%*********************************************************************
%*********************************************************************

%developed by: Mehrnoush  Masihpour 25.12.2009
%Magnetic Communication Link Budget _ GUI_ gets the value of each parameter
%and return the value of received power also the graphs of received power vs.
%different system parameters
%file name: nfmicGUI.m
% sub_program: NFMIC.m
```

```
clc
clear all
close all

% **************************************************************************
% Fiure for GUI
figure1 = figure('Color',[0.7 0.9 0.99],'Name',...
'  Magnetic Induction Communication Power Equation_(C)Copyright Reserved
    (University of Technology, Sydney, Australia)__Mehrnoush Masihpour',...
      'NumberTitle','off','Position',[25 50 900 650]);


% input initial data  ***************************************************

txp=23;
qr0=1000;
qt0=1000;
reff0=0.98;
teff0=0.98;
rmu0=10000;
tmu0=10000;
tr=2.5;
rr=1.5;
d=50;
mu0=log10(4.*pi.* 10.^-7);
qr=log10(qr0);
qt=log10(qt0);
reff=log10(reff0);
teff=log10(teff0);
rmu=log10(rmu0)-mu0;
tmu=log10(tmu0)-mu0;
mu0=log10(4.*pi.* 10.^-7);


% text box and variable box for each parameter ****************************
Text_1=uicontrol(gcf,'Style','text','Position',[25 570 150 20],...
    'String','TX_radius [cm] ','FontSize',12,...
    'HorizontalAlignment','left','BackgroundColor',[.7 0.8 1]);
Edit_1=uicontrol(gcf,'Style','edit','Position',[175 570 50 20],...
    'String',tr,'FontSize',12,'BackgroundColor',[1 1 1],'Callback',...
    'tr=str2num(get(Edit_1,''String''));');


Text_2=uicontrol(gcf,'Style','text','Position',[25 550 150 20],...
    'String','RX_radius [cm] ','FontSize',12,...
    'HorizontalAlignment','left','BackgroundColor',[.9 0.9 1]);
Edit_2=uicontrol(gcf,'Style','edit','Position',[175 550 50 20],...
    'String',rr,'FontSize',12,'BackgroundColor',[1 1 1],'Callback',...
    'rr=str2num(get(Edit_2,''String''));');


Text_3=uicontrol(gcf,'Style','text','Position',[25 530 150 20],...
    'String','TX_Power [dBm] ','FontSize',12,...
```

```
    'HorizontalAlignment','left','BackgroundColor',[.7 0.8 1]);
Edit_3=uicontrol(gcf,'Style','edit','Position',[175 530 50 20],...
    'String',txp,'FontSize',12,'BackgroundColor',[1 1 1],'Callback',...
    'txp=str2num(get(Edit_3,''String''));');


Text_4=uicontrol(gcf,'Style','text','Position',[25 510 150 20],...
    'String','Distance [cm] ','FontSize',12,...
    'HorizontalAlignment','left','BackgroundColor',[.9 0.9 1]);
Edit_4=uicontrol(gcf,'Style','edit','Position',[175 510 50 20],...
    'String',d,'FontSize',12,'BackgroundColor',[1 1 1],'Callback',...
    'd=str2num(get(Edit_4,''String''));');


Text_5=uicontrol(gcf,'Style','text','Position',[25 490 150 20],...
    'String','TX_QuaityFactor ','FontSize',12,...
    'HorizontalAlignment','left','BackgroundColor',[.7 0.8 1]);
Edit_5=uicontrol(gcf,'Style','edit','Position',[175 490 50 20],...
    'String',qt0,'FontSize',12,'BackgroundColor',[1 1 1],'Callback',...
    'qt0=str2num(get(Edit_5,''String''));');


Text_6=uicontrol(gcf,'Style','text','Position',[25 470 150 20],...
    'String','RX_QuaityFactor','FontSize',12,...
    'HorizontalAlignment','left','BackgroundColor',[.9 0.9 1]);
Edit_6=uicontrol(gcf,'Style','edit','Position',[175 470 50 20],...
    'String',qr0,'FontSize',12,'BackgroundColor',[1 1 1],'Callback',...
    'qr0=str2num(get(Edit_6,''String''));');


Text_7=uicontrol(gcf,'Style','text','Position',[25 450 150 20],...
    'String','TX_permeability','FontSize',12,...
    'HorizontalAlignment','left','BackgroundColor',[.7 0.8 1]);
Edit_7=uicontrol(gcf,'Style','edit','Position',[175 450 50 20],...
    'String',tmu0,'FontSize',12,'BackgroundColor',[1 1 1],'Callback',...
    'tmu0=str2num(get(Edit_7,''String''));');


Text_8=uicontrol(gcf,'Style','text','Position',[25 430 150 20],...
    'String','RX_permeability','FontSize',12,...
    'HorizontalAlignment','left','BackgroundColor',[.9 0.9 1]);
Edit_8=uicontrol(gcf,'Style','edit','Position',[175 430 50 20],...
    'String',rmu0,'FontSize',12,'BackgroundColor',[1 1 1],'Callback',...
    'rmu0=str2num(get(Edit_8,''String''));');


Text_9=uicontrol(gcf,'Style','text','Position',[25 410 150 20],...
    'String','TX_Efficiency','FontSize',12,...
    'HorizontalAlignment','left','BackgroundColor',[.7 0.8 1]);
Edit_9=uicontrol(gcf,'Style','edit','Position',[175 410 50 20],...
    'String',teff0,'FontSize',12,'BackgroundColor',[1 1 1],'Callback',...
    'teff0=str2num(get(Edit_9,''String''));');
```

```
Text_10=uicontrol(gcf,'Style','text','Position',[25 390 150 20],...
    'String','RX_Efficiency','FontSize',12,...
    'HorizontalAlignment','left','BackgroundColor',[.9 0.9 1]);
Edit_10=uicontrol(gcf,'Style','edit','Position',[175 390 50 20],...
    'String',reff0,'FontSize',12,'BackgroundColor',[1 1 1],'Callback',...
    'reff0=str2num(get(Edit_10,''String''));');

%RADIO BUTTON_ showing the graphs for air and ferrite core coils  ************
Option=1;


radiobutton1 = uicontrol(gcf,'Style','radiobutton','Value',0,'Position',
                [10 350 230 20],...
    'Min',0,'Max',1,'BackgroundColor',[1 .7 0.8],...
    'String','RX_power vs RX_radius','FontName','Arial',...
    'FontSize',12,'FontWeight','normal','HorizontalAlignment','left',
    'CallBack','x=1; Option=1; NFMIC');

radiobutton2 = uicontrol(gcf,'Style','radiobutton','Value',0,'Position',
                [10 325 230 20],...
    'Min',0,'Max',1,'BackgroundColor',[1 .7 0.8],...
    'String','RX_power vs TX_radius','FontName','Arial',...
    'FontSize',12,'FontWeight','normal','HorizontalAlignment','left',
    'CallBack','x=1; Option=2; NFMIC');

radiobutton3 = uicontrol(gcf,'Style','radiobutton','Value',0,'Position',
                [10 300 230 20],...
    'Min',0,'Max',1,'BackgroundColor',[1 .7 0.8],...
    'String','RX_power vs TX_power','FontName','Arial',...
    'FontSize',12,'FontWeight','normal','HorizontalAlignment','left',
    'CallBack','x=1; Option=3; NFMIC');

radiobutton4 = uicontrol(gcf,'Style','radiobutton','Value',0,'Position',
                [10 275 230 20],...
    'Min',0,'Max',1,'BackgroundColor',[1 .7 0.8],...
    'String','RX_power vs Distance','FontName','Arial',...
    'FontSize',12,'FontWeight','normal','HorizontalAlignment','left',
    'CallBack','x=1; Option=4; NFMIC');

radiobutton5 = uicontrol(gcf,'Style','radiobutton','Value',0,'Position',
                [10 250 230 20],...
    'Min',0,'Max',1,'BackgroundColor',[1 .7 0.8],...
    'String','RX_power vs TX_Q-Factor','FontName','Arial',...
    'FontSize',12,'FontWeight','normal','HorizontalAlignment','left',
    'CallBack','x=1; Option=5; NFMIC');

radiobutton6 = uicontrol(gcf,'Style','radiobutton','Value',0,'Position',
                [10 225 230 20],...
    'Min',0,'Max',1,'BackgroundColor',[1 .7 0.8],...
    'String','RX_power vs RX_Q-Factor','FontName','Arial',...
    'FontSize',12,'FontWeight','normal','HorizontalAlignment','left',
    'CallBack','x=1; Option=6; NFMIC');
```

```
radiobutton7 = uicontrol(gcf,'Style','radiobutton','Value',0,'Position',
               [10 200 230 20],...
    'Min',0,'Max',1,'BackgroundColor',[1 .7 0.8],...
    'String','RX_power vs TX_permeability','FontName','Arial',...
    'FontSize',12,'FontWeight','normal','HorizontalAlignment','left',
    'CallBack','x=1; Option=7; NFMIC');

radiobutton8 = uicontrol(gcf,'Style','radiobutton','Value',0,'Position',
               [10 175 230 20],...
    'Min',0,'Max',1,'BackgroundColor',[1 .7 0.8],...
    'String','RX_power vs RX_permeability','FontName','Arial',...
    'FontSize',12,'FontWeight','normal','HorizontalAlignment','left',
    'CallBack','x=1; Option=8; NFMIC');

radiobutton9 = uicontrol(gcf,'Style','radiobutton','Value',0,'Position',
               [10 150 230 20],...
    'Min',0,'Max',1,'BackgroundColor',[1 .7 0.8],...
    'String','RX_power vs TX_efficiency','FontName','Arial',...
    'FontSize',12,'FontWeight','normal','HorizontalAlignment','left',
    'CallBack','x=1; Option=9; NFMIC');

radiobutton10 = uicontrol(gcf,'Style','radiobutton','Value',0,'Position',
               [10 125 230 20],...
    'Min',0,'Max',1,'BackgroundColor',[1 .7 0.8],...
    'String','RX_power vs RX_efficiency','FontName','Arial',...
    'FontSize',12,'FontWeight','normal','HorizontalAlignment','left',
    'CallBack','x=1; Option=10; NFMIC');

radiobutton11 = uicontrol(gcf,'Style','radiobutton','Value',0,'Position',
               [1 90 250 20],...
    'Min',0,'Max',1,'BackgroundColor',[1 .7 0.8],...
    'String','RX_power vs RX&TX_efficiency','FontName','Arial',...
    'FontSize',12,'FontWeight','normal','HorizontalAlignment','left',
    'CallBack','x=1; Option=11; NFMIC');

radiobutton12 = uicontrol(gcf,'Style','radiobutton','Value',0,'Position',
               [1 65 250 20],...
    'Min',0,'Max',1,'BackgroundColor',[1 .7 0.8],...
    'String','RX_power vs RX&TX_permeability','FontName','Arial',...
    'FontSize',12,'FontWeight','normal','HorizontalAlignment','left',
    'CallBack','x=1; Option=12; NFMIC');

radiobutton13 = uicontrol(gcf,'Style','radiobutton','Value',0,'Position',
               [1 40 250 20],...
    'Min',0,'Max',1,'BackgroundColor',[1 .7 0.8],...
    'String','RX_power vs RX&TX_Q-Factor','FontName','Arial',...
    'FontSize',12,'FontWeight','normal','HorizontalAlignment','left',
    'CallBack','x=1; Option=13; NFMIC');
%CREATING THE LINE    ****************************************************
line1=uicontrol(gcf,'Style','text','Position',[250 20 3 606],...
    'String',' ','BackgroundColor',[1 0 0]);
%TEXT BOXES    ****************************************************
Text_11=uicontrol(gcf,'Style','text','Position',[10 600 230 25],...
```

```
      'String','Initial Input Data',...
      'FontSize',14,'HorizontalAlignment',...
      'center','BackgroundColor',[.8 .9 .5]);

%Axis    ********************************************************************
plot1=subplot('Position',[0.35 0.45 0.63 0.45]); axis on;

%**********************************************************************************
%CLOSE BUTTON
p3 = uicontrol(gcf,'Style','pushbutton','Position',[160 10 70 25],...
      'String','Close','BackgroundColor',[.8 .9 .5],'FontName','Arial',...
      'FontSize',14,'FontWeight','normal',...
      'HorizontalAlignment','center','CallBack','close');
%CALCULATION BUTTON (FERRITE)
p5 = uicontrol(gcf,'Style','pushbutton','Position',[280 200 600 35],
      'BackgroundColor',[.8 .9 .5],...
      'String','To calculate the power (dBm) at the receiver click here please',
      'FontName','Arial',...
      'FontSize',14,'FontWeight','normal',...
      'HorizontalAlignment','center','CallBack','x=2; NFMIC');
hb99 = uicontrol('Style','pushbutton','position',[80 10 70 25],
            'FontName','Arial',...
            'FontSize',14,'FontWeight','normal','BackgroundColor',[.8 .9 .5],
            'HorizontalAlignment','left',...
            'String','Back');
            set(hb99,'Callback','chaingui')
%**********************************************************************************
%**********************************************************************************
%                              THE END
%**********************************************************************************
%**********************************************************************************


%developed by: Mehrnoush  Masihpour 12.12.2009
%file name: air.m

mu0=log10(4.*pi.* 10.^-7);
axis on
if f==1
      subplot('Position',[0.38 0.49 0.6 0.37]); axis on;
    if st==1
          %Distance as a function of transmitter coil radius
Ptx=23; % power at the transmitter
Prx=0; % power at the receiver
qur=1000;
qut=1000;
Qrx=log10(qut);  % Quality factor of the transmitter
Qtx=log10(qur);  % Quality factor of the receiver
eff1=0.98;
eff2=0.98;
EFICIENCYtx=log10(eff1);  %Efficiency of the transmitter
EFICIENCYrx=log10(eff2);  %Efficiency of the receiver {
mu0=log10(4.*pi.* 10.^-7);
```

```
radius_tx=(0:0.005:8);        % radius of the transmitter coil {cm}
radius_rx =1.5;% radius of the receiver coil {cm}
distance_pr=10.^((Ptx+mu0+mu0+Qtx+Qrx+EFICIENCYtx+EFICIENCYrx+30.
             *log10(radius_rx.*radius_tx)-Prx)./60).*1.5;
distance_del=10.^(-0.5.*log10(1+(radius_tx./distance_pr).^2));
distance_c= distance_pr.*distance_del;
line11=0:0.05:8;
%straight line for comparison between the ghraph trend and a liniar pattern
line22=5.*line11;
plot(radius_tx,distance_c,'-',radius_tx,distance_pr,'--')
%plot of distance as a function of transmitter ciol radius
xlabel('Transmitter Coil Radius (cm)')
ylabel('Distance (cm)')
legend('with delta d consideration','without delta d consideration',2)

    end
if st==2
        %Distance as a function of receiver coil radius
TxP=23; % power at the transmitter {dBm}
RxP=0; % power at the receiver {dBm}
qur=1000;
qut=1000;
RxQ=log10(qut);  % Quality factor of the transmitter
TxQ=log10(qur);  % Quality factor of the receiver
eff1=0.98;
eff2=0.98;
TxE=log10(eff1);  %Efficiency of the transmitter
RxE=log10(eff2);  %Efficiency of the receiver
TXr=2.5;
 % radius of the transmitter coil {cm}
RXr=(0.001:0.005:7);% radius of the receiver coil {cm}
distance_pr=10.^((TxP+mu0+mu0+TxQ+mu0+mu0+RxQ+TxE+RxE+30.
             *log10(RXr.*TXr)-RxP)./60).*1.5;
distance_del=10.^(-0.5.*log10(1+(TXr./distance_pr).^2));
distance_c= distance_pr.*distance_del;
line11=0:0.05:7;
%straight line for comparison between the ghraph trend and a liniar pattern
line22=7.*line11;
plot(RXr,distance_c,'-',RXr,distance_pr,'--')
%plot of distance as a function of receiver ciol radius
xlabel('Receiver Coil Radius (cm)')
ylabel('Distance (cm)')
legend('with delta d consideration' ,'without delta d consideration',2)
%text(4.55,22,'(X=4.55,Y=22)');

end

if st==3
        %Distance as a function of transmission power
Prx=0; % power at the transmitter {dBm}
Ptx=(0:0.05:70); % power at the receiver {dBm}
qur=1000;
qut=1000;
```

```
Qrx=log10(qut); % Quality factor of the transmitter
Qtx=log10(qur); % Quality factor of the receiver {
 eff1=0.98;
eff2=0.98;
EFICIENCYrx=log10(eff1); %Efficiency of the transmitter
EFICIENCYtx=log10(eff2); %Efficiency of the receiver
radius_tx=2.5 ;    % radius of the transmitter coil {cm}
radius_rx =1.5;% radius of the receiver coil {cm}
distance_pr=10.^((Ptx+mu0+mu0+Qtx+Qrx+EFICIENCYtx+EFICIENCYrx+30.
            *log10(radius_rx.*radius_tx)-Prx)./60).*1.5;
distance_del=10.^(-0.5.*log10(1+(radius_tx./distance_pr).^2));
distance_c= distance_pr.*distance_del;
plot(Ptx,distance_c,'-',Ptx,distance_pr,'--')
%plot of distance as a function of transmission power
xlabel('Transmission Power(dBm)')
ylabel('Distance (cm)')
legend('with delta d consideration','without delta d consideration',2)

end
if st==4

        %Distance as a function of receiver power
Ptx=23; % power at the transmitter
Prx=(0:0.05:23); % power at the receiver
qur=1000;
qut=1000;
Qrx=log10(qut); % Quality factor of the transmitter
Qtx=log10(qur); % Quality factor of the receiver
eff1=0.98;
eff2=0.98;
EFICIENCYrx=log10(eff1); %Efficiency of the transmitter
EFICIENCYtx=log10(eff2); %Efficiency of the receiver
radius_tx=2.5  ;    % radius of the transmitter coil {cm}
radius_rx =1.5;% radius of the receiver coil {cm}
distance_pr=10.^((Ptx+mu0+mu0+Qtx+Qrx+EFICIENCYtx+EFICIENCYrx+30.
             *log10(radius_rx.*radius_tx)-Prx)./60).*1.5;
distance_del=10.^(-0.5.*log10(1+(radius_tx./distance_pr).^2));
distance_c= distance_pr.*distance_del;
plot(Prx,distance_c,'-',Prx,distance_pr,'--')
%plot of distance as a function of reception power
xlabel('Received Power(dBm)')
ylabel('Distance(cm)')
legend('with delta d consideration','without delta d consideration')

end
if st==5
        % Distance as a function of efficiency of receiver and transmitter coil
Ptx=23; % power at the transmitter {dBm}
Prx=0; % power at the receiver {dBm}
qur=1000;
qut=1000;
Qrx=log10(qut); % Quality factor of the transmitter
Qtx=log10(qur); % Quality factor of the receiver
```

```
eff1=0.01:0.05:1
eff2=0.01:0.05:1;
EFICIENCYtx=log10(eff1);
%Efficiency of the transmitter EFICIENCYrx=log10(eff2);
%Efficiency of the receiver
radius_tx=2.5;      % radius of the transmitter coil {cm}
radius_rx =1.5;% radius of the receiver coil {cm}
distance_pr=10.^((Ptx+mu0+mu0+Qtx+Qrx+EFICIENCYtx+EFICIENCYrx+30.
            *log10(radius_rx.*radius_tx)-Prx)./60).*1.5;
distance_del=10.^(-0.5.*log10(1+(radius_tx./distance_pr).^2));
distance_c= distance_pr.*distance_del;
plot(eff1,distance_c,'-',eff2,distance_pr,'--')
%plot of distance as a function of receiver coil efficiency
xlabel('Receiver and Transmitter Coil Efficiency')
ylabel('Distance (cm)')
legend('with delta d consideration','without delta d consideration',2)

end
if st==6
        % Distance as a function of efficiency of receiver coil
Ptx=23; % power at the transmitter {dBm}
Prx=0; % power at the receiver {dBm}
qur=1000;
qut=1000;
Qrx=log10(qut); % Quality factor of the transmitter
Qtx=log10(qur); % Quality factor of the receiver
eff1=0.98;
eff2=0.01:0.05:1;
EFICIENCYtx=log10(eff1);
%Efficiency of the transmitter EFICIENCYrx=log10(eff2);
%Efficiency of the receiver
radius_tx=2.5;      % radius of the transmitter coil {cm}
radius_rx =1.5;% radius of the receiver coil {cm}
distance_pr=10.^((Ptx+mu0+mu0+Qtx+Qrx+EFICIENCYtx+EFICIENCYrx+30.
            *log10(radius_rx.*radius_tx)-Prx)./60).*1.5;
distance_del=10.^(-0.5.*log10(1+(radius_tx./distance_pr).^2));
distance_c= distance_pr.*distance_del;
plot(eff2,distance_c,'-',eff2,distance_pr,'--')
%plot of distance as a function of receiver coil efficiency
xlabel('Receiver Coil Efficiency')
ylabel('Distance (cm)')
legend('with delta d consideration' ,'without delta d consideration',2)
end
if st==7
  %Distance as a function of efficiency of transmitter coil
Ptx=23; % power at the transmitter {dBm}
Prx=0; % power at the receiver {dBm}
qur=1000;
qut=1000;
Qrx=log10(qut); % Quality factor of the transmitter
Qtx=log10(qur); % Quality factor of the receiver
eff2=0.98;
eff1=0.01:0.05:1;
```

```
EFICIENCYtx=10.*log10(eff1);  %Efficiency of the transmitter
EFICIENCYrx=10.*log10(eff2);  %Efficiency of the receiver
radius_tx=2.5  ;   % radius of the transmitter coil {cm}
radius_rx =1.5;% radius of the receiver coil {cm}
distance_pr=10.^((Ptx+mu0+mu0+Qtx+Qrx+EFICIENCYtx+EFICIENCYrx+30.
           *log10(radius_rx.*radius_tx)-Prx)./60).*1.5;
distance_del=10.^(-0.5.*log10(1+(radius_tx./distance_pr).^2));
distance_c= distance_pr.*distance_del;
plot(eff1,distance_c,'-',eff1,distance_pr,'--')
%plot of distance as a function of transmitter ciol efficiency
xlabel('Transmitter Coil Efficiency (dB)')
ylabel('Distance (cm)')
legend('with delta d consideration','without delta d consideration',2)

end
if st==8
    %Distance as a function of Quality factor of receiver and transmitter coil
Ptx=23; % power at the transmitter
 Prx=0; % power at the receiver
qut=0.01:0.05:1000;
qur=0.01:0.05:1000;
Qrx=log10(qut);  % Quality factor of the transmitter
Qtx=log10(qur);  % Quality factor of the receiver
eff1=0.98;
eff2=0.98;
EFICIENCYtx=log10(eff1);  %Efficiency of the transmitter
EFICIENCYrx=log10(eff2);  %Efficiency of the receiver
radius_tx=2.5 ;   % radius of the transmitter coil {cm}
radius_rx =1.5;% radius of the receiver coil {cm}
distance_pr=10.^((Ptx+mu0+mu0+Qtx+Qrx+EFICIENCYtx+EFICIENCYrx+30.
           *log10(radius_rx.*radius_tx)-Prx)./60).*1.5;
distance_del=10.^(-0.5.*log10(1+(radius_tx./distance_pr).^2));
distance_c= distance_pr.*distance_del;
plot(qur,distance_c,'-',qur,distance_pr,'--')
%plot of distance as a function of Quality factor of receiver coil
xlabel('Receiver and Transmitter Coil Quality Factor')
ylabel('Distance (cm)')
legend('with delta d consideration' ,'without delta d consideration',2)

end
if st==9
     %Distance as a function of Quality factor of transmitter coil
Ptx=23; % power at the transmitter {dBm}
Prx=0; % power at the receiver {dBm}
qut=0.01:0.05:1000;
qur=1000;
Qrx=log10(qut);  % Quality factor of the transmitter
Qtx=log10(qur);  % Quality factor of the receiver
eff1=0.98;
eff2=0.98;
EFICIENCYtx=log10(eff1);  %Efficiency of the transmitter
EFICIENCYrx=log10(eff2);  %Efficiency of the receiver
radius_tx=2.5;     % radius of the transmitter coil {cm}
```

```
radius_rx =1.5;% radius of the receiver coil {cm}
distance_pr=10.^((Ptx+mu0+mu0+Qtx+Qrx+EFICIENCYtx+EFICIENCYrx+30.
            *log10(radius_rx.*radius_tx)-Prx)./60).*1.5;
distance_del=10.^(-0.5.*log10(1+(radius_tx./distance_pr).^2));
distance_c= distance_pr.*distance_del;
plot(qut,distance_c,'-',qut,distance_pr,'--');
%plot of distance as a function of  Quality factor of transmitter coil
xlabel('Transmitter Coil Quality Factor')
ylabel('Distance (cm)')
legend('with delta d consideration','without delta d consideration',2)


end
if st==10

%Distance as a function of Quality factor of receiver coil
Ptx=23; % power at the transmitter
Prx=0; % power at the receiver
qur=0.01:0.05:1000;
qut=1000;
Qrx=log10(qut);  % Quality factor of the transmitter
Qtx=log10(qur);  % Quality factor of the receiver
eff1=0.98;
eff2=0.98;
EFICIENCYtx=log10(eff1);  %Efficiency of the transmitter
EFICIENCYrx=log10(eff2);  %Efficiency of the receiver
radius_tx=2.5 ; % radius of the transmitter coil {cm}
radius_rx =1.5;% radius of the receiver coil {cm}
distance_pr=10.^((Ptx+mu0+mu0+Qtx+Qrx+EFICIENCYtx+EFICIENCYrx+30.
            *log10(radius_rx.*radius_tx)-Prx)./60).*1.5;
distance_del=10.^(-0.5.*log10(1+(radius_tx./distance_pr).^2));
distance_c= distance_pr.*distance_del;
plot(qur,distance_c,'-',qur,distance_pr,'--')
%plot of distance as a function of Quality factor of receiver coil
xlabel('Receiver Coil Quality Factor')
ylabel('Distance  (cm)')
legend('with delta d consideration','without delta d consideration',2)

end


elseif f==2
    subplot('Position',[0.38 0.06 0.6 0.33]); axis on;
    if sr==1
%Distance as a function of transmitter coil radius (ferrite core coil)
Ptx_dBm=23; % power at the transmitter {dBm}
Prx_dBm=0; % power at the receiver {dBm}
qur=1000;
qut=1000;
Qrx_dB=log10(qut);  % Quality factor of the transmitter
Qtx_dB=log10(qur);  % Quality factor of the receiver
eff1=0.98;
eff2=0.98;
EFICIENCYtx_dB=log10(eff1);  %Efficiency of the transmitter
```

```
EFICIENCYrx_dB=log10(eff2); %Efficiency of the receiver
mu0=log10(4.*pi.* 10.^-7);
mu1p=10000;
mu1rel=log10(mu1p);
MU_tx=mu1rel-mu0;
mu2p=10000;
mu2rel=log10(mu2p);
MU_rx=mu2rel-mu0;
radius_tx_cm=(0:0.005:8); % radius of the transmitter coil {cm}
radius_rx_cm =1.5;% radius of the receiver coil {cm}
distance_prime=10.^((Ptx_dBm+Qtx_dB+Qrx_dB+EFICIENCYtx_dB+EFICIENCYrx_dB+30.
            *log10(radius_rx_cm.*radius_tx_cm)-Prx_dBm+MU_tx+MU_rx)./60).
            *1.5;
distance_delta=10.^(-0.5.*log10(1+(radius_tx_cm./distance_prime).^2));
distance_cm= distance_prime.*distance_delta;
line1=0:0.05:8;
line2=30.*line1;
plot(radius_tx_cm,distance_cm,'-',radius_tx_cm,distance_prime,'--');
%plot of distance as a function of transmitter ciol radius
xlabel('Transmitter Coil Radius (cm)')
ylabel('Distance (cm)')
legend('with delta d consideration','without delta d consideration',2)

    end

if sr==2
%Distance as a function of receiver coil radius(ferrite core coil)
Ptx_dBm=23; % power at the transmitter {dBm}
Prx_dBm=0; % power at the receiver {dBm}
qur=1000;
qut=1000;
Qrx=log10(qut);  % Quality factor of the transmitter
Qtx=log10(qur);  % Quality factor of the receiver
mu0=log10(4.*pi.* 10.^-7);
eff1=0.98;
eff2=0.98;
EFICIENCYtx_dB=log10(eff1);   %Efficiency of the transmitter
EFICIENCYrx_dB=log10(eff2);   %Efficiency of the receiver
mu1p=10000;                 %range of the permeability of the transmitting coil
mu1rel=log10(mu1p);
MU_tx=mu1rel-mu0;
mu2p=10000;
mu2rel=log10(mu2p);
MU_rx=mu2rel-mu0;
radius_tx_cm=2.5; % radius of the transmitter coil {cm}
radius_rx_cm=(0:0.005:7);% radius of the receiver coil {cm}
distance_prime=10.^((Ptx_dBm+Qtx_dB+Qrx_dB+EFICIENCYtx_dB+EFICIENCYrx_dB+30.
            *log10(radius_rx_cm.*radius_tx_cm)-Prx_dBm+MU_tx+MU_rx)./60).
            *1.5;
distance_delta=10.^(-0.5.*log10(1+(radius_tx_cm./distance_prime).^2));
distance_cm= distance_prime.*distance_delta;
line1=0:0.05:7;
line2=30.*line1;
```

```
plot(radius_rx_cm,distance_cm,'-',radius_rx_cm,distance_prime,'--')
%plot of distance as a function of receiver ciol radius
xlabel('Receiver Coil Radius (cm)')
ylabel('Distance (cm)')
legend('with delta d consideration','without delta d consideration',2)


end
if sr==3

%Distance as a function of transmission power (ferrite core coil)
Prx_dBm=0; % power at the transmitter
Ptx_dBm=(0.001:0.05:40); % power at the receiver
qur=1000;
qut=1000;
Qrx=log10(qut); % Quality factor of the transmitter
Qtx=log10(qur); % Quality factor of the receiver
mu0=log10(4.*pi.* 10.^-7);
eff1=0.98;
eff2=0.98;
EFICIENCYtx_dB=log10(eff1); %Efficiency of the transmitter
EFICIENCYrx_dB=log10(eff2); %Efficiency of the receiver
mu1p=10000;                 %range of the permeability of the transmitting coil
mu1rel=log10(mu1p);
MU_tx=mu1rel-mu0;
mu2p=10000;
mu2rel=log10(mu2p);
MU_rx=mu2rel-mu0;
radius_tx_cm=2.5 ;     % radius of the transmitter coil {cm}
radius_rx_cm =1.5;% radius of the receiver coil {cm}
distance_prime=10.^((Ptx_dBm+Qtx_dB+Qrx_dB+EFICIENCYtx_dB+EFICIENCYrx_dB+30.
              *log10(radius_rx_cm.*radius_tx_cm)-Prx_dBm+MU_tx+MU_rx)./60).
              *1.5;
distance_delta=10.^(-0.5.*log10(1+(radius_tx_cm./distance_prime).^2));
distance_cm= distance_prime.*distance_delta;
plot(Ptx_dBm,distance_cm,'-',Ptx_dBm,distance_prime,'--')
%plot of distance as a function of transmission power
xlabel('Transmission Power(dBm)')
ylabel('Distance (cm)')
legend('with delta d consideration','without delta d consideration',2)


end
if sr==4

%Distance as a function of receiver power (ferrite core coil)
Ptx_dBm=23; % power at the transmitter {dBm}
Prx_dBm=(0.001:0.05:23); % power at the receiver {dBm}
qur=1000;
qut=1000;
Qrx=log10(qut); % Quality factor of the transmitter
Qtx=log10(qur); % Quality factor of the receiver
mu0=log10(4.*pi.* 10.^-7);
eff1=0.98;
```

```
eff2=0.98;
EFICIENCYtx_dB=log10(eff1);
%Efficiency of the transmitter EFICIENCYrx_dB=log10(eff2);
%Efficiency of the receiver
mu1p=10000;
mu1rel=log10(mu1p);
MU_tx=mu1rel-mu0;
mu2p=10000; % permeability of the receiver coil
mu2rel=log10(mu2p);
MU_rx=mu2rel-mu0;
radius_tx_cm=2.5  ;  % radius of the transmitter coil {cm}
radius_rx_cm =1.5;% radius of the receiver coil {cm}
distance_prime=10.^((Ptx_dBm+Qtx_dB+Qrx_dB+EFICIENCYtx_dB+EFICIENCYrx_dB+30.
              .*log10(radius_rx_cm.*radius_tx_cm)-Prx_dBm+MU_tx+MU_rx)./60).
              .*1.5;
distance_delta=10.^(-0.5.*log10(1+(radius_tx_cm./distance_prime).^2));
distance_cm= distance_prime.*distance_delta;
plot(Prx_dBm,distance_cm,'-',Prx_dBm,distance_prime,'--')
%plot of distance as a function of reception power
xlabel('Received Power(dBm)')
ylabel('Distance (cm)')
legend('with delta d consideration','without delta d consideration')



end
if sr==5
% Distance as a function of efficiency of receiver and transmitter coil
   (ferrite core coil)
Ptx_dBm=23; % power at the transmitter
Prx_dBm=0; % power at the receiver
qur=1000;
qut=1000;
Qrx=log10(qut);  % Quality factor of the transmitter
Qtx=log10(qur);  % Quality factor of the receiver
mu0=log10(4.*pi.* 10.^-7);
eff1=0.01:0.005:1;
eff2=0.01:0.005:1;
EFICIENCYtx_dB=log10(eff1);  %Efficiency of the transmitter
EFICIENCYrx_dB=log10(eff2);  %Efficiency of the receiver
mu1p=10000;  %range of the permeability of the transmitting coil
mu1rel=log10(mu1p);
MU_tx=mu1rel-mu0;
mu2p=10000; %range of the permeability of the receiver coil
mu2rel=log10(mu2p);
MU_rx=mu2rel-mu0;
radius_tx_cm=2.5;     % radius of the transmitter coil {cm}
radius_rx_cm =1.5;% radius of the receiver coil {cm}
distance_prime=10.^((Ptx_dBm+Qtx_dB+Qrx_dB+EFICIENCYtx_dB+EFICIENCYrx_dB+30.
              .*log10(radius_rx_cm.*radius_tx_cm)-Prx_dBm+MU_tx+MU_rx)./60).
              .*1.5;
distance_delta=10.^(-0.5.*log10(1+(radius_tx_cm./distance_prime).^2));
distance_cm= distance_prime.*distance_delta;
```

```
plot(eff2,distance_cm,'-',eff2,distance_prime,'--')
%plot of distance as a function of receiver coil efficiency
xlabel('Receiver and Transmitter Coil Efficiency')
ylabel('Distance (cm)')
legend('with delta d consideration','without delta d consideration',2)


end
if sr==6
% Distance as a function of efficiency of receiver coil(ferrite core coil)
Ptx_dBm=23; % power at the transmitter {dBm}
Prx_dBm=0; % power at the receiver {dBm}
qur=1000;
qut=1000;
Qrx=log10(qut); % Quality factor of the transmitter
Qtx=log10(qur); % Quality factor of the receiver
mu0=log10(4.*pi.* 10.^-7);
eff1=0.98;
eff2=0.01:0.005:1;
EFICIENCYtx_dB=log10(eff1);  %Efficiency of the transmitter
EFICIENCYrx_dB=log10(eff2); %Efficiency of the receiver
mu1p=10000; %range of the permeability of the transmitting coil
mu1rel=log10(mu1p);
MU_tx=mu1rel-mu0;
mu2p=10000; %range of the permeability of the receiver coil
mu2rel=log10(mu2p);
MU_rx=mu2rel-mu0;
radius_tx_cm=2.5;     % radius of the transmitter coil {cm}
radius_rx_cm =1.5;% radius of the receiver coil {cm}
distance_prime=10.^((Ptx_dBm+Qtx_dB+Qrx_dB+EFICIENCYtx_dB+EFICIENCYrx_dB+30.
               *log10(radius_rx_cm.*radius_tx_cm)-Prx_dBm+MU_tx+MU_rx)./60).
               *1.5;
distance_delta=10.^(-0.5.*log10(1+(radius_tx_cm./distance_prime).^2));
distance_cm= distance_prime.*distance_delta;
plot(eff2,distance_cm,eff2,distance_prime,'--')
%plot of distance as a function of receiver coil efficiency
xlabel('Receiver Coil Efficiency')
ylabel('Distance (cm)')

legend('with delta d consideration','without delta d consideration',2)

end
 if sr==7
%Distance as a function of efficiency of transmitter coil (ferrite core coil)
Ptx_dBm=23; % power at the transmitter
Prx_dBm=0; % power at the receiver
qur=1000;
qut=1000;
Qrx=log10(qut); % Quality factor of the transmitter
Qtx=log10(qur); % Quality factor of the receiver
mu0=log10(4.*pi.* 10.^-7);
eff21=0.98;
eff1=0.01:0.005:1;
```

```
EFICIENCYtx_dB=log10(eff1);  %Efficiency of the transmitter
EFICIENCYrx_dB=log10(eff2);  %Efficiency of the receiver
mu1p=10000;                       %range of the permeability of the transmitting coil
mu1rel=log10(mu1p);
MU_tx=mu1rel-mu0;
mu2p=10000; %range of the permeability of the receiver coil
mu2rel=log10(mu2p);
MU_rx=mu2rel-mu0;
radius_tx_cm=2.5  ; % radius of the transmitter coil {cm}
radius_rx_cm =1.5;% radius of the receiver coil {cm}
distance_prime=10.^((Ptx_dBm+Qtx_dB+Qrx_dB+EFICIENCYtx_dB+EFICIENCYrx_dB+30.
               *log10(radius_rx_cm.*radius_tx_cm)-Prx_dBm+MU_tx+MU_rx)./60).
               *1.5;
distance_delta=10.^(-0.5.*log10(1+(radius_tx_cm./distance_prime).^2));
distance_cm= distance_prime.*distance_delta;
plot(eff1,distance_cm,'-',eff1,distance_prime,'--')
%plot of distance as a function of transmitter ciol efficiency
xlabel('Transmitter Coil Efficiency')
ylabel('Distance(cm)')

legend('with delta d consideration','without delta d consideration',2)


end
if sr==8
 %Distance as a function of Quality factor of receiver and transmitter coil
Ptx_dBm=23; % power at the transmitter {dBm}
Prx_dBm=0; % power at the receiver {dBm}
qur=0.0001:0.005:1000;
qut=0.0001:0.005:1000;
mu0=log10(4.*pi.* 10.^-7);
Qrx_dB=log10(qut);  % Quality factor of the transmitter
Qtx_dB=log10(qur);  % Quality factor of the receiver
mu1p=10000; %range of the permeability of the transmitting coil
mu1rel=log10(mu1p);
MU_tx=mu1rel-mu0;
mu2p=10000;   %range of the permeability of the receiver coil
mu2rel=log10(mu2p);
MU_rx=mu2rel-mu0;
eff2=0.98;
eff1=0.98;
EFICIENCYtx_dB=log10(eff1);  %Efficiency of the transmitter
EFICIENCYrx_dB=log10(eff2);  %Efficiency of the receiver
radius_tx_cm=2.5 ; % radius of the transmitter coil {cm}
radius_rx_cm =1.5;% radius of the receiver coil {cm}
distance_prime=10.^((Ptx_dBm+Qtx_dB+Qrx_dB+EFICIENCYtx_dB+EFICIENCYrx_dB+30.
               *log10(radius_rx_cm.*radius_tx_cm)-Prx_dBm+MU_rx+MU_tx)./60).
               *1.5;
distance_delta=10.^(-0.5.*log10(1+(radius_tx_cm./distance_prime).^2));
distance_cm= distance_prime.*distance_delta;
plot(qur,distance_cm,'-',qur,distance_prime,'--')
%plot of distance as a function of Quality factor of receiver coil
xlabel('Receiver and Transmitter Coil Quality Factor')
```

```
ylabel('Distance (cm)')
legend('with delta d consideration','without delta d consideration',2)


end
if sr==9
 %Distance as a function of Quality factor of transmitter coil
 (ferrite core coil)
Ptx_dBm=23; % power at the transmitter {dBm}
Prx_dBm=0; % power at the receiver
qur=1000;
qut=0.0001:0.005:1000;
Qrx_dB=log10(qut); % Quality factor of the transmitter
Qtx_dB=log10(qur); % Quality factor of the receiver
eff2=0.98;
eff1=0.98;
mu0=log10(4.*pi.* 10.^-7);
EFICIENCYtx_dB=log10(eff1);
%Efficiency of the transmitter EFICIENCYrx_dB=log10(eff2);
%Efficiency of the receiver
mu1p=10000;  %range of the permeability of the transmitting coil
mu1rel=log10(mu1p);
MU_tx=mu1rel-mu0;
mu2p=10000;  %range of the permeability of the receiver coil
mu2rel=log10(mu2p);
MU_rx=mu2rel-mu0;
radius_tx_cm=2.5;      % radius of the transmitter coil {cm}
radius_rx_cm =1.5;% radius of the receiver coil {cm}
distance_prime=10.^((Ptx_dBm+Qtx_dB+Qrx_dB+EFICIENCYtx_dB+EFICIENCYrx_dB+30.
               .*log10(radius_rx_cm.*radius_tx_cm)-Prx_dBm+MU_tx+MU_rx)./60).
               .*1.5;
distance_delta=10.^(-0.5.*log10(1+(radius_tx_cm./distance_prime).^2));
distance_cm= distance_prime.*distance_delta;
plot(qut,distance_cm,'-',qut,distance_prime,'--');
%plot of distance as a function of  Quality factor of transmitter coil
xlabel('Transmitter Coil Quality Factor ')
ylabel('Distance (cm)')

legend('with delta d consideration','without delta d consideration',2)


end
if sr==10
  %Distance as a function of Quality factor of receiver coil
  (ferrite core coil)
Ptx_dBm=23; % power at the transmitter
Prx_dBm=0; % power at the receiver
qur=0.0001:0.005:1000;
qut=1000;
Qrx_dB=log10(qut);  % Quality factor of the transmitter
Qtx_dB=log10(qur);  % Quality factor of the receiver
mu0=log10(4.*pi.* 10.^-7);
eff2=0.98;
```

```
eff1=0.98;
EFICIENCYtx_dB=log10(eff1); %Efficiency of the transmitter
EFICIENCYrx_dB=log10(eff2); %Efficiency of the receiver
mu1p=10000;   %range of the permeability of the transmitting coil
mu1rel=log10(mu1p);
MU_tx=mu1rel-mu0;

mu2p=10000;      %range of the permeability of the receiver coil
mu2rel=log10(mu2p);
MU_rx=mu2rel-mu0;radius_tx_cm=2.5 ; % radius of the transmitter coil {cm}
radius_rx_cm =1.5;% radius of the receiver coil {cm}
distance_prime=10.^((Ptx_dBm+Qtx_dB+Qrx_dB+EFICIENCYtx_dB+EFICIENCYrx_dB+30.
              *log10(radius_rx_cm.*radius_tx_cm)-Prx_dBm+MU_tx+MU_rx)./60).
              *1.5;
distance_delta=10.^(-0.5.*log10(1+(radius_tx_cm./distance_prime).^2));
distance_cm= distance_prime.*distance_delta;
plot(qur,distance_cm,'-',qur,distance_prime,'--')
%plot of distance as a function of Quality factor of receiver coil
xlabel('Receiver Coil Quality Factor')
ylabel('Distance (cm)')
legend('with delta d consideration','without delta d consideration',2)



end
if sr==11
%Distance as a function of transmitter and receiver coil core permeability
 (ferrite core coil)
Prx_dBm=0; % power at the receiver

Ptx_dBm=23;
qur=1000;
qut=1000;
mu0=log10(4.*pi.* 10.^-7);
Qrx=log10(qut);  % Quality factor of the transmitter
Qtx=log10(qur);  % Quality factor of the receiver
eff2=0.98;
eff1=0.98;
EFICIENCYtx_dB=log10(eff1);  %Efficiency of the transmitter
EFICIENCYrx_dB=log10(eff2); %Efficiency of the receiver
mu1p=1:0.005:10000; %range of the permeability of the transmitting coil
mu1rel=log10(mu1p);
MU_tx=mu1rel-mu0;

mu2p=1:0.005:10000; %range of the permeability of the receiver coil
mu2rel=log10(mu2p);
MU_rx=mu2rel-mu0;

radius_tx_cm=2.5;  % radius of the transmitter coil {cm}
radius_rx_cm =1.5;% radius of the receiver coil {cm}
distance_prime=10.^((Ptx_dBm+Qtx_dB+Qrx_dB+EFICIENCYtx_dB+EFICIENCYrx_dB+30.
              *log10(radius_rx_cm.*radius_tx_cm)-Prx_dBm+MU_tx+MU_rx)./60).
              *1.5;
distance_delta=10.^(-0.5.*log10(1+(radius_tx_cm./distance_prime).^2));
```

```
distance_cm= distance_prime.*distance_delta;
plot(mu1p,distance_cm,'-',mu1p,distance_prime,'--');
%plot of distance as a function of transmitter ciol radius
xlabel('Transmitter and Receiver Coil Relative Permeability')
ylabel('Distance (cm)')

legend('with delta d consideration','without delta d consideration',2)


end

if sr==12
Prx_dBm=0;
Ptx_dBm=23;% power at the transmitter {dBm}
mu0=log10(4.*pi.* 10.^-7);
qur=1000;
qut=1000;
Qrx=log10(qut);  % Quality factor of the transmitter
Qtx=log10(qur);  % Quality factor of the receiver
eff2=0.98;
eff1=0.98;
EFICIENCYtx_dB=log10(eff1);
%Efficiency of the transmitter EFICIENCYrx_dB=log10(eff2);
%Efficiency of the receiver
mu1p=1:0.005:10000; %range of the permeability of the transmitting coil
mu1rel=log10(mu1p);
MU_tx=mu1rel-mu0;
mu2p=10000; %range of the permeability of the receiver coil
mu2rel=log10(mu2p);
MU_rx=mu2rel-mu0;
radius_tx_cm=2.5; % radius of the transmitter coil {cm}
radius_rx_cm =1.5;% radius of the receiver coil {cm}
distance_prime=10.^((Ptx_dBm+Qtx_dB+Qrx_dB+EFICIENCYtx_dB+EFICIENCYrx_dB+30.
              *log10(radius_rx_cm.*radius_tx_cm)-Prx_dBm+MU_tx+MU_rx)./60).
              *1.5;
distance_delta=10.^(-0.5.*log10(1+(radius_tx_cm./distance_prime).^2));
distance_cm= distance_prime.*distance_delta;
plot(mu1p,distance_cm,'-',mu1p,distance_prime,'--');
%plot of distance as a function of transmitter ciol radius
xlabel('Transmitter Coil relative Permeability')
ylabel('Distance(cm)')

legend('with delta d consideration','without delta d consideration',2)


end
  if sr==13
Prx_dBm=0; % power at the receiver {dBm}
Ptx_dBm=23;
mu0=log10(4.*pi.* 10.^-7);
qur=1000;
qut=1000;
Qrx=10.*log10(qut);  % Quality factor of the transmitter
```

```
Qtx=10.*log10(qur);  % Quality factor of the receiver)
eff2=0.98;
eff1=0.98;
EFICIENCYtx_dB=log10(eff1);  %Efficiency of the transmitter
EFICIENCYrx_dB=log10(eff2);  %Efficiency of the receiver }
mu1p=10000;           %range of the permeability of the transmitting coil
mu1rel=log10(mu1p);
MU_tx=mu1rel-mu0;

mu2p=1:0.005:10000;
mu2rel=log10(mu2p);
MU_rx=mu2rel-mu0;
radius_tx_cm=2.5;        % radius of the transmitter coil {cm}
radius_rx_cm =1.5;% radius of the receiver coil {cm}
distance_prime=10.^((Ptx_dBm+Qtx_dB+Qrx_dB+EFICIENCYtx_dB+EFICIENCYrx_dB+30.
            *log10(radius_rx_cm.*radius_tx_cm)-Prx_dBm+MU_tx+MU_rx)./60).
            *1.5;
distance_delta=10.^(-0.5.*log10(1+(radius_tx_cm./distance_prime).^2));
distance_cm= distance_prime.*distance_delta;
plot(mu2p,distance_cm,'-',mu2p,distance_prime,'--');
%plot of distance as a function of transmitter ciol radius
xlabel('Receiver Coil relative Permeability')
ylabel('Distance (cm)')
legend('with delta d consideration','without delta d consideration',2)


    end

end
 hb = uicontrol('Style','pushbutton','position',[80 160 100 30],
       'FontName','Arial',...
          'FontSize',14,'FontWeight','normal',...
                'String','Reset');
          set(hb,'Callback','twogragh')


%********************************************************************
%********************************************************************
%                        THE END
%********************************************************************
%********************************************************************

%developed by: Mehrnoush  Masihpour 18.12.2009
%file name: minoo.m

if x==1 % ---------------------------------------------------
    if Option==1
        set (radiobutton1, 'Value',1);
        set (radiobutton2, 'Value',0);
        set (radiobutton3, 'Value',0);
        set (radiobutton4, 'Value',0);
        set (radiobutton5, 'Value',0);
        set (radiobutton6, 'Value',0);
```

```
set (radiobutton7, 'Value',0);
set (radiobutton8, 'Value',0);
set (radiobutton9, 'Value',0);
set (radiobutton10, 'Value',0);
set (radiobutton11, 'Value',0);
set (radiobutton12, 'Value',0);
set (radiobutton13, 'Value',0);
TXQ1=log10(TXQ);
RXQ1=log10(RXQ);
TXeff1=log10(TXeff);
RXeff1=log10(RXeff);
TXmu1=log10(TXmu);
RXmu1=log10(RXmu);
mu0=log10(4.*pi.* 10.^-7);
r1=0.0001:0.005:8;        %range of transmitter coil radius



d1=10.^((mu0+TXp+TXQ1+RXQ1+TXeff1+RXeff1+30.
    *log10(RXr.*r1)-RXp)./60).*1.5;
d2=10.^(-0.5.*log10(1+(r1./d1).^2));
da=d1.*d2;
subplot('Position',[0.35 0.55 0.63 0.35]); axis on;
plot(r1,da);
xlabel('Transmitter Coil Radius (cm)')
ylabel('Distance (cm)')

d11=10.^((RXmu1+TXmu1+TXp+TXQ1+RXQ1+TXeff1+RXeff1+30.
    *log10(RXr.*r1)-RXp)./60).*1.5;
d22=10.^(-0.5.*log10(1+(r1./d11).^2));
df= d11.*d22;
subplot('Position',[0.35 0.05 0.63 0.35]); axis on;
plot(r1,df);
xlabel('Transmitter Coil Radius (cm)')
ylabel('Distance (cm)')

elseif Option==2
    set (radiobutton1, 'Value',0);
    set (radiobutton2, 'Value',1);
    set (radiobutton3, 'Value',0);
    set (radiobutton4, 'Value',0);
    set (radiobutton5, 'Value',0);
    set (radiobutton6, 'Value',0);
    set (radiobutton7, 'Value',0);
    set (radiobutton8, 'Value',0);
    set (radiobutton9, 'Value',0);
    set (radiobutton10, 'Value',0);
    set (radiobutton11, 'Value',0);
    set (radiobutton12, 'Value',0);
    set (radiobutton13, 'Value',0);
```

```
        r2=0.0001:0.005:7;        %range of receiver coil radius

        d1=10.^((mu0+mu0+TXp+TXQ1+RXQ1+TXeff1+RXeff1+30.
           *log10(r2.*TXr)-RXp)./60).*1.5;
        d2=10.^(-0.5.*log10(1+(TXr./d1).^2));
        da= d1.*d2;
        subplot('Position',[0.35 0.55 0.63 0.35]); axis on;
        plot(r2,da);
        xlabel('Receiver Coil Radius (cm)')
        ylabel('Distance (cm)')

        d11=10.^((RXmu1+TXmu1+TXp+TXQ1+RXQ1+TXeff1+RXeff1+30.
           *log10(r2.*TXr)-RXp)./60).*1.5;
        d22=10.^(-0.5.*log10(1+(TXr./d11).^2));
        df= d11.*d22;
        subplot('Position',[0.35 0.05 0.63 0.35]); axis on;
        plot(r2,df);
        xlabel('Receiver Coil Radius (cm)')
        ylabel('Distance (cm)')


elseif Option==3
        set (radiobutton1, 'Value',0);
        set (radiobutton2, 'Value',0);
        set (radiobutton3, 'Value',1);
        set (radiobutton4, 'Value',0);
        set (radiobutton5, 'Value',0);
        set (radiobutton6, 'Value',0);
        set (radiobutton7, 'Value',0);
        set (radiobutton8, 'Value',0);
        set (radiobutton9, 'Value',0);
        set (radiobutton10, 'Value',0);
        set (radiobutton11, 'Value',0);
        set (radiobutton12, 'Value',0);
        set (radiobutton13, 'Value',0);

        p1=0.0001:0.005:25;           %transmission power range

        subplot('Position',[0.35 0.55 0.63 0.35]); axis on;
        d1=10.^((mu0+mu0+p1+TXQ1+RXQ1+TXeff1+RXeff1+30.
           *log10(RXr.*TXr)-RXp)./60).*1.5;
        d2=10.^(-0.5.*log10(1+(TXr./d1).^2));
        da= d1.*d2;
        plot(p1,da);
        xlabel('Transmission Power (dBm)')
        ylabel('Distance (cm)')

        subplot('Position',[0.35 0.05 0.63 0.35]); axis on;
        d11=10.^((RXmu1+TXmu1+p1+TXQ1+RXQ1+TXeff1+RXeff1+30.
           *log10(RXr.*TXr)-RXp)./60).*1.5;
        d22=10.^(-0.5.*log10(1+(TXr./d11).^2));
        df= d11.*d22;
        plot(p1,df);
```

```
        xlabel('Transmission Power (dBm)')
        ylabel('Distance (cm)')

elseif Option==4
        set (radiobutton1, 'Value',0);
        set (radiobutton2, 'Value',0);
        set (radiobutton3, 'Value',0);
        set (radiobutton4, 'Value',1);
        set (radiobutton5, 'Value',0);
        set (radiobutton6, 'Value',0);
        set (radiobutton7, 'Value',0);
        set (radiobutton8, 'Value',0);
        set (radiobutton9, 'Value',0);
        set (radiobutton10, 'Value',0);
        set (radiobutton11, 'Value',0);
        set (radiobutton12, 'Value',0);
        set (radiobutton13, 'Value',0);

        p2=0.0001:0.005:20;            %range of the power at the receiver

        subplot('Position',[0.35 0.55 0.63 0.35]); axis on;
        d1=10.^((mu0+mu0+TXp+TXQ1+RXQ1+TXeff1+RXeff1+30.
            *log10(RXr.*TXr)-p2)./60).*1.5;
        d2=10.^(-0.5.*log10(1+(TXr./d1).^2));
        da= d1.*d2;
        plot(p2,da);
        xlabel('Power at the Receiver (dBm)')
        ylabel('Distance (cm)')

        subplot('Position',[0.35 0.05 0.63 0.35]); axis on;
        d11=10.^((RXmu1+TXmu1+TXp+TXQ1+RXQ1+TXeff1+RXeff1+30.
            *log10(RXr.*TXr)-p2)./60).*1.5;
        d22=10.^(-0.5.*log10(1+(TXr./d11).^2));
        df= d11.*d22;
        plot(p2,df);
        xlabel('Power at the Receiver (dBm)')
        ylabel('Distance (cm)')

elseif Option==5
        set (radiobutton1, 'Value',0);
        set (radiobutton2, 'Value',0);
        set (radiobutton3, 'Value',0);
        set (radiobutton4, 'Value',0);
        set (radiobutton5, 'Value',1);
        set (radiobutton6, 'Value',0);
        set (radiobutton7, 'Value',0);
        set (radiobutton8, 'Value',0);
        set (radiobutton9, 'Value',0);
        set (radiobutton10, 'Value',0);
        set (radiobutton11, 'Value',0);
        set (radiobutton12, 'Value',0);
        set (radiobutton13, 'Value',0);
```

```
TXQ1=log10(TXQ);
RXQ1=log10(RXQ);
TXeff1=log10(TXeff);
RXeff1=log10(RXeff);
TXmu1=log10(TXmu);
RXmu1=log10(RXmu);
mu0=log10(4.*pi.* 10.^-7);
q1p=0.0001:0.005:1000;
q1=log10(q1p)    ;
subplot('Position',[0.35 0.55 0.63 0.35]); axis on;
d1=10.^((mu0+mu0+TXp+q1+RXQ1+TXeff1+RXeff1+30.
    *log10(RXr.*TXr)-RXp)./60).*1.5;
d2=10.^(-0.5.*log10(1+(TXr./d1).^2));
da= d1.*d2;
plot(q1p,da);
xlabel('Transmitter Coil Quality Factor ')
ylabel('Distance (cm)')

subplot('Position',[0.35 0.05 0.63 0.35]); axis on;
d11=10.^((RXmu1+TXmu1+TXp+q1+RXQ1+TXeff1+RXeff1+30.
    *log10(RXr.*TXr)-RXp)./60).*1.5;
d22=10.^(-0.5.*log10(1+(TXr./d11).^2));
df= d11.*d22;
plot(q1p,df);
xlabel('Transmitter Coil Quality Factor')
ylabel('Distance (cm)')

elseif Option==6
    set (radiobutton1, 'Value',0);
    set (radiobutton2, 'Value',0);
    set (radiobutton3, 'Value',0);
    set (radiobutton4, 'Value',0);
    set (radiobutton5, 'Value',0);
    set (radiobutton6, 'Value',1);
    set (radiobutton7, 'Value',0);
    set (radiobutton8, 'Value',0);
    set (radiobutton9, 'Value',0);
    set (radiobutton10, 'Value',0);
    set (radiobutton11, 'Value',0);
    set (radiobutton12, 'Value',0);
    set (radiobutton13, 'Value',0);

    TXQ1=log10(TXQ);
    RXQ1=log10(RXQ);
    TXeff1=log10(TXeff);
    RXeff1=log10(RXeff);
    TXmu1=log10(TXmu);
    RXmu1=log10(RXmu);
    mu0=log10(4.*pi.* 10.^-7);
    q2p=0.0001:0.005:10000;
    q2=log10(q2p) ;
    subplot('Position',[0.35 0.55 0.63 0.35]); axis on;
    d1=10.^((mu0+mu0+TXp+TXQ1+q2+TXeff1+RXeff1+30.
```

```
         *log10(RXr.*TXr)-RXp)./60).*1.5;
    d2=10.^(-0.5.*log10(1+(TXr./d1).^2));
    da= d1.*d2;
    plot(q2p,da);
    xlabel('Receiver Coil Quality Factor')
    ylabel('Distance (cm)')

    subplot('Position',[0.35 0.05 0.63 0.35]); axis on;
    d11=10.^((RXmu1+TXmu1+TXp+TXQ1+q2+TXeff1+RXeff1+30.
         *log10(RXr.*TXr)-RXp)./60).*1.5;
    d22=10.^(-0.5.*log10(1+(TXr./d11).^2));
    df= d11.*d22;
    plot(q2p,df);
    xlabel('Receiver Coil Quality Factor')
    ylabel('Distance (cm)')

elseif Option==7
    set (radiobutton1, 'Value',0);
    set (radiobutton2, 'Value',0);
    set (radiobutton3, 'Value',0);
    set (radiobutton4, 'Value',0);
    set (radiobutton5, 'Value',0);
    set (radiobutton6, 'Value',0);
    set (radiobutton7, 'Value',1);
    set (radiobutton8, 'Value',0);
    set (radiobutton9, 'Value',0);
    set (radiobutton10, 'Value',0);
    set (radiobutton11, 'Value',0);
    set (radiobutton12, 'Value',0);
    set (radiobutton13, 'Value',0);


    TXQ1=log10(TXQ);
    RXQ1=log10(RXQ);
    TXeff1=log10(TXeff);
    RXeff1=log10(RXeff);
    TXmu1=log10(TXmu);
    RXmu1=log10(RXmu);
    mu0=log10(4.*pi.* 10.^-7);
    mu1p=1:0.005:10000;
    mu1rel=log10(mu1p);
    mu1=mu1rel-mu0;
    plot2=subplot('Position',[0.35 0.55 0.63 0.35]); axis off;
    delete(plot2);

    subplot('Position',[0.35 0.05 0.63 0.35]); axis on;
    d11=10.^((mu0+mu0+RXmu1+mu1+TXp+TXQ1+RXQ1+TXeff1+RXeff1+30.
         *log10(RXr.*TXr)-RXp)./60).*1.5;
    d22=10.^(-0.5.*log10(1+(TXr./d11).^2));
    df= d11.*d22;
    plot(mu1p,df);
    xlabel('Transmitter Coil Relative permeability')
    ylabel('Distance (cm)')
```

```
elseif Option==8
    set (radiobutton1, 'Value',0);
    set (radiobutton2, 'Value',0);
    set (radiobutton3, 'Value',0);
    set (radiobutton4, 'Value',0);
    set (radiobutton5, 'Value',0);
    set (radiobutton6, 'Value',0);
    set (radiobutton7, 'Value',0);
    set (radiobutton8, 'Value',1);
    set (radiobutton9, 'Value',0);
    set (radiobutton10, 'Value',0);
    set (radiobutton11, 'Value',0);
    set (radiobutton12, 'Value',0);
    set (radiobutton13, 'Value',0);

    TXQ1=log10(TXQ);
    RXQ1=log10(RXQ);
    TXeff1=log10(TXeff);
    RXeff1=log10(RXeff);
    TXmu1=log10(TXmu);
    RXmu1=log10(RXmu);
    mu0=log10(4.*pi.* 10.^-7);
    mu2p=1:0.005:1000;
    mu2rel=log10(mu2p);
    mu2=mu2rel-mu0;
    plot2=subplot('Position',[0.35 0.55 0.63 0.35]); axis off;
    delete(plot2);
    subplot('Position',[0.35 0.05 0.63 0.35]); axis on;
    d11=10.^((mu2+TXmu1+TXp+TXQ1+RXQ1+TXeff1+RXeff1+30.
        *log10(RXr.*TXr)-RXp)./60).*1.5;
    d22=10.^(-0.5.*log10(1+(TXr./d11).^2));
    df= d11.*d22;
    plot(mu2p,df);
    xlabel('receiver Coil Relative permeability')
    ylabel('Distance (cm)')


elseif Option==9
    set (radiobutton1, 'Value',0);
    set (radiobutton2, 'Value',0);
    set (radiobutton3, 'Value',0);
    set (radiobutton4, 'Value',0);
    set (radiobutton5, 'Value',0);
    set (radiobutton6, 'Value',0);
    set (radiobutton7, 'Value',0);
    set (radiobutton8, 'Value',0);
    set (radiobutton9, 'Value',1);
    set (radiobutton10, 'Value',0);
    set (radiobutton11, 'Value',0);
    set (radiobutton12, 'Value',0);
    set (radiobutton13, 'Value',0);
```

```
eff1p=0.01:.005:1;          %range of the transmitter coil efficiency
eff1=log10(eff1p);
subplot('Position',[0.35 0.55 0.63 0.35]); axis on;
d1=10.^((mu0+mu0+TXp+TXQ1+RXQ1+eff1+RXeff1+30.
   .*log10(RXr.*TXr)-RXp)./60).*1.5;
d2=10.^(-0.5.*log10(1+(TXr./d1).^2));
da= d1.*d2;
plot(eff1p,da);
xlabel('Transmitter Coil Efficiency ')
ylabel('Distance (cm)')


subplot('Position',[0.35 0.05 0.63 0.35]); axis on;
d11=10.^((RXmu1+TXmu1+TXp+TXQ1+RXQ1+eff1+RXeff1+30.
    .*log10(RXr.*TXr)-RXp)./60).*1.5;
d22=10.^(-0.5.*log10(1+(TXr./d11).^2));
df= d11.*d22;
plot(eff1p,df);
xlabel('Transmitter Coil Efficiency ')
ylabel('Distance (cm)')


elseif Option==10
    set (radiobutton1, 'Value',0);
    set (radiobutton2, 'Value',0);
    set (radiobutton3, 'Value',0);
    set (radiobutton4, 'Value',0);
    set (radiobutton5, 'Value',0);
    set (radiobutton6, 'Value',0);
    set (radiobutton7, 'Value',0);
    set (radiobutton8, 'Value',0);
    set (radiobutton9, 'Value',0);
    set (radiobutton10, 'Value',1);
    set (radiobutton11, 'Value',0);
    set (radiobutton12, 'Value',0);
    set (radiobutton13, 'Value',0);

    eff2p=0.01:0.005:1;
    eff2=log10(eff2p);
    subplot('Position',[0.35 0.55 0.63 0.35]); axis on;
    d1=10.^((mu0+mu0+TXp+TXQ1+RXQ1+TXeff1+eff2+30.
       .*log10(RXr.*TXr)-RXp)./60).*1.5;
    d2=10.^(-0.5.*log10(1+(TXr./d1).^2));
    da= d1.*d2;
    plot(eff2p,da);
    xlabel('Receiver Coil Efficiency')
    ylabel('Distance (cm)')

    subplot('Position',[0.35 0.05 0.63 0.35]); axis on;
    d11=10.^((RXmu1+TXmu1+TXp+TXQ1+RXQ1+TXeff1+eff2+30.
        .*log10(RXr.*TXr)-RXp)./60).*1.5;
    d22=10.^(-0.5.*log10(1+(TXr./d11).^2));
    df= d11.*d22;
    plot(eff2p,df);
```

```
        xlabel('Receiver Coil Efficiency')
        ylabel('Distance (cm)')

elseif Option==11
        set (radiobutton1, 'Value',0);
        set (radiobutton2, 'Value',0);
        set (radiobutton3, 'Value',0);
        set (radiobutton4, 'Value',0);
        set (radiobutton5, 'Value',0);
        set (radiobutton6, 'Value',0);
        set (radiobutton7, 'Value',0);
        set (radiobutton8, 'Value',0);
        set (radiobutton9, 'Value',0);
        set (radiobutton10, 'Value',0);
        set (radiobutton11, 'Value',1);
        set (radiobutton12, 'Value',0);
        set (radiobutton13, 'Value',0);

        eff11p=0.01:0.005:1;
        eff12p=0.01:0.005:1;
        eff11=log10(eff11p);
        eff12=log10(eff12p);

        subplot('Position',[0.35 0.55 0.63 0.35]); axis on;
        d1=10.^((mu0+mu0+TXp+TXQ1+RXQ1+eff11+eff12+30.
            *log10(RXr.*TXr)-RXp)./60).*1.5;
        d2=10.^(-0.5.*log10(1+(TXr./d1).^2));
        da= d1.*d2;
        plot(eff12,da);
        xlabel('Receiver and Transmitter Coil Efficiency')
        ylabel('Distance (cm)')

        subplot('Position',[0.35 0.05 0.63 0.35]); axis on;
        d11=10.^((RXmu1+TXmu1+TXp+TXQ1+RXQ1+eff11+eff12+30.
            *log10(RXr.*TXr)-RXp)./60).*1.5;
        d22=10.^(-0.5.*log10(1+(TXr./d11).^2));
        df= d11.*d22;
        plot(eff12,df);
        xlabel('Receiver and Transmitter Coil Efficiency')
        ylabel('Distance (cm)')

elseif Option==12
        set (radiobutton1, 'Value',0);
        set (radiobutton2, 'Value',0);
        set (radiobutton3, 'Value',0);
        set (radiobutton4, 'Value',0);
        set (radiobutton5, 'Value',0);
        set (radiobutton6, 'Value',0);
        set (radiobutton7, 'Value',0);
        set (radiobutton8, 'Value',0);
        set (radiobutton9, 'Value',0);
        set (radiobutton10, 'Value',0);
        set (radiobutton11, 'Value',0);
```

```
        set (radiobutton12, 'Value',1);
        set (radiobutton13, 'Value',0);


        TXQ1=log10(TXQ);
        RXQ1=log10(RXQ);
        TXeff1=log10(TXeff);
        RXeff1=log10(RXeff);
        TXmu1=log10(TXmu);
        RXmu1=log10(RXmu);
        mu0=log10(4.*pi.* 10.^-7);
        mu11p=1:0.005:10000;
        mu12p=1:0.005:10000;
        mu11rel=log10(mu11p);
        mu12rel=log10(mu12p);
        mu11=mu11rel-mu0;
        mu12=mu12rel-mu0;
        plot2=subplot('Position',[0.35 0.55 0.63 0.35]); axis on;
        delete(plot2);

        subplot('Position',[0.35 0.05 0.63 0.35]); axis on;
        d11=10.^((mu11+mu12+TXp+TXQ1+RXQ1+TXeff1+RXeff1+30.
            *log10(RXr.*TXr)-RXp)./60).*1.5;
        d22=10.^(-0.5.*log10(1+(TXr./d11).^2));
        df= d11.*d22;
        plot(mu11p,df);
        xlabel('Receiver and Transmitter Coil Relative Permeability')
        ylabel('Distance (cm)')

elseif Option==13
        set (radiobutton1, 'Value',0);
        set (radiobutton2, 'Value',0);
        set (radiobutton3, 'Value',0);
        set (radiobutton4, 'Value',0);
        set (radiobutton5, 'Value',0);
        set (radiobutton6, 'Value',0);
        set (radiobutton7, 'Value',0);
        set (radiobutton8, 'Value',0);
        set (radiobutton9, 'Value',0);
        set (radiobutton10, 'Value',0);
        set (radiobutton11, 'Value',0);
        set (radiobutton12, 'Value',0);
        set (radiobutton13, 'Value',1);

        q11p=0.001:0.005:1000;
        %range of the transmitter coil quality factor
        q12p=0.001:0.005:1000;
        %range of the receiver coil quality factor(identical to the
         transmitter's)

        q11=log10(q11p);
        q12=log10(q12p);
```

```
        subplot('Position',[0.35 0.55 0.63 0.35]); axis on;
        d1=10.^((mu0+mu0+TXp+q11+q12+TXeff1+RXeff1+30.
           *log10(RXr.*TXr)-RXp)./60).*1.5;
        d2=10.^(-0.5.*log10(1+(TXr./d1).^2));
        da= d1.*d2;
        plot(q11p,da);
        xlabel('Receiver and Transmitter Coil Quaity Factor')
        ylabel('Distance (cm)')

        subplot('Position',[0.35 0.05 0.63 0.35]); axis on;
        d11=10.^((RXmu1+TXmu1+TXp+q11+q12+TXeff1+RXeff1+30.
           *log10(RXr.*TXr)-RXp)./60).*1.5;
        d22=10.^(-0.5.*log10(1+(TXr./d11).^2));
        df= d11.*d22;
        plot(q11p,df);
        xlabel('Receiver and Transmitter Coil Quality Factor')
        ylabel('Distance (cm)')

    end
% DISTANCE CALCULATION FOR AIR CORE COIL *****************
elseif x==2
        d1=10.^((mu0+mu0+TXp+RXQ1+TXQ1+TXeff1+RXeff1+30.
           *log10(RXr.*TXr)-RXp)./60).*1.5;
        d2=10.^(-0.5.*log10(1+(TXr./d1).^2));
        distancea= d1.*d2;

        pText1=uicontrol(gcf,'Style','text','Position',[460 610 150 25],...
        'String',distancea,'FontSize',12,'FontWeight','bold',...
        'HorizontalAlignment','center','FontName','Arial',
        'BackgroundColor',[1 1 .5]);

% DISTANCE CALCULATION FOT FERRITE CORE COIL**************
elseif x==3
        d11=10.^((RXmu1+TXmu1+TXp+RXQ1+TXQ1+TXeff1+RXeff1+30.
           *log10(RXr.*TXr)-RXp)./60).*1.5;
        d22=10.^(-0.5.*log10(1+(TXr./d11).^2));
        distancef= d11.*d22;

        pText2=uicontrol(gcf,'Style','text','Position',[460 290 150 25],...
        'String',distancef,'FontSize',12,'FontWeight','bold',...
        'HorizontalAlignment','center','FontName','Arial',
        'BackgroundColor',[1 1 .5]);
end

% RESET BUTTON ***********************************************************

        hb = uicontrol('Style','pushbutton','position',[5 10 70 25],
              'FontName','Arial',...
        'FontSize',14,'FontWeight','normal','BackgroundColor',[.8 1 1],...
        'String','Reset');
        set(hb,'Callback','minoogui')
%***********************************************************************
%***********************************************************************
```

```
%                          THE END
%*********************************************************************
%*********************************************************************


%developed by: Mehrnoush  Masihpour 23.12.2009
%file name: NFMIC.m

if x==1
    if Option==1
        set (radiobutton1, 'Value',1);
        set (radiobutton2, 'Value',0);
        set (radiobutton3, 'Value',0);
        set (radiobutton4, 'Value',0);
        set (radiobutton5, 'Value',0);
        set (radiobutton6, 'Value',0);
        set (radiobutton7, 'Value',0);
        set (radiobutton8, 'Value',0);
        set (radiobutton9, 'Value',0);
        set (radiobutton10, 'Value',0);
        set (radiobutton11, 'Value',0);
        set (radiobutton12, 'Value',0);
        set (radiobutton13, 'Value',0);
rr1=0.001:0.005:8;
rxp=txp+qt+qr+teff+reff+tmu+rmu+10.*log10(9.85.*((tr.*rr1).
    /((d.^2)+(tr.^2))).^3);
subplot('Position',[0.35 0.45 0.63 0.45]);
plot(rr1,rxp)
xlabel('receiver coil radius (cm)')
ylabel('Power at the Receiver (dBm)')
title('Power at the receiver vs receiver coil radius')


    elseif Option==2
        set (radiobutton1, 'Value',0);
        set (radiobutton2, 'Value',1);
        set (radiobutton3, 'Value',0);
        set (radiobutton4, 'Value',0);
        set (radiobutton5, 'Value',0);
        set (radiobutton6, 'Value',0);
        set (radiobutton7, 'Value',0);
        set (radiobutton8, 'Value',0);
        set (radiobutton9, 'Value',0);
        set (radiobutton10, 'Value',0);
        set (radiobutton11, 'Value',0);
        set (radiobutton12, 'Value',0);
        set (radiobutton13, 'Value',0);
tr1=0.001:0.005:8;
rxp=txp+qt+qr+teff+reff+tmu+rmu+10.*log10(9.85.*((tr1.*rr).
    /((d.^2)+(tr1.^2))).^3);
subplot('Position',[0.35 0.45 0.63 0.45]);
plot(tr1,rxp)
```

```
xlabel('transmitter coil radius (cm)')
ylabel('Power at the Receiver (dBm)')
title('Power at the receiver vs transmitter coil radius')


    elseif Option==3
        set (radiobutton1, 'Value',0);
        set (radiobutton2, 'Value',0);
        set (radiobutton3, 'Value',1);
        set (radiobutton4, 'Value',0);
        set (radiobutton5, 'Value',0);
        set (radiobutton6, 'Value',0);
        set (radiobutton7, 'Value',0);
        set (radiobutton8, 'Value',0);
        set (radiobutton9, 'Value',0);
        set (radiobutton10, 'Value',0);
        set (radiobutton11, 'Value',0);
        set (radiobutton12, 'Value',0);
        set (radiobutton13, 'Value',0);
txp1=0.001:0.005:25;
rxp=txp1+qt+qr+teff+reff+tmu+rmu+10.*log10(9.85.*((tr.*rr).
    /((d.^2)+(tr.^2))).^3);
subplot('Position',[0.35 0.45 0.63 0.45]);
plot(txp1,rxp)
xlabel('Transmission power (dBm)')
ylabel('Power at the Receiver (dBm)')
title('Power at the receiver vs transmission power')


    elseif Option==4
        set (radiobutton1, 'Value',0);
        set (radiobutton2, 'Value',0);
        set (radiobutton3, 'Value',0);
        set (radiobutton4, 'Value',1);
        set (radiobutton5, 'Value',0);
        set (radiobutton6, 'Value',0);
        set (radiobutton7, 'Value',0);
        set (radiobutton8, 'Value',0);
        set (radiobutton9, 'Value',0);
        set (radiobutton10, 'Value',0);
        set (radiobutton11, 'Value',0);
        set (radiobutton12, 'Value',0);
        set (radiobutton13, 'Value',0);
d1=0.0001:0.005:500;
rxp=txp+qt+qr+teff+reff+tmu+rmu+10.*log10(9.85.*((tr.*rr).
    /((d1.^2)+(tr.^2))).^3);
subplot('Position',[0.35 0.45 0.63 0.45]);
plot(d1,rxp)
xlabel('Communication Range (cm)')
ylabel('Power at the Receiver (dBm)')
title('Power at the receiver vs distance')
```

```
    elseif Option==5
        set (radiobutton1, 'Value',0);
        set (radiobutton2, 'Value',0);
        set (radiobutton3, 'Value',0);
        set (radiobutton4, 'Value',0);
        set (radiobutton5, 'Value',1);
        set (radiobutton6, 'Value',0);
        set (radiobutton7, 'Value',0);
        set (radiobutton8, 'Value',0);
        set (radiobutton9, 'Value',0);
        set (radiobutton10, 'Value',0);
        set (radiobutton11, 'Value',0);
        set (radiobutton12, 'Value',0);
        set (radiobutton13, 'Value',0);
qtt=0.1:0.005:1000;
qt1=log10(qtt);
rxp=txp+qt1+qr+teff+reff+tmu+rmu+10.*log10(9.85.*((tr.*rr).
    /((d.^2)+(tr.^2))).^3);
subplot('Position',[0.35 0.45 0.63 0.45]);
plot(qtt,rxp)
xlabel('Quality Factor of the transitter coil')
ylabel('Power at the Receiver (dBm)')
title('Power at the receiver vs Quality factor of the transmitter coil')


    elseif Option==6
    set (radiobutton1, 'Value',0);
        set (radiobutton2, 'Value',0);
        set (radiobutton3, 'Value',0);
        set (radiobutton4, 'Value',0);
        set (radiobutton5, 'Value',0);
        set (radiobutton6, 'Value',1);
        set (radiobutton7, 'Value',0);
        set (radiobutton8, 'Value',0);
        set (radiobutton9, 'Value',0);
        set (radiobutton10, 'Value',0);
        set (radiobutton11, 'Value',0);
        set (radiobutton12, 'Value',0);
        set (radiobutton13, 'Value',0);
qrr=0.1:0.005:1000;
qr1=log10(qrr);
rxp=txp+qt+qr1+teff+reff+tmu+rmu+10.*log10(9.85.*((tr.*rr).
    /((d.^2)+(tr.^2))).^3);
subplot('Position',[0.35 0.45 0.63 0.45]);
plot(qrr,rxp)
xlabel('Quality Factor of the receiver coil')
ylabel('Power at the Receiver (dBm)')
title('Power at the receiver vs Quality factor of the receiver coil')


    elseif Option==7
        set (radiobutton1, 'Value',0);
```

```
          set (radiobutton2, 'Value',0);
          set (radiobutton3, 'Value',0);
          set (radiobutton4, 'Value',0);
          set (radiobutton5, 'Value',0);
          set (radiobutton6, 'Value',0);
          set (radiobutton7, 'Value',1);
          set (radiobutton8, 'Value',0);
          set (radiobutton9, 'Value',0);
          set (radiobutton10, 'Value',0);
          set (radiobutton11, 'Value',0);
          set (radiobutton12, 'Value',0);
          set (radiobutton13, 'Value',0);
tmuu=0.1:0.005:10000;
tmu1=log10(tmuu);
rxp=txp+qt+qr+teff+reff+tmu1+rmu+10.*log10(9.85.*((tr.*rr).
    /((d.^2)+(tr.^2))).^3);
subplot('Position',[0.35 0.45 0.63 0.45]);
plot(tmuu,rxp)
xlabel('transmitter coil Relative permeability')
ylabel('Power at the Receiver (dBm)')
%legend('Power at the receiver vs transmitter coil permeability')
title('Power at the receiver vs transmitter coil permeability')




    elseif Option==8
    set (radiobutton1, 'Value',0);
          set (radiobutton2, 'Value',0);
          set (radiobutton3, 'Value',0);
          set (radiobutton4, 'Value',0);
          set (radiobutton5, 'Value',0);
          set (radiobutton6, 'Value',0);
          set (radiobutton7, 'Value',0);
          set (radiobutton8, 'Value',1);
          set (radiobutton9, 'Value',0);
          set (radiobutton10, 'Value',0);
          set (radiobutton11, 'Value',0);
          set (radiobutton12, 'Value',0);
          set (radiobutton13, 'Value',0);
rmuu=0.1:0.005:10000;
rmu1=log10(rmuu);
rxp=txp+qt+qr+teff+reff+tmu+rmu1+10.*log10(9.85.*((tr.*rr).
    /((d.^2)+(tr.^2))).^3);
subplot('Position',[0.35 0.45 0.63 0.45]);
plot(rmuu,rxp)
xlabel('receiver coil Relative permeability')
ylabel('Power at the Receiver (dBm)')
title('Power at the receiver vs receiver coil permeability')

    elseif Option==9
          set (radiobutton1, 'Value',0);
          set (radiobutton2, 'Value',0);
```

```
             set (radiobutton3, 'Value',0);
             set (radiobutton4, 'Value',0);
             set (radiobutton5, 'Value',0);
             set (radiobutton6, 'Value',0);
             set (radiobutton7, 'Value',0);
             set (radiobutton8, 'Value',0);
             set (radiobutton9, 'Value',1);
             set (radiobutton10, 'Value',0);
             set (radiobutton11, 'Value',0);
             set (radiobutton12, 'Value',0);
             set (radiobutton13, 'Value',0);
tefff=0.0:0.05:1;
teff1=log10(tefff);
rxp=txp+qt+qr+teff1+reff+tmu+rmu+10.*log10(9.85.*((tr.*rr).
    /((d.^2)+(tr.^2))).^3);
subplot('Position',[0.35 0.45 0.63 0.45]);
plot(tefff,rxp)
xlabel('transmitter coil efficiency')
ylabel('Power at the Receiver (dBm)')
title('Power at the receiver vs transmitter coil efficiency')




    elseif Option==10
    set (radiobutton1, 'Value',0);
        set (radiobutton2, 'Value',0);
        set (radiobutton3, 'Value',0);
        set (radiobutton4, 'Value',0);
        set (radiobutton5, 'Value',0);
        set (radiobutton6, 'Value',0);
        set (radiobutton7, 'Value',0);
        set (radiobutton8, 'Value',0);
        set (radiobutton9, 'Value',0);
        set (radiobutton10, 'Value',1);
        set (radiobutton11, 'Value',0);
        set (radiobutton12, 'Value',0);
        set (radiobutton13, 'Value',0);
refff=0.1:0.05:1;
reff1=log10(refff);
rxp=txp+qt+qr+teff+reff1+tmu+rmu+10.*log10(9.85.*((tr.*rr).
    /((d.^2)+(tr.^2))).^3);
subplot('Position',[0.35 0.45 0.63 0.45]);
plot(refff,rxp)
xlabel('receiver coil efficiency')
ylabel('Power at the Receiver (dBm)')
title('Power at the receiver vs receiver coil efficiency')




    elseif Option==11
        set (radiobutton1, 'Value',0);
        set (radiobutton2, 'Value',0);
        set (radiobutton3, 'Value',0);
```

```
        set (radiobutton4, 'Value',0);
        set (radiobutton5, 'Value',0);
        set (radiobutton6, 'Value',0);
        set (radiobutton7, 'Value',0);
        set (radiobutton8, 'Value',0);
        set (radiobutton9, 'Value',0);
        set (radiobutton10, 'Value',0);
        set (radiobutton11, 'Value',1);
        set (radiobutton12, 'Value',0);
        set (radiobutton13, 'Value',0);
refff=0.1:0.05:1;
tefff=.1:0.05:1;

reff1=log10(tefff);
teff1=log10(tefff);
rxp=txp+qt+qr+teff1+reff1+tmu+rmu+10.*log10(9.85.*((tr.*rr).
    /((d.^2)+(tr.^2))).^3);
subplot('Position',[0.35 0.45 0.63 0.45]);
plot(refff,rxp)
xlabel('receiver and transmitter coil efficiency')
ylabel('Power at the Receiver (dBm)')
title('Power at the receiver vs receiver and transmitter coil efficiency')


    elseif Option==12
        set (radiobutton1, 'Value',0);
        set (radiobutton2, 'Value',0);
        set (radiobutton3, 'Value',0);
        set (radiobutton4, 'Value',0);
        set (radiobutton5, 'Value',0);
        set (radiobutton6, 'Value',0);
        set (radiobutton7, 'Value',0);
        set (radiobutton8, 'Value',0);
        set (radiobutton9, 'Value',0);
        set (radiobutton10, 'Value',0);
        set (radiobutton11, 'Value',0);
        set (radiobutton12, 'Value',1);
        set (radiobutton13, 'Value',0);
mu0=log10(4.*pi.* 10.^-7);
rmuu=1:0.5:10000;
tmuu=1:0.5:10000;
rmu1=log10(rmuu)-mu0;
tmu1=log10(tmuu)-mu0;
rxp=txp+qt+qr+teff+reff+tmu1+rmu1+10.*log10(9.85.*((tr.*rr).
    /((d.^2)+(tr.^2))).^3);
subplot('Position',[0.35 0.45 0.63 0.45]);
plot(rmuu,rxp)
xlabel('receiver and transmitter  coil permeability')
ylabel('Power at the Receiver (dBm)')
title('Power at the receiver vs receiver and transmitter coil permeability')
```

```
    elseif Option==13
        set (radiobutton1, 'Value',0);
        set (radiobutton2, 'Value',0);
        set (radiobutton3, 'Value',0);
        set (radiobutton4, 'Value',0);
        set (radiobutton5, 'Value',0);
        set (radiobutton6, 'Value',0);
        set (radiobutton7, 'Value',0);
        set (radiobutton8, 'Value',0);
        set (radiobutton9, 'Value',0);
        set (radiobutton10, 'Value',0);
        set (radiobutton11, 'Value',0);
        set (radiobutton12, 'Value',0);
        set (radiobutton13, 'Value',1);
qrr=0.1:0.5:1000;
qtt=0.1:0.5:1000;
qr1=log10(qrr);
qt1=log10(qtt);
rxp=txp+qt1+qr1+teff+reff+tmu+rmu+10.*log10(9.85.*((tr.*rr).
    /((d.^2)+(tr.^2))).^3);
subplot('Position',[0.35 0.45 0.63 0.45]);
plot(qrr,rxp)
xlabel('Quality Factor of the receiver and transmitter coil')
ylabel('Power at the Receiver (dBm)')
title('Power at the receiver vs Quality factor of the receiver and
        transmitter coil')
    end
elseif x==2
        rxp=txp+qt+qr+teff+reff+tmu+rmu+10.*log10(9.85.*((tr.*rr).
            /((d.^2)+(tr.^2))).^3);
        rxp_mW=10.^(rxp./100);

        pText2=uicontrol(gcf,'Style','text','Position',[280 100 500 60],...
        'String','The power at the receiver is = (dBm)','FontSize',14,
        'FontWeight','bold',...
        'HorizontalAlignment','center','FontName','Arial','BackgroundColor',
        [0.7 0.9 0.99]);
        pText212=uicontrol(gcf,'Style','text','Position',[780 100 100 60],...
        'String',rxp,'FontSize',14,'FontWeight','bold',...
        'HorizontalAlignment','left','FontName','Arial','BackgroundColor',
        [0.7 0.9 0.99]);

        pText33=uicontrol(gcf,'Style','text','Position',[280 40 500 60],...
        'String','The power at the receiver is = (mW)','FontSize',14,
        'FontWeight','bold',...
        'HorizontalAlignment','center','FontName','Arial','BackgroundColor',
        [0.7 0.9 0.99]);
        pText313=uicontrol(gcf,'Style','text','Position',[780 40 100 60],...
        'String',rxp_mW,'FontSize',14,'FontWeight','bold',...
        'HorizontalAlignment','left','FontName','Arial','BackgroundColor',
        [0.7 0.9 0.99]);
end
% RESET BUTTON ************************************************************
```

```
hb = uicontrol('Style','pushbutton','position',[5 10 70 25],
    'FontName','Arial',...
'FontSize',14,'FontWeight','normal','BackgroundColor',[.8 .9 .5],...
'String','Reset');
set(hb,'Callback','NFMICgui')
```

```
%********************************************************************
%                           THE END
%********************************************************************
```

Index

3-port splitters, 256

Agbinya, 343
Agbinya-Masihpour, 345
a solenoid, 85
AC resistance, 86, 87
Agbinya, 224, 225, 227, 228,
        232, 237
air cores, 80
Akyildiz, 225, 228–230, 232,
        233, 237
AM models, 232
AM2, 230, 234
amplitude shift keying (ASK), 159,
        164
angular misalignment, 79, 81, 82, 314
antenna coils, 57, 302, 303, 345
anti-symmetric modes, 258
ASK, 159–161, 164, 166, 177, 179

biological tissues, 187
Biomedical Monitoring, 269
biomedical monitoring, 274
Biphase encoding, 154
Blood pressure management, 321
blood pressure management,
        275, 321
Bluetooth, 23, 270, 271, 274
body area networks, 45, 135,
        172, 195
boundary condition, 246, 260
BPSQ, 154
Bragg gratings, 248

bubble factors, 39, 45

CET, 336, 338, 339
circuit configuration modulator, 166
cochlear implants, 326, 327
communication bubble, 39, 44,
        45, 273
conductance, 87, 277
conductivity, 85, 86, 115, 312, 313
conjugate match, 75
Contact-less payment, 269
Contactless Energy Transfer (CET),
        336
Costas loop, 171, 172, 175
coupling coefficient, 30, 32, 33, 40,
        43, 45, 47, 53, 74, 103,
        106, 108, 113, 135, 138,
        142, 163, 198–200, 203,
        213, 221, 222, 307, 309,
        311, 324, 329
coupling of energy, 63
cross talk, 85, 146
crosstalk, 103, 104, 133, 139, 142,
        146, 201, 203, 206

dipole antenna, 3
directional coupler, 257
dispersion equation, 258, 263, 266
dispersion equations, 258, 260
distance bubble factor, 45, 48

ECMA-340 standard, 283
ECMA-352, 284

J. Ihyeh Agbinya (PhD), Principles of Inductive Near Field Communications for Internet of Things, 385–389.
© 2011 *River Publishers. All rights reserved.*